Structure and Functions of Amine Oxidases

Editor

Bruno Mondovì
Professor of Applied Biochemistry
Institute of Applied Biochemistry
University of Rome and CNR
Center for Molecular Biology
Rome, Italy

CRC Press
Taylor & Francis Group
Boca Raton London New York

CRC Press is an imprint of the
Taylor & Francis Group, an **informa** business

First published 1985 by CRC Press
Taylor & Francis Group
6000 Broken Sound Parkway NW, Suite 300
Boca Raton, FL 33487-2742

Reissued 2018 by CRC Press

Library of Congress Cataloging in Publication Data
Main entry under title:

Structure and functions of amine oxidases.

Bibliography: p.
Includes index.
1. Amine oxidase. I. Mondovì, Bruno.
QP603.M6S77 1985 574.19'258 85-4240
ISBN 0-8493-5869-8

A Library of Congress record exists under LC control number: 85004240

ISBN 13: 978-1-315-89785-1 (hbk)
ISBN 13: 978-1-351-07695-1 (ebk)

Visit the Taylor & Francis Web site at http://www.taylorandfrancis.com and the
CRC Press Web site at http://www.crcpress.com

DEDICATION

This book is dedicated to the memory of Professor Eraldo Antonini, famous biochemist and friend, prematurely deceased on March 19th, 1983.

ACKNOWLEDGMENT

Part of this book has been compiled under the auspices of the Italian National Research Council.

PREFACE

This book has been compiled in the hopes that it may be of use to those students and researchers in the fields of medicine, biological sciences, and chemistry who are interested in the structure and functions of amine oxidases.

As a general discussion on the historical background will be made in the introductory chapter, I will limit myself here to briefly describing the general plan of this book and the terminology employed herein.

During recent years more extensive literature on amine oxidases has been available concerning the basic as well as the applied points of view. Unfortunately, there is often much confusion, to the detriment of unspecialized scientists, regarding the identification of these enzymes, both structurally and functionally. As recently stated by Zeller, "anarchy is reigning supreme in the nomenclature of amine oxidases."[1]

A first attempt to classify amine oxidases on the basis of their sensitivity towards specific inhibitors or with respect to their affinity towards particular substrates could not be satisfactorily considered adequate, for neither the known inhibitors are very specific, nor do these enzymes show a strict substrate specificity.[2-4]

It was then decided to operate a first general division according to the nature of the prosthetic group contained in the enzyme molecule. On the basis of this criterion two classes of amine oxidases can be described: one containing FAD and the other having copper and a second organic cofactor, as prosthetic groups. Thus, the first distinction to make for an adequate nomenclature appears to be that between the FAD- and the copper-dependent amine oxidases. However, it is not always possible to be so limited; therefore, amine oxidases, and particularly the copper-dependent enzymes, must in some cases be subdivided in relation to their substrate specificity.

After examining many possibilities and accepting many helpful suggestions from the authors, the different enzymes have been abbreviated by using the symbols reported in Table 1. The mitochondrial monoamine oxidase (MAO) and the cytosolic polyamine oxidase (PAO) are both FAD-dependent enzymes. The copper-dependent enzymes have been abbreviated AO. In the latter group, the abbreviation for the tissue of origin, when well established, has been placed before the symbol. For example, the connective tissue copper-dependent amine oxidase is cited as CTAO.

When strictly necessary, the initial of the name of the best substrate has been placed before the symbol; such is the case of those copper-dependent animal enzymes which oxidize diamines and histamine, and of those which oxidize peptide-bound lysine in collagen and elastin. Consequently, the names diamine oxidase (DAO) and lysyl oxidase (LAO) have been adopted. Actually, DAO and LAO, from both a historical and a physiopathological point of view, play a well-defined role and, especially in the case of DAO, the change of the name and symbol creates only more confusion. As an example, two copper-dependent enzymes are present in most mammals' plasma: one oxidizes monoamines and polyamines, and the other diamines.

To avoid confusion with the FAD-dependent polyamine oxidase (PAO), the plasma enzymes have been referred to as serum amine oxidase (SAO) and serum diamine oxidase (SDAO).

If one would like to be more concise when describing an amine oxidase of a particular animal or plant, the initial of the animal's (or plant's) name can be written before the initial(s) of the tissue of origin, followed by the abbreviated symbol of the enzyme. So, SKDAO, HSAO, HPDAO, LSAO, and BALAO refer to swine kidney DAO, human serum AO, human placental DAO, lentil seedlings AO, and bovine aorta LAO, respectively.

The role of copper- and FAD-dependent amine oxidases in medicine has been particularly stressed. Throughout the different chapters of this book attention has been focused on the

Table 1
ABBREVIATIONS

ANAO	= *Aspergillus niger* amine oxidase	5-HTP	= 5-Hydroxytryptophan
AO	= Amine oxidase (copper-dependent)	IAA	= Imidazoleacetic acid
BALAO	= Bovine aorta lysyl oxidase	LAO	= Lysyl oxidase
BAPN	= β-Amino propionitrile	LSAO	= Lentil (*Lens esculenta*) seedlings amine oxidase
BSAO	= Bovine serum (plasma) amine oxidase	MAO	= Monoamine oxidase (mitochon drial, (FAD-dependent)
CTAO	= Connective tissue amine oxidase		
DA	= Dopamine	MHPG	= 3-Methoxy-4-hydroxyphenylglycol
DAO	= Diamine oxidase	ODC	= Ornithine decarboxylase
DDC	= Diethyldithiocarbamate	PAO	= Polyamine oxidase
DFMO	= α-Difluoromethylornithine	PCA	= *Para*-Chloroamphetamine
E	= Epinephrine	PCPA	= *para*-Chlorophenylalanine
EGF	= Epidermal growth factor	PEA	= Phenylethylamine
ELAO	= *Euphorbia latex* amine oxidase	PLP	= Pyridoxal-5-phosphate
FAD	= Flavin-adenin-dinucleotide	PSAO	= Pea (*Pisum sativum*) seedlings amine oxidase
GABA	= γ-Aminobutyric acid		
HDC	= Histidine decarboxylase	SAO	= Serum (plasma) amine oxidase
HFC	= Histamine forming capacity	SDAO	= Serum (plasma) diamine oxidase
HKDAO	= Human kidney diamine oxidase	SKDAO	= Swine kidney diamine oxidase
HPDAO	= Human placental diamine oxidase	SSAO	= Swine serum (plasma) amine oxidase
HSAO	= Human serum amine oxidase		
HTC	= Hepatoma tissue culture	SSCRAO	= Semicarbazide-sensitive-clorgy-line-resistant amine oxidase
HVA	= Homovanillic acid		
5-HIAA	= 5-Hydroxyindoleacetic acid	TCP	= Tranylcypromine
5-HT	= 5-Hydroxytriptamine	VLAO	= *Vicia faba* leaves amine oxidase

important catalytic properties of amine oxidases, though their role in both physiological and pathological processes has also been emphasized.

A good portion of this book has been devoted to the copper-dependent enzymes, these being the more numerous. The chapter dealing with serum amine oxidases also focuses attention on their catalytic mechanism, as these enzymes have been studied in greater depth.

As the presentation of topics whose experimental basis is rapidly developing is likely to stimulate the reader's interest, many bibliographic references have been included.

Readers could find this book poor, as far as many topics are dealt with in a relativley little space, but we believe it essential to trace the background of our present knowledge in the field of amine oxidases, stressing the future outlook of research on these enzymes, for they are becoming more and more important in general and medical biochemistry.

In closing, I would like to express my sincerest thanks to all the authors who worked so hard to make this book possible.

REFERENCES

1. **Zeller, E. A.,** Classification and nomenclature of monoamine oxidases and other amine oxidsases, in *Monoamine Oxidase: Structure, Function and Altered Functions,* Singer, T. P., Korff, R. W., and Murphy, D. L., Eds., Academic Press, New York, 1979, 531.
2. **Zeller, E. A.,** Oxidation of amines, in *The Enzymes,* Vol. 2, (Part I), 1st ed., Summer, J. B. and Myrbäck, K., Eds., Academic Press, New York, 1951, 536.
3. **Malström, B. G., Andreasson, L. E., and Reinhammar, B.,** Copper containing oxidases and superoxide dismutase, in *The Enzymes,* Vol. 12, 3rd ed., Boyer, P. D., Ed., Academic Press, New York, 1975, 507.
4. **Blaschko, H.,** Amine oxidase, in *The Enzymes,* Vol. 8, Boyer, P. D., Lardy, H., and Myrbäck, K., Eds. Academic Press, New York, 1963, 337.

THE EDITOR

Bruno Mondovì, M.D., Professor of Applied Biochemistry at the University of Rome, received his early training at the University of Rome where he obtained the M.D. degree in 1951. In 1958 he obtained the *Libera Docenza* (the Italian equivalent of the Ph.D.) in Biological Chemistry and in 1961 in Applied Biochemistry.

In 1952, Dr. Mondovì began his academic career as Assistant Professor in Biological Chemistry at the University of Rome. He turned his attention to the study of sulpur-containing amino acid metabolism up to 1960. His current interests include a cancer research project concerning the biochemical basis of tumor cell selective sensitivity to heat. Since 1960 he has been largely engaged in research on the structure and function of copper-containing proteins, and particularly amine oxidases. In 1961 he was appointed Associate Professor in Applied Biochemistry at the same University. Winner of an open competition, he became Full Professor of Applied Biochemistry at the Faculty of Pharmacy, University of Rome, in 1970. Since 1971 he has been chief of the Institute of Applied Biochemistry at the University of Rome.

Bruno Mondovì is a member of the American Chemical Society, the New York Academy of Science, the European Association for Cancer Research, the International Society of Magnetic Resonance, the International Clinical Hyperthermia Society, the Italian Society of Cancerology, and the Italian Biophysic and Molecular Biology Society. He is also a member of the Managing Committee of the Italian Biochemical Society.

Dr. Mondovì has attended several international symposia as an invited speaker. He was also the Italian organizer of international symposia on the structure and function of copper proteins in 1971, 1974, 1976, and 1979.

Dr. Mondovì is the author of more than 230 papers and has contributed to 5 books. He has also edited 1 book and was co-editor of another. He is cited in *Who's who in the World*, 1980.

CONTRIBUTORS

Maria P. Argento-Cerù
Associate Professor
Biology Department
Faculty of Science
2nd University of Rome
Rome, Italy

Francesco Autuori
Professor
Biology Department
Faculty of Science
2nd University of Rome
Rome, Italy

Uriel Bachrach, M.Sc., Ph.D.
Professor
Department of Molecular Biology
Hebrew University
Hadassah Medical School
Jerusalem, Israel

W. G. Bardsley, Ph.D., D.Sc.
Senior Lecturer in Biochemistry
Department of Obstetrics and Gynecology
St. Mary's Hospital
University of Manchester
Manchester, England

Stephen B. Baylin, M.D.
Associate Professor of Oncology and Medicine
Division of Endocrine Oncology
Johns Hopkins University School of Medicine
Baltimore, Maryland

Tadeusz Bieganski, M.D., Ph.D.
Associate Professor
Department of Biogenic Amines
Polish Academy of Sciences
Lodz, Poland

Giuseppe Bombardieri, M.D.
Associate Professor of Hepatology
Institute of Clinical Medicine
Universita Cattolica del S. Cuore
Rome, Italy

Franca Buffoni
Professor
Department of Preclinical and Clinical Pharmacology
University of Florence
Florence, Italy

Maria Alfonsina Desiderio, Sc.D.
Researcher
Institute of General Pathology
University of Milan
Milan, Italy

Madeleine Ennis
Department of Theoretical Surgery
Centre of Operative Medicine I
University of Marburg
Marburg, West Germany

Alessandro Finazzi-Agrò
Professor of Enzymology
Faculty of Medicine
University of Rome II
Rome, Italy

Giovanni Floris
Associate Professor
Department of Biochemistry
University of Cagliari
Cagliari, Italy

Wieslawa A. Fogel, Ph.D.
Assistant Professor
Department of Biogenic Amines
Polish Academy of Sciences
Lodz, Poland

Christopher J. Fowler, M.A., Ph.D.
Adjunct Professor of Human Biochemical Pharmacology
Research and Development Laboratories
Astra Lakemedel AB
Sodertalje, Sweden

Anna Giartosio
Professor
Institute of Appliled Biochemistry
Biochimica Applicata
University of Rome La Sapienza
Rome, Italy

V. Z. Gorkin
Professor
Institute of Biological and Medical Chemistry
Academy of Medical Sciences
Moscow, U.S.S.R.

Rudolf Hesterberg
Surgical Clinic Centre of Operative Medicine I
University of Marburg
Marburg, West Germany

Mary Ellen Kitler, Ph.D., F.C.P.
Manager
Experimental Therapeutics
Zyma SA
Nyon, Switzerland

Donald M. Kuhn
Section on Biochemical Pharmacology
N.H.L.B.I.
National Institutes of Health
Building 10, 7N262
Bethesda, Maryland

Hidehiko Kumagai, Ph.D.
Associate Professor
Department of Food Science and Technology
Faculty of Agriculture
Kyoto University
Kyoto, Japan

Jurgen Kusche
Division of Experimental Surgery and Biochemistry
Department of Surgery
University of Cologne
Cologne, West Germany

Wilfried Lorenz
Department of Theoretical Surgery
Centre of Operative Medicine I
University of Marburg
Marburg, West Germany

Gordon D. Luk, M.D.
Associate Professor of Medicine and Oncology
Johns Hopkins University School of Medicine
Baltimore, Maryland

Czeslaw Maśliński, M.D., Ph.D.
Professor
Department of Biogenic Amines
Polish Academy of Sciences
Lodz, Poland

A. Milani, M.D.
Institute of Patologia Medica
Universita Cattolica del Sacro Cuore
Rome, Italy

Bruno Mondovì
Professor of Applied Biochemistry
Institute of Applied Biochemistry
University of Rome La Sapienza
and
Center for Molecular Biology
Rome, Italy

D. L. Murphy
Section on Clinical Neuropharmacology,
Laboratory of Clinical Science — NIMH,
NIH Clinical Center,
Building 10/3S229,
Bethesda, Maryland

Antonio Perin, M.D., Ph.D.
Professor of General Pathology
Institute of General Pathology
University of Milan
Milan, Italy

Gosta Pettersson
Biochemistry 1
Chemical Center
Lund, Sweden

Pierluigi Riccio
Institute of Applied Biochemistry
University of Rome La Sapienza
Rome, Italy

Augusto Rinaldi
Director and Professor
Department of Biochemistry
University of Cagliari
Cagliari, Italy

Lodovico Rossi, M.D.
Associate Professor of Medical Therapy
Institute of Medical Pathology
Universita Cattolica S. Cuore
Rome, Italy

Giuseppe Rotilio, M.D., Ph.D.
Professor of Biochemistry
Department of Biology
2nd University of Rome
Rome, Italy

Jan Sääf, M.Sc., Dr. Med. Sc.
Researcher
Department of Psychiatry
St. Goran's Hospital
Stockholm, Sweden

Jorg Sattler
Department of Theoretical Surgery
Centre of Operative Medicine I
University of Marburg
Marburg, West Germany

Nikolaus Seiler, Ph.D.
Professor
Merrell-Dow Research Institute
Strasbourg, France

Angela Sessa, M.D.
Researcher
Institute of General Pathology
Center for Research in Cell Pathology
Milan, Italy

Thomas P. Singer, Ph.D.
Professor
Department of Pharmaceutical Chemistry
Biochemistry, and Biophysics
University of California at San Francisco
and
Head, Molecular Biology Division
Veterans Administration Medical Center
San Francisco, California

Carl-Detmar Stahlknecht
Department of Theoretical Surgery
Center of Operative Medicine I
University of Marburg
Marburg, West Germany

Celia White Tabor
Laboratory of Biochemical Pharmacology
National Institute of Arthritis, Diabetes,
and Digestive and Kidney Diseases
National Institutes of Health
Bethesda, Maryland

Anthony Ka Lung Tan, M.S.
Graduate Student
Department of Biochemistry
University of Hawaii at Manoa
Honolulu, Hawaii

Hideaki Yamada, D.A.
Professor
Department of Agricultural Chemistry
Kyoto University
Kyoto, Japan

Kerry T. Yasunobu
Professor
Department of Biochemistry-Biophysics
University of Hawaii
Honolulu, Hawaii

Moussa B. H. Youdim
Professor and Chairman
Department of Pharmacology
Technion
Rappaport Institute
Haifa, Israel

TABLE OF CONTENTS

Chapter 1

INTRODUCTION

Herbert Tabor

The history of investigations on amine oxidases (AO) represents a striking interplay of several different disciplines, each using different techniques and approaches. In addition to basic biochemical investigations, the pharmacological activities of the various substrates led to many pharmacologic and clinical studies.

There have been a number of reviews of the early history of the amine oxidases.[1-8] The first in vitro study was the demonstration in 1928 of the presence of tyramine oxidase in liver preparations by Mary Hare (later Mary Bernheim). Several years later (1937), Blaschko and colleagues showed that the pharmacologically active catecholamines were substrates of the same monoamine oxidase (MAO) preparation. Subsequent studies showed that MAO preparations also oxidized 5-hydroxytryptamine and a variety of other monoamines.

These MAO preparations were associated with the mitochondria, and early attempts to purify the enzymes were unsuccessful. Consequently, all of the earlier work had to be carried out on impure preparations, with resultant difficulties in interpretation of many of the experiments. It was not until much later, when techniques became available for the solubilization of mitochondrial enzymes, that MAO preparations were solubilized and purified.

Shortly after the initial study on tyramine oxidase, histaminase was reported in lung by Best, and later a diamine oxidase (DAO) that oxidized putrescine and cadaverine was reported in hog kidney by Zeller. Zeller showed that one enzyme oxidized both histamine and diamines (such as putrescine), and he named the enzyme ''diamine oxidase''. Since DAO was a soluble enzyme, it was partially purified from hog kidney in 1951, and the identity of histaminase and DAO in this tissue was confirmed. The classification of AOs into MAOs and DAOs was, in general, accepted, although there was some controversy about whether histaminase and DAO were identical in all tissues.

Those working on these enzymes felt that the AOs must have significant physiological importance since they were concerned with the oxidation and inactivation of a number of very potent pharmacologic agents. However, there was little general interest in these enzymes until the dramatic and accidental discovery in the early 1950s* of the relationship of AOs to mental disorders. As is well known, the observation was made that tuberculosis patients treated with iproniazid exhibited a number of behavioral side effects; this correlation, plus the findings of Zeller, Barsky, and Berman, that iproniazid inhibited MAO, led to an explosive interest in MAO. A very large number of studies were carried out on the synthesis and action of new inhibitors, on amine concentration and metabolism in nervous tissues, and on the possible involvement of amines in a variety of mental disorders. These studies are still going on and have been reported in several recent symposia.[9-13]

In addition to the work directed specifically towards problems related to neurological function and mental disorders, there have been an increasing number of basic biochemical studies on the nature of the enzymes involved. Amine oxidase activities have been demonstrated in a wide variety of animal and plant tissues, as well as in various bacteria and fungi. A number of these enzymes have been purified and partly characterized.

* S. S. Kety, in his preface to Reference 9 summarizes this aspect of the history of MAO and points out how this story is an excellent example of how scientific discoveries are made and how fundamental discoveries lead to practical therapies. As a criticism of targeted research, he wondered whether any current committee involved in programing research on mental illness would have funded Mary Hare's study of a liver enzyme with no known relation to nervous disease at the time.

Studies on the substrate and inhibitor specificity of these preparations showed that the simple classification into MAOs and DAOs is incorrect; indeed, the relative rates of activity towards monoamines, diamines, and polyamines differ in all preparations.* The complexity increased even further as new substrates (such as acetylspermidine [see Chapter 3] and lysine [see Chapter 8]) and new enzyme sources were tested. Similarly, the different preparations are affected to different degrees by various inhibitors. Even enzyme preparations from different tissues in the same organism or from the same tissue in different species are very different. In some cases it seems likely that a single tissue may have two different types of activity, as "monoamine oxidase A" and "monoamine oxidase B". An even more complicated picture emerges from the studies of Gorkin (see Chapter 19), in which the substrate specificity of a given enzyme can be changed by sulfhydryl oxidation. Other work has indicated that changes can be induced in the enzyme by the isolation procedures and by the presence or absence of phospholipids. Although some of these effects result from the use of impure preparations, the same picture has been found with homogeneous preparations and are clearly related to the characteristics of the enzyme protein.

The observation that each preparation has its own substrate and inhibitor specificity is most unusual. In other systems, enzymes from closely related sources usually have only relatively minor differences and usually have comparable substrate and inhibitor specificities. The availability of a series of closely related enzyme proteins with different substrate specificities affords an excellent opportunity for future investigators to correlate primary and secondary protein structure with specific enzyme activities. It would be of great interest to know whether these functionally related enzymes are related structurally.

From the biochemical point of view, one of the most interesting aspects of the AOs is the nature of the cofactors present and the mechanism of their action. Most AOs have flavin or metal cofactors. However, in general, the role of these cofactors in the mechanism of electron transfer is not clear and, in some of the AOs, the nature of the cofactors present is unknown or uncertain.

One of these cofactors is flavin adenine dinucleotide, which was shown to be present in mitochondrial MAO in covalent linkage as 8-α-cysteinyl-FAD. Extensive studies have been carried out on the nature of the interaction between this flavin cofactor and a variety of mechanism-based ("suicide") inhibitors.[9-13] In another study, both FAD and a heme were found to be cofactors for *Serratia* spermidine dehydrogenase (see Chapter 5).

A number of amine oxidases (AOs) do not contain flavin cofactors, but do contain stoichiometric amounts of copper, which are needed for activity.[9-13] Included in this group are AOs from *Aspergillus niger*, pea seedlings, bovine and pig plasma, and pig kidney. The stoichiometry, visible spectra, and EPR spectra of these enzymes have been extensively studied, but there is still complete uncertainty on the role of copper in the mechanism of electron transfer from the substrate to oxygen. Another area of uncertainty concerns the nature and function of the carbonyl group, which is found in many of the AOs and which accounts for the inhibition caused by carbonyl-binding reagents.

In summary, the articles in this volume, plus those in the many recent symposia, clearly indicate the many developments in this area in recent years. However, they also indicate that even though these enzymes have been studied for over 50 years, much more basic biochemical work is needed before we can understand the mechanism of action, physiological significance, and interrelationship of these functionally related series of enzymes. It seems likely that the present availability of methods for the amplification of enzyme proteins and of improved techniques for obtaining amino acid and gene sequences will markedly facilitate further studies.

* The problem of developing a satisfactory nomenclature for the AOs has not been solved and is discussed at greater length by Bachrach (Chapter 2 in this volume.)

ADDENDUM

Since this article was written, evidence has been presented that several copper-containing AOs contain a pyrroloquinoline quinone cofactor.[14]

REFERENCES

1. **Blaschko, H.,** Amine oxidase and amine metabolism, *Pharmacol. Rev.*, 4, 415, 1952.
2. **Blaschko, H.,** Enzymic oxidation of amines, *Br. Med. Bull.*, 9, 46, 1953.
3. **Blaschko, H.,** Amine oxidase, in *The Enzymes*, Vol. 8, 2nd ed., Boyer, P. D., Lardy, H., and Myrbäck, K., Eds., Academic Press, New York, 1963, 337.
4. **Buffoni, F.,** Histaminase and related amine oxidases, *Pharmacol. Rev.*, 18, 1163, 1966.
5. **Tabor, H.,** Metabolic studies on histidine, histamine, and related imidazoles, *Pharmacol. Rev.*, 6, 299, 1954.
6. **Tabor, H. and Tabor, C. W.,** Spermidine, spermine and related amines, *Pharmacol. Rev.*, 16, 245, 1964.
7. **Zeller, E. A.,** Oxidation of amines, in *The Enzymes*, Vol. 2, 1st ed., Boyer, P. D., Ed., Academic Press, New York, 1951, 536.
8. **Zeller, E. A.,** Diamine oxidases, in *The Enzymes*, Vol. 8, 2nd ed., Boyer, P. D., Lardy, H., and Myrbäck, K., Ed., Academic Press, New York, 1963, 314.
9. **Wolstenholme, G. E. W. and Knight, J., Eds.,** *Monoamine Oxidase and Its Inhibition*, Ciba Found. Symp. 39, Elsevier/North-Holland, Amsterdam, 1976, 415 pp.
10. **Youdim, M. B. H. and Paykel, E. S., Eds.,** *Monoamine Oxidase Inhibitors*, John Wiley & Sons, Chichester, 1981, 214 pp.
11. **Kamijo, K., Usdin, E., and Nagatsu, T., Eds.,** *Monoamine Oxidase: Basic and Clinical Frontiers*, Excerpta Medica, Amsterdam, 1982, 378 pp.
12. **Malmström, B. G., Andréasson, L. E., and Reinhammar, B.,** Copper-containing oxidases and superoxide dismutase, in *The Enzymes*, Vol. 12, 3rd ed., Boyer, P. D., Ed., Academic Press, New York, 1975, 511; see also Chapters 11 and 12 in this volume.
13. **Singer, T. P., von Korff, R. W., and Murphy, D. L., Eds.,** *Monoamine Oxidase: Structure, Function, and Altered Functions*, Academic Press, New York, 1979, 557 pp.; see also Chapter 21 in this volume.
14. **Ameyama, M. et al.,** *Agric Biol. Chem.*, 48, 561, 1984; Lobenstein-Verbeek, C. L. et al., *FEBS Lett.*, 170, 305, 1984.

Chapter 2

COPPER AMINE OXIDASES AND AMINES AS REGULATORS OF CELLULAR PROCESSES

Uriel Bachrach

TABLE OF CONTENTS

I. INTRODUCTION

It has been well established that polyamines and diamines play an important role in controlling growth and differentiation processes.[1-5] Cellular polyamine levels may be regulated by their synthesis, excretion, or degradation.[6-7] While the biosynthetic enzymes, ornithine decarboxylase and *S*-adenosyl-L-methionine decarboxylase have been well characterized,[8-11] the nature and properties of the catabolic enzymes are only poorly understood.

II. AMINE OXIDASES

The oxidation of amines, diamines, and polyamines is catalyzed by amine oxidases,[12] the classification of which may be based on the following criteria:

1. Substrate specificity
2. Mode of oxidation
3. Sensitivity to inhibitors
4. Structure of enzyme and cofactor requirement

The first three criteria are based on phenotypic properties of the enzyme. These properties are not absolute and may depend on environmental factors, or on the procedures employed for their assay. The fourth criterion, on the other hand, is much more characteristic and is based on the intrinsic properties of each of the respective enzymes.

A. Substrate Specificity

The easiest way to classify amine oxidases (AOs) could be based on their substrate specificities. AOs may thus be divided into monoamine oxidases (MAOs), diamine oxidases (DAOs), and polyamine oxidases (PAOs), according to their ability to catalyze the oxidation of monoamines, diamines, or polyamines, respectively. Indeed, the original division into two groups, mono- and diamine oxidases,[13] was based on this approach.

It soon became apparent[14] that this is an oversimplification, as most enzymes lack strict substrate specificity. Thus, the classical DAO also uses histamine as substrate and the enzyme has also been designated as histaminase (Table 1).

The classical plasma amine oxidase (SAO) oxidizes aromatic amines (such as benzylamine) and peptidyl lysine, in addition to polyamines.[15] When more sensitive analytic tools are used, the spectrum of substrates is even wider. Recent studies have shown that immobilized SAO attacks spermine, spermidine, cadaverine, and methylamine at the relative rate of 100, 20, and 4%, respectively.[16]

The separation of AOs according to their substrate specificity was even more confused when Gorkin and Tatyanenko[17] reported that the substrate specificity of mitochondrial MAO can be altered by treating the enzyme with peroxides of oleic acid. The enzyme, which usually oxidizes only monoamines, acquired a new substrate specificity and also oxidized polyamines (Table 1) (see Chapter 19).

Unfortunately, this is not the only example representing changes in substrate specificity by chemical modification of the enzyme. Okada et al.[18] treated DAO from *Micrococcus rubens* with 1-ethyl-3-(3-dimethylaminopropyl) carbodiimide. The modified enzyme lost most of its capacity to oxidize putrescine, but acquired the ability to oxidize monoamines such as methylamines. Monoamines are not oxidized by the native enzyme.

Table 1 summarizes some of the properties of various AOs. It may be seen that a MAO from human serum oxidizes mono- and polyamines. Most of the known DAOs catalyze the oxidation of diamines and histamine, while the dog small intestinal DAO also oxidizes benzylamine (Table 1). Human pregnancy DAOs (purified 50-fold) oxidize diamines, po-

Table 1
SUBSTRATE SPECIFICITY OF AMINE OXIDASES

Enzyme	Monoamines	Diamines	Polyamines	Other	Ref.
Monoamine oxidase					
Mitochondria, rat liver	+				17
Mitochondria, transformed	+		+		17
Human serum	+ +		+		19
Brain	+ +			Acetyl putrescine	20
Diamine oxidase					
Hog liver		+ +		Histamine	13
Pig small intestine	+	+ +		Histamine	21
Dog small intestine	+	+ +		Benzylamine	
Mouse liver		+ +	+		22
Amine oxidase					
Bovine serum			+ +	N^8-Acetyl spermidine	23
				Aromatic amines	
				Peptidyl lysine	24
Chick aorta			+ +	Peptidyl lysine	25
Dental pulp			+ +	Aromatic amines	26
Serum albumin		+ + +	+	Peptidyl lysine	27
Liver			+ +	N^1-Acetyl spermidine	28
Human seminal plasma		+ +	+ + +		29
Pregnancy serum		+ +	+ + +	Acetyl putrescine	30
Human placenta	+ +	+ +		Benzylamine	31

lyamines, and their acetyl derivatives.[30] It is very likely that more than one enzyme is present in the sera of pregnant women. Fetal bovine serum AO indeed contains a mixture of DAO and PAO (see Chapters 14 and 18).

Bacteria also exhibit AO activity (Table 2). In this system, enzymatic activities can be modulated by changing the composition of the growth medium. Since many of the enzymes are adaptive ones, their activity can be induced by adding the respective amine to the growth medium. Thus, *Serratia marcescens* can degrade spermidine or spermine when either amine is added to the growth medium.[34] Plant enzymes also lack substrate specificity, as illustrated in Table 3.

It may, therefore, be concluded that it would be rather difficult to characterize AOs by their substrate specificity.

B. Mode of Oxidation

Amine oxidases could also be characterized by their mechanism of oxidation. Most of the enzymes catalyze the oxidative deamination of terminal primary amino groups. Such enzymes include the classical SKDAO, SAO, and various bacterial enzymes (Figure 1). Another group of enzymes catalyze the oxidation of secondary amino groups leading to the cleavage of the polyamine molecule. This group of enzymes can be further subdivided by the nature of the oxidation product. Thus, liver PAO yields diaminobutane,[51] while various bacterial enzymes give rise to diaminopropane.

Plant AOs oxidize polyamines differently,[51] and yield cyclic compounds such as 1,3-aminopropyl-pyrroline (Figure 2).

These findings thus suggest that characterization of oxidation products would not lead to a reasonable enzyme taxonomy. The pathways are either very general or results may be complicated by secondary reactions.

C. Sensitivity to Inhibitors

Zeller[13] first proposed the use of inhibitor specificities as one of the criteria for AO classification. This view was also adopted by Malström et al.,[12] who defined MAOs as

Table 2
AMINE OXIDASES IN MICROORGANISMS

Source	Substrate: Putrescine	Spermidine	Spermine	Ref.
Pseudomonas aeruginosa		+ +	+ +	32
Corynebacterium pseudodiphtheriticum	+ +			
Mycobacterium smegmatis	+ +	+ +	+ +	33
Serretia marcescens	−	+ +	(+)	34
Micrococcus rubens	+ +	+ +		35
Neisseria perflava		+ +	+ +	36
Hemophilus parainfluenzae		+ +	+ +	37
Penicillium PO-1	+ +	+ +	+ +	38
Penicillium chrysogenum	+ +	+ +	+ +	39
Candida boidinii	+ +			40
Mucor fragilis		−	+ +	41
Fusarium oxysporum		+	+ +	
Cylindrocarpon didymum		+ +	+ +	
Aspergillus terreus		+ +	+ +	42
Aspergillus niger	+ +			43
Anacystis nidulans		+ +	+ +	44

Table 3
AMINE OXIDASES IN PLANTS

Source	Substrate: Putrescine	Spermidine	Spermine	Ref.
Pea seedlings	+ + +	+ +	−	45
Barley leaves		+ +	+	46
Oat leaves		+ +	+ +	47
Soy bean seedlings	+ + +	+	−	48
Maize shoots		+ +	+ +	49
Euphorbia latex	+ +			50

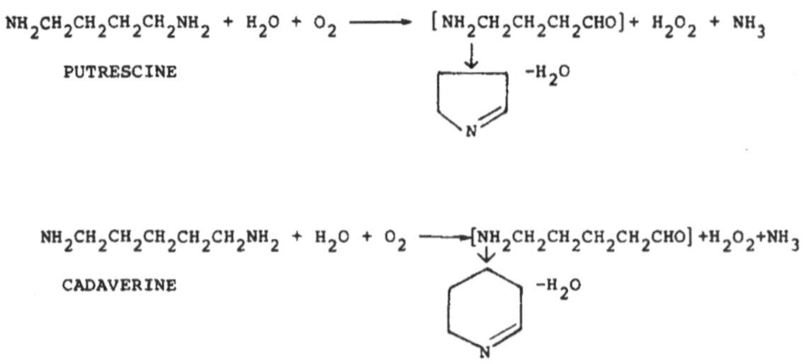

FIGURE 1. Oxidation of diamines by hog kidney oxidase.

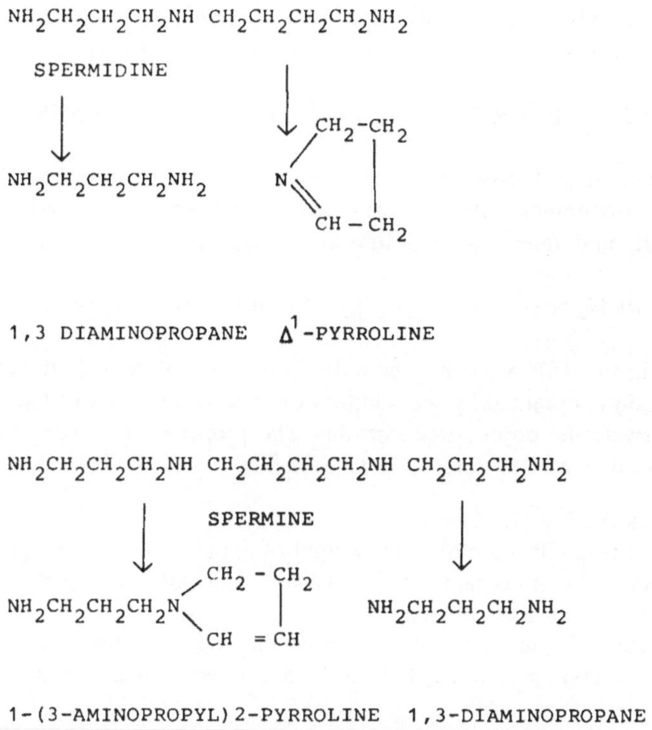

FIGURE 2. Oxidation of diamines and polyamines by pea seedling oxidase.

enzymes which are inhibited by 2-phenylcyclopropylamine. DAOs, on the other hand, are inhibited by hydrazine and semicarbazide. In many cases, the specificity of the sensitivity to the drugs is not absolute. Thus, aminoguanidine, which was regarded as a typical specific inhibitor of DAO, also inhibits the oxidation of polyamines. Rojansky et al.[52] reported that injecting aminoguanidine into pregnant rats led to an increase in urinary diamines and polyamines. Seiler et al.[53] extended these studies and demonstrated the inhibition of AO of newborn calf serum by aminoguanidine. It should be noted, however, that PAO required 5×10^{-5} *M* aminoguanidine for 50% inhibition. SKDAO, on the other hand, was inhibited by the drug at a concentration of 7×10^{-7} *M*. Seiler et al.[53] also reported that DAO of the small intestine of mice is inhibited by aminoacetonitrile, which was regarded as a typical inhibitor of SAO.[54]

It thus appears that AOs cannot be characterized by their sensitivity to inhibitors (see Chapter 13 and 21).

D. Structure of Enzyme and Cofactors

The above-described data imply that new approaches must be used for a better nomenclature of the AOs. The structure of the enzyme and its cofactor requirement may be an alternative basis for nomenclature. Many of the AOs contain copper ions. These include:

- Bovine plasma AO[55]
- Pig plasma AO[56]
- Pig kidney DAO[57]
- Pea seedling DAO[58]
- *Aspergillus niger* AO[43]
- *Euphorbia latex* DAO[50]

In addition, the following enzymes also appear to contain Cu^{++}: chick bone tissue AO,[25] pig liver mitochondria AOs,[59] human placental enzymes,[60] and soybean seedlings.[48]

III. COPPER-CONTAINING AMINE OXIDASES

The copper-containing AOs are enzymes which catalyze the oxidative deamination of mono-, di-, and polyamines with the formation of stoichiometric amounts of aldehydes, hydrogen peroxide, and ammonia according to the equation:

$$RCH_2NH_2 + O_2 + H_2O \rightarrow RCHO + H_2O_2 + NH_3$$

The copper ions in the AOs are firmly bound. On removal of copper, the enzymic activity is lost. Reactivation is obtained by the addition of suitable amounts of Cu^{++}. Other metal ions do not reactivate the copper-free enzyme. The properties of some copper-containing AOs are given below (see Chapters 6, 7, 10, and 12).

A. Bovine Plasma Amine Oxidase

The crystalline enzyme has a molecular weight of about 170,000 and apparently consists of two identical subunits. It contains one to two molecules of Cu^{++} per enzyme molecule, and a pyridoxal phosphate-like substance.[55]

The enzyme catalyzes the oxidation of spermidine and spermine. Benzylamine and butylamine, N^8-monoacetyl spermidine, but not N^1-acetylspermidine, serve as substrate[23] (see Chapter 10).

B. Pig Plasma Amine Oxidase

The crystalline enzyme has a molecular weight of approximately 195,000 and contains two molecules of Cu^{++} per enzyme molecule.

The enzyme catalyzes the oxidation of monoamine, histamine, and benzylamine. Diamines and polyamines are not metabolized.[56]

C. Pig Kidney Diamine Oxidase

The purified enzyme has a molecular weight of approximately 170,000 and contains two molecules of Cu^{++} and one molecule of pyridoxal phosphate-like substance per mole of enzyme. It catalyzes the oxidation of diamines and histamine[61] (see Chapter 7).

D. Pea Seedling Diamine Oxidase

This enzyme contains 0.08 to 0.09% Cu^{++}, firmly bound to the protein. It catalyzes the oxidation of putrescine, cadaverine, lysine, and benzylamine[62] (see Chapter 6).

E. *Aspergillus niger* Amine Oxidase

The crystallized enzyme has a molecular weight of 273,000 and contains three molecules of Cu^{++} per enzyme molecule. It also contains a pyridoxal-like material.[63]

The enzyme oxidizes monoamines, 2 to 6 carbon chains, benzylamine, and histamine and aliphatic diamines (see Chapter 3).

F. *Euphorbia latex* Amine Oxidase

The enzyme is purified 144-fold, has a molecular weight of 144,000, and is composed of two identical subunits. It contains two Cu^{++} ions per dimer. The enzyme oxidizes putrescine and cadaverine, but not histamine, spermidine, or spermine[50] (see Chapter 6).

IV. REGULATION OF CELLULAR FUNCTIONS

Amine oxidases (AOs) catalyze the oxidation of amines or polyamines and thereby regulate their cellular levels. Growth and differentiation processes are closely related to cellular

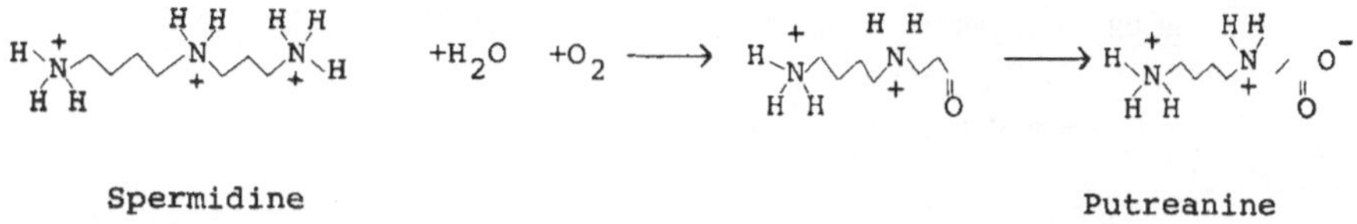

FIGURE 3. Formation of putreanine from spermidine.

polyamine levels; hence, these vital processes may be regulated by modulating cellular amine levels. In many cases, the oxidation of the amines is followed by additional enzymatic reactions, yielding compounds of physiological significance. Thus, in the brain, putrescine is converted into γ-aminobutyric acid.[64] This has also been demonstrated during the process of differentiation of chick embryo retina. During the early stages of development, putrescine is synthesized. Upon differentiation, when growth is arrested, the diamine is converted into γ-aminobutyric acid,[65] which serves as a neurotransmitter and also inhibits protein synthesis. The formation of β-alanine from the aminopropyl moiety of spermidine by bacteria is another example of the conversion of a polyamine or diamine into an amino acid.[66,67]

A similar conversion is the formation of putreanine from spermidine[68] or spermic acid (Figure 3) from spermine.[69] The initial step of these reactions is catalyzed by the respective AO. The aldehydes thus formed are further oxidized by aldehyde dehydrogenases. These two enzymes, namely, the AO and the aldehyde dehydrogenase, may be located in different tissues or in separate cellular compartments.[53,70] The distribution of the various enzymes in the different tissues may lead to the accumulation of the products at specific regions of the brain or liver. Although little is known about the biological function of putreanine, isoputreanine, or spermic acid, it is obvious that these derivatives lack the cationic nature which is so typical of diamines or polyamines. Therefore, their association with nucleic acids or membrane constituents such as phospholipids would be altered. It is therefore most likely that the conversion of diamines into the two respective amino acids may have a far-reaching consequence as to the structure of nucleic acids and on the behavior of the cellular membrane.

In many cases, the primary products of amine oxidation, namely the aldehydes, are not further metabolized. These aminoaldehydes may have a regulatory function (see below).

Very often, the oxidation of amines is studied in in vivo experiments and the enzymes which catalyze the reactions are not well characterized. Moreover, even in vitro studies are sometimes based on reactions catalyzed by crude enzymes. These preparations may contain a mixture of enzymes which are not well defined. Evidence is available, however, that at least the following reactions are catalyzed by copper-containing AOs:

1. Formation of aminoaldehydes by PAO
2. Oxidation of diamines by DAO
3. Formation of cyclic derivatives in plants by AOs
4. Oxidation of peptidyl lysine

In addition, perturbation of AOs has been associated with the following pathological conditions (these may also be due to copper-containing enzymes):

1. Cancer
2. Uremia
3. Behavior
4. Pregnancy

A. Formation of Aminoaldehydes

The presence of an enzyme in sheep and bovine sera which catalyzes the oxidative deamination of polyamines was first reported by Hirsch.[71] This enzyme has been purified

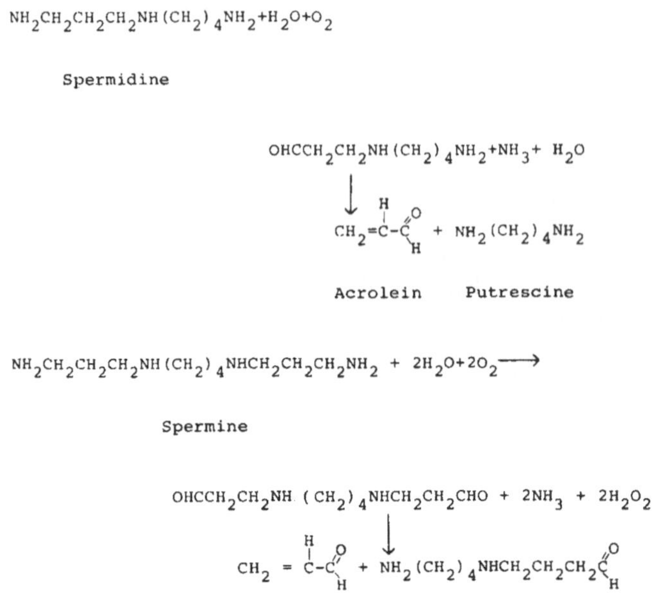

FIGURE 4. Formation of acrolein from polyamines.

and crystallized from bovine plasma and has been shown to act on spermidine and spermine as follows[72] (Figure 4).

The aldehydes produced were characterized by reduction with sodium borohydride and compared with the corresponding synthetic amino alcohol.[72] The synthesis of the diacetal derivatives of the amino aldehydes has been described, and their biological activity has been documented.[73] The primary oxidation products of the polyamines are rather labile, and may undergo secondary spontaneous degradation processes.[74,75] Acrolein, which is one of these secondary degradation products,[76] is cytotoxic, but its mode of action and spectrum of activity differ from those of the aminoaldehydes.[77] The latter are known to interact with nucleotides[78] or with DNA,[79] impairing biological activity such as genetic transformation of *B. subtilis*[80] and of T4 DNA.[81] The interaction of oxidized polyamines with DNA also results in conformational changes which are manifested by stabilization against thermal denaturation and enzymatic degradation of the polynucleotide chains.[82] Hirsch and Dubos were the first to demonstrate the effect of oxidized polyamines or tubercle bacilli.[83] These findings were subsequently confirmed and extended to include other bacteria such as *Staphylococcus aureus* and *Escherichia coli*.[84] Oxidized polyamines are also toxic to parasites such as *Trypanosoma equiperdum*,[85] coliphages,[81] animal,[86] and plant viruses.[87] Oxidized polyamines are also toxic to tumor cells.[88]

It is difficult to ascertain whether or not oxidized polyamines play a role in defending a host against infection. Hirsch and Dubos[83] suggested that natural resistance to infection by tubercle bacilli may be due to the presence of both polyamines and the oxidase in the infection-resistant lung.

Immunoregulation is another important biological function which has been attributed to polyamines and their oxidases. According to Byrd et al.,[89] polyamines have a reversible suppressive effect on in vitro parameters of immunity. When murine lymphocytes were incubated with irradiated DAB/2 spleen cells in the presence of polyamines and bovine serum, the synthesis of DNA was significantly reduced, and the cytolytic response was also impaired. Spermine and spermidine without bovine serum were not inhibitory. It was concluded that the polyamines were converted into an inhibitory oxidation product, or alternatively, triggered the release of an active molecule present in the bovine serum. This inhibitory phenomenon may be related to endogenous tissue-specific inhibitors of cell proliferation described as "chalones". The chalone concept deals with the growth regulation of normal tissues, but there is some evidence that tumor cells may synthesize and respond to the chalones of their tissue origin.[90] Barfod[91] described the properties of such a chalone from JB-1 ascites tumors and identified spermine as one of its components. It has been shown that this chalone arrests JB-1 cells at the G1-S border and in the G2 phase. Gonzales and Verly[92] also isolated an inhibitor specific for mammary cells and suggested the involvement of a polyamine or its oxidation product in the inhibitory process. Similarly, a basic polypeptide, which presumably contains polyamines, has been isolated from fibroblasts and lymphocytes. The association of chalones with polyamines and the inhibition of lymphocyte proliferation by polyamines and their oxidation products were the subject of recent reviews by Barfod[91] and Gaugas.[93]

Gaugas and Curzen[94] tested the possibility that polyamines or their oxidation products are involved in the survival of the fetoplacental unit despite an immunological incompatibility of the mother. They found that polyamines interact with pregnancy serum and elicit potent suppression of lymphocyte transformation in vitro. It is of special interest that the maternal part of the placenta has a remarkably high amine oxidase (AO) activity, unlike the fetal part, which contains low enzyme activity.[95] It is conceivable that this distribution of enzyme and substrate leads to the formation of a "safety belt" around the fetoplacental unit and prevents the formation of undesired immunological responses against the fetus. The recent findings of Ishizuka et al.,[96] that polyamines inhibit the blastogenesis of concanavalin-A stimulated lymphocytes, is in line with the above-described data.

B. Oxidation of Diamines

Numerous studies indicate that the activity of DAOs increases in rapidly growing tissues and particularly in tumors.[97,98] In human and rat gestation, there is a striking increase in blood DAO activity.[99,100] In baby hamster[101] and in cultured glioma cells,[102] DAO activity increases in early phases of growth. Desiderio et al.[103] have recently demonstrated the increase in DAO activity in regenerating liver and kidney undergoing compensatory hypertrophy. It has also been demonstrated that DAO activity is elevated in hypertrophic heart of spontaneously hypertensive rats.[104] The enzyme was aminoguanidine sensitive, but it has not been established whether or not it contained copper. Mondovì et al.[105] have recently described the effect of immobilized DAO on the growth of Ehrlich ascites cells in Swiss mice. Further details on the involvement of DAOs in cancer and in pregnancy are given below.

C. Formation of Cyclic Derivatives in Plants

The formation of pyrroline and its derivatives during polyamine oxidation by plants may be of physiological significance. A number of plant alkaloids consist of pyrrole or pyrrolidine. These include nicotine, which constitutes 95% of the alkaloids of *Nicotina tabacum*, and strychnine, which is an alkaloid containing a pyrrolidine group. It has indeed been demonstrated that radioactive putrescine is incorporated into nicotine[106] or hyoscyamine in *Atropa belladona* and *Datura stramonium*.[107,108] Hygrine, on the other hand, is formed during the oxidation of *N*-methylputrescine by plant DAO.[109] Likewise, piperidine may be formed during the oxidation of cadaverine. Radioactive cadaverine was indeed shown to be incorporated into lupinine,[110] sparteine,[111] lupenine,[112] matrine,[113] and anabasine.[114]

D. Oxidation of Peptidyl Lysine

The formation of cross-links in elastin and collagen may also be mediated by copper-containing AOs. Traub and Piez[115] found that lysine and hydroxylysine residues in these proteins are oxidized to the respective aldehydes by AOs. These products undergo spontaneous aldol and aldimine condensations to yield molecular cross-links. The reported activity of copper-containing AOs in connective tissues may lead to the formation of similar peptidyl lysine cross-links[25] (see Chapters 7 and 8). It is also possible that the synthesis of the so-called polyamine conjugates is catalyzed by AOs. Thus, Chan et al.[116] described the occurrence of polyamine-containing peptides in human amniotic fluid throughout gestation. Putrescine-bound peptides were mainly of molecular weight between 1000 to 10,000, while spermidine was bound to larger peptides. One should consider the possibility that these conjugates are the result of aldol or aldimine condensations between oxidized polyamines or oxidized lysine residues of the peptides. The formation of Schiff bases between oxidation products and amino groups cannot be ruled out.

Copper-containing enzymes may also be involved in the syndromes described in the sections below.

1. Cancer

The relationship between AO activity and neoplastic growth has been discussed extensively in a recent excellent review by Scalabrino and Ferioli[117] (see Chapter 17). They correctly pointed out that DAO in various tissues oxidizes not only diamines, but also histamine. The extensive data in the review may be summarized as follows.

Baylin et al.[118] first reported an increase in DAO activity in the sera and autopsy material of patients with medullary carcinoma. It has been suggested that measuring the activity of the enzyme in the serum may be useful for the detection of metastases and for evaluating the efficacy of surgical treatment.[119] Indeed, after removal of a localized tumor, enzyme activity returned to normal levels, but remained elevated in the sera of patients with metastases. It appears that this phenomenon is not general and the sera of patients with occult thyroid tumors do not exhibit a consistent elevated DAO activity.[120] Baylin et al.[118] demonstrated that the elevated activity of the enzyme in the serum originates from the tumor which excretes DAO. It has also been demonstrated, by administration of aminoguanidine, that the turnover of the enzyme in the serum is approximately 2 days.

Subsequently, Lin et al.[121] found elevated DAO activities in the ascites fluids from cancer of the ovary, breast, stomach, colon, and lung. Lin et al.[122] measured DAO activity in more than 400 effusion fluids collected from 162 cancer patients and found a good correlation between increased DAO activity and the production of carcinoembryonic antigens in patients with cancer of the colon and stomach. On the other hand, Küsche et al.[123] refuted the generality of the phenomenon and showed that DAO activity was not elevated in the GI tract of patients with adenocarcinoma. At present, the optimism of using DAO as a prognostic marker for small cell carcinoma of the lung remains doubtful.

In their review, Scalabrino and Ferioli[117] also summarized experiments demonstrating alterations in DAO activities in in vitro studies. Quash et al.[101] reported that 24-hr-old normal rat kidney cells have about half the DAO activity of their counterparts transformed by Rous sarcoma virus. Perin et al.[124] also found elevated DAO activity in 4-DAB hepatoma and Yoshida ascites hepatoma cells. It has been speculated[102] that the increase in DAO activity coincides with elevation in ODC activity, which is typical for neoplastic growth. This has been demonstrated for cultured glioma cells[102] and other brain tumors.[98] Moreover, Mondovì et al.[98] provided evidence for a relationship between AO activity and the degree of malignancy in astrocytomas. Again, this elevated enzyme activity is not general for brain tumors, and medulloblastoma, which is one of the most malignant tumors, showed low AO activity.

It may thus be concluded that increased DAO activity is not typical for all cancer cells. Its monitoring may have a limited prognostic value for certain cancer cells at definite stages of growth. It should also be stressed that results depend on the analytical procedures employed. It should be remembered that the primary oxidation product, namely, Δ^1-pyrroline, may undergo secondary metabolism and thus introduce a serious error if the isotope method[125] is employed.

2. Uremia

The levels of both aliphatic and aromatic amines have been shown to be elevated in the plasma of patients with renal failure.[126] It has been postulated that certain amines may contribute to the toxicity of the uremia. Reduction in AO activity may cause the accumulation of the toxic amines which could contribute to the pathogenesis of the uremic syndrome.[127]

Wang et al.[128] determined the activities of mono- and diamine oxidase in the tissues of uremic rats using benzylamine and putrescine as substrates. They showed that DAO activity in uremic rats was decreased in kidney, increased in plasma, and unchanged in liver and muscle of the uremic rats. MAO activity was reduced in both kidney and muscle. It is likely that the decrease in AO activity in the kidney might cause the accumulation of amines in uremic urine. It should also be noted that the activity of rabbit kidney AO is inhibited in vitro by urea, in a concentration similar to that present in human uremic plasma.[129] Hence, in vivo, high urea concentrations may also have inhibitory effects on AO activity.

At present, the kidney DAO of uremic patients has not been purified and characterized. Yet, if it resembles the classical SKDAO, then the physiopathology of uremia may be associated with a copper enzyme. (See also a review by Campbell et al.[130])

3. Behavior

Biogenic amines play a crucial role in brain physiology, and AO, (mainly MAO) activity might affect behavior. These enzymes degrade various biogenic amines and their metabolites, and thus regulate the concentration and synthesis of these neurotransmitter amines in presynaptic terminals.[131] Early studies indicated that MAO activity is reduced in some psychiatric patients and their close relatives.[132] More recently, low platelet MAO and SAO activities were associated with psychological characteristics such as increased sensation seeking.[133]

Sostek et al.[134] examined the relationship between AO activity and behavior in infants. Enzyme activities were determined in the cord blood after separating platelets from the plasma, using benzylamine as substrate. It has been demonstrated that infants with low MAO were more motorically active than those with higher enzyme activity. The AO has not been purified and characterized. Yet, if we accept that it resembles other SAOs, then we may assume that we are dealing with a copper-containing enzyme (see Chapters 22 and 23).

4. Pregnancy

Pregnancy in the human is accompanied by an increase in the serum levels of DAO.[15] The activity of this enzyme increases 500- to 1000-fold and reaches a plateau at about 16 weeks of gestation.[135] Maternal serum DAO assays were suggested for the determination of pregnancy.[136] Moreover, Elmfors et al.[137] proposed the use of AO assays for the determination of gestational age and the predicted date of delivery. It is very likely that this enzyme originates from the decidual elements of the placenta. For further details, see recent reviews by Henningsson et al.[138] (see Chapters 14, 17, and 18).

V. AMINE OXIDASES AS ANALYTICAL TOOLS

A. Diamines

Diamines may be oxidized to pyrroline by hog kidney DAO or by bacterial enzymes. *o*-Aminobenzaldehyde reacts with Δ^1-pyrroline to yield 2,3-trimethylene-1,2-dihydroquinazolium, which can be measured spectrophotometrically at 435 nm (E_{435} = 1.860 mol^{-1} Em^{-1}). The method permits the determination of 5 to 10 nmol of the diamine. Recently, Sakamoto and Samejima[139] prepared antibodies against 2,3-trimethylene-4-quinazolone. The latter was obtained by oxidation of 2,3-trimethylene-1,2-dihydroquinozolium with chromium trioxide. This permitted a development of an enzyme immunoassay as follows: Δ^1-pyrroline formed during the oxidation of putrescine was treated with chromium trioxide and the oxidation product was then extracted with benzene. The benzene layer was then treated with HCl and the extract was neutralized and incubated with immobilized anti-2,3-trimethylene-4-quinazolone antibodies. The quantity of free antibody was then assayed by adding peroxidase bound to 2,3-trimethylene-4-quinazolone and determining the luminescence. This method, which is quite laborious, permits the detection of 15 to 20 pmol of putrescine.

B. Polyamines

Spermidine and spermine are stoichimetrically oxidized by plasma AO to aminoaldehydes. These end products can be assayed by coupling with resorcinol and measuring the fluorescence at 520 nm (excitation at 366 nm). This method has the sensitivity of 200 nmol.[140] Another method is based on the formation of a blue color by incubating with *N*-methyl-2-benzothiazolone hydrazone. This method has the sensitivity of 3 to 5 nmol.[141] Other methods are based on the assay of hydrogen peroxide formed during the oxidation. This can be accomplished by adding 4-aminoantipyrine, phenol and peroxidase,[41] *p*-hydroxyphenylacetic acid and peroxidase,[142] indigo disulfonate and peroxidase,[142] *N,N*-dimethyaniline, aminoantipyrine, and peroxidase.[38] Immobilized beef plasma AO may be used for this procedure.[16] Bacterial enzymes may be used instead of plasma AO. Thus, spermidine is oxidized by *Serratia marcescens* to yield Δ^1-pyrroline.[141] This product can be assayed as described above. *Penicillium chrysogenum* IFO 4266 and *Aspergillus terreus* IFO 6346 also possess PAO activity,[39] while putrescine oxidase activity was detected in extracts of *Micrococcus rubens*.[39] These enzymes permit the detection of 10 nmol of polyamines. It is very likely that the enzymatic procedures could be improved and perfected. They eventually may be used as a simple and sensitive analytical tool for the detection and quantitation of di- and polyamines.

REFERENCES

1. **Tabor, H. and Tabor, C. W.,** Spermidine, spermine and related amines, *Pharmacol. Rev.*, 16, 245, 1964.
2. **Cohen, S. S.,** *Introduction to the Polyamines*, Prentice-Hall, Englewood Cliffs, N.J., 1971.
3. **Bachrach, U.,** *Function of Naturally Occurring Polyamines*, Academic Press, New York, 1973.
4. **Jänne, J., Raina, A., and Siimes, M.,** Spermidine and spermine in rat tissues at different ages, *Acta Physiol. Scand.*, 62, 352, 1964.
5. **Heby, O.,** Role of polyamines in the control of cell proliferation and differentiation, *Differentiation*, 19, 1, 1981.
6. **Russell, D. and Snyder, S. H.,** Amine synthesis in rapidly growing tissues: ornithine decarboxylase activity in regenerating rat liver, chick embryo, and various tumors, *Proc. Natl. Acad. Sci. U.S.A.*, 60, 1420, 1968.
7. **Melvin, M. A. L. and Keir, H. M.,** Excretion of polyamines from mammalian cells in culture, in *Polyamines in Biochemical Research*, Gaugas, J. M., Ed., John Wiley & Sons, New York, 1980, 363.
8. **Bachrach, U.,** The induction of ornithine decarboxylase in normal and neoplastic cells, in *Polyamines in Biochemical Research*, Gaugas, J. M., Ed., John Wiley & Sons, New York, 1980, 81.
9. **Russell, D. H., Ed.,** *Polyamines in Normal and Neoplastic Cells*, Raven Press, New York, 1973.
10. **Russell, D. H. and Durie, B. G. M.,** *Polyamines as Biochemical Markers of Normal and Malignant Growth*, Raven Press, New York, 1978.
11. **Zappia, V., Cacciapuoti, G., Pontoni, G., Della Ragione, F., and Carteni-Farina, M.,** Mechanism of enzymatic propylamine transfer reactions, in *Advances in Polyamine Research*, Vol. 3, Caldarera, C. M., Zappia, V., and Bachrach, U., Eds., Raven Press, New York, 1981, 39.
12. **Malmström, B. G., Andréasson, L-E., and Reinhammar, B.,** Copper-containing oxidases and superoxide dismutex, in *The Enzymes*, Vol. 12, Boyer, P. D., Ed., Academic Press, New York, 1975, 507.
13. **Zeller, E. A.,** Diamine-oxidase, *Adv. Enzymol.*, 2, 93, 1942.
14. **Blaschko, H.,** Amine oxidase and amine metabolism, *Pharmacol. Rev.*, 4, 415, 1952.
15. **Kapeller-Adler, R.,** *Amine Oxidases and Methods for their Study*, Wiley-Interscience, New York, 1970.
16. **Isobe, K., Tani, Y., Yamada, H., and Hiromi, K.,** Determination of polyamines with immobilized beef plasma amine oxidase, *Agric. Biol. Chem.*, 44, 615, 1980.
17. **Gorkin, V. Z. and Tatyanenko, L. V.,** "Transformation" of mitochondrial monoamine oxidase into a diamine oxidase-like enzyme *in vitro*, *Biochem. Biophys. Res. Commun.*, 27, 613, 1967.
18. **Okada, M., Kawashima, S., and Imahori, K.,** Mode of inactivation of putrescine oxidase by 1-ethyl-3(3-dimethylaminopropyl) carbodiimide or metal ions, *J. Biochem.*, 88, 481, 1980.
19. **McEwen, C. M., Jr. and Cohen, J. D.,** An amine oxidase in normal human serum, *J. Lab. Clin. Med.*, 62, 766, 1963.

20. **Seiler, N. and Al-Therib, M. J.**, Putrescine catabolism in mammalian brain, *Biochem. J.*, 144, 29, 1974.
21. **Kusche, J., Schmidt, J., Schmidt, A., and Lorenz, W.**, Diamine oxidases in the small intestine of rabbits, dogs and pigs: separation from soluble monoamine oxidases and substrate specificity of the enzymes, *Agents Actions*, 5, 440, 1975.
22. **Syatkin, S. P. and Galev, Yu, V.**, Oxidation of putrescine, spermidine and spermine by mouse liver diamine oxidase, *Biokhimiya*, 42, 783, 1977.
23. **Tabor, C. W., Tabor, H., and Rosenthal, S. M.**, Purification of amine oxidase from beef plasma, *J. Biol. Chem.*, 208, 645, 1954.
24. **Oda, O., Manabe, T., and Okuyama, T.**, Oxidation of ϵ-amino group of lysyl peptides by bovine serum amine oxidase, *J. Biochem.*, 89, 1317, 1981.
25. **Rucker, R. B. and O'Dell, B. L.**, Connective tissue amine oxidase. I. Purification of bovine aorta amine oxidase and its comparison with plasma amine oxidase, *Biochim. Biophys. Acta*, 235, 32, 1971.
26. **Nakono, G., Harada, M., and Nagatsu, T.**, Purification and properties of an amine oxidase in bovine dental pulp and its comparison with serum amine oxidase, *Biochim. Biophys. Acta*, 341, 366, 1974.
27. **Al-Naji, A. S. and Clarke, D. E.**, Amine oxidase in commercial preparations of bovine serum albumin, *Life Sci.*, 32, 635, 1983.
28. **Hölttä, E.**, Oxidation of spermidine and spermine in rat liver: purification and properties of polyamine oxidase, *Biochemistry*, 16, 91, 1971.
29. **Jänne, J., Hölttä, E., Haaranen, P., and Elfving, K.**, Polyamines and polyamine metabolizing enzyme activities in human semen, *Clin. Chim. Acta*, 48, 393, 1973.
30. **Gahl, W. A. and Pitot, H. C.**, Acetylated polyamines as substrates for human pregnancy serum diamine oxidase, *Life Sci.*, 29, 2177, 1981.
31. **Bardsley, W. G., Crabbe, J. C., and Scott, I. V.**, The amine oxidase of human placenta and pregnancy plasma, *Biochem. J.*, 139, 169, 1974.
32. **Razin, S., Gery, I., and Bachrach, U.**, The degradation of natural polyamines and diamines by bacteria, *Biochem. J.*, 71, 551, 1959.
33. **Roulet, F. and Zeller, E. A.**, Über die Diamin-Oxydase der Smegmabazillen, *Helv. Clin. Acta*, 28, 1326, 1945.
34. **Bachrach, U.**, Spermidine oxidase from *Serratia marcercens*, *J. Biol. Chem.*, 237, 3443, 1962.
35. **Yamada, H., Tanaka, A., and Ogata, K.**, Putrescine oxidase of *Micrococcus rubens*, *Mem. Res. Inst. Food Sci.*, 26, 1, 1965.
36. **Weaver, R. H. and Herbst, E. J.**, The oxidation of polyamines by *Neisseria perflava*, *J. Biol. Chem.*, 231, 647, 1958.
37. **Weaver, R. H. and Herbst, E. J.**, Metabolism of diamines and polyamines in microorganisms, *J. Biol. Chem.*, 231, 637, 1958.
38. **Kobayashi, Y. and Horikoshi, K.**, Production of extracellular polyamine oxidase by Penicillum sp. No PO-1, *Agric. Biol. Chem.*, 45, 2943, 1981.
39. **Isobe, K., Tani, Y., and Yamada, H.**, Differential determination procedure for putrescine, spermidine and spermine with polyamine oxidase from fungi and putrescine oxidase, *Agric. Biol. Chem.*, 45, 727, 1981.
40. **Haywood, G. W. and Large, P. J.**, Microbial oxidation of amines, *Biochem. J.*, 199, 187, 1981.
41. **Yamada, H., Isobe, K., and Tani, Y.**, Oxidation of polyamines by fungal enzymes, *Agric. Biol. Chem.*, 44, 2469, 1980.
42. **Isobe, K., Tani, Y., and Yamada, H.**, Polyamine oxidase of fungi. IV. Kinetic properties of fungal polyamine oxidases and their application to differential determination of spermine and spermidine, *Agric. Biol. Chem.*, 44, 2955, 1980.
43. **Yamada, H., Adachi, O., and Ogata, K.**, Amine oxidases of microorganisms. I. Formation of amine oxidases by fungi, *Agric. Biol. Chem.*, 29, 117, 1965.
44. **Ramakrishna, S., Guarino, L., and Cohen, S. S.**, Polyamines of *Anacystis nidulans* and metabolism of exogenous spermidine and spermine, *J. Bacteriol.*, 134, 744, 1978.
45. **Kenten, R. H. and Mann, P. J. G.**, The oxidation of amines by extracts of pea seedlings, *Biochem. J.*, 50, 360, 1951.
46. **Smith, T. A. and Bickley, D. A.**, Further properties of the polyamine oxidase of barley leaves, *Phytochemistry*, 13, 2437, 1974.
47. **Smith, T. A.**, Polyamine oxidase from barley and oats, *Phytochemistry*, 15, 633, 1976.
48. **Suzuki, Y.**, Some properties of amine oxidase of soybean seedlings, *Plant Cell Physiol.*, 14, 413, 1973.
49. **Hirasawa, E. and Suzuki, Y.**, Purification and properties of a polyamine oxidase from Zea mays, *Phytochemistry*, 14, 99, 1975.
50. **Rinaldi, A., Floris, G., and Finazzi-Agrò, A.**, Purification and properties of diamine oxidase from *Euphorbia latex*, *Eur. J. Biochem.*, 127, 417, 1982.
51. **Bachrach, U.**, Oxidation of polyamines and diamines, in *Polyamines in Biology and Medicine*, Morris, D. R. and Marton, L. J., Eds., Marcel Dekker, New York, 1981, 151.

52. **Rojansky, M., Neufeld, E., and Chayen, R.,** Excretion of polyamines by the pregnant rat following inhibition of diamine oxidase, *Biochem. Biophys. Acta*, 586, 1, 1979.
53. **Seiler, H., Knödgen, B., Bink, G., Sarhan, S., and Bolkenius, F.,** Diamine oxidase and polyamine catabolism, in *Advances in Polyamine Research*, Vol. 4, Bachrach, U., Kaye, A., and Chayen, R., Eds., Raven Press, New York, 1983, 135.
54. **Riceberg, L. J., Simon, M., Van Vunakis, H., and Abeles, R. H.,** Effects of aminoacetonitrite, an amine oxidase inhibitor, on mescaline metabolism in the rabbit, *Biochem. Pharmacol.*, 24, 119, 1975.
55. **Yamada, H. and Yasunobu, K. T.,** Monoamine oxidase. I. Purification, crystallization and properties of plasma monoamine oxidase, *J. Biol. Chem.*, 237, 1511, 1962.
56. **Buffoni, F. and Blaschko, H.,** Benzylamine oxidase and histaminase. Studies of a crystalline preparation of the amine oxidase of pig plasma, *Biochem. J.*, 89, 111p, 1963.
57. **Mondovì, B., Rotilio, G., Costa, M. T., Finazzi-Agrò, A., Chiancone, E., Hansen, R. E., and Beinert, H.,** Diamine oxidase from pig kidney. Improved purifications and properties, *J. Biol. Chem.*, 242, 1160, 1967.
58. **Werle, E., Trautschold, I., and Aures, D.,** Reinigung und Charakterisierung der Diamin-Oxydase aus Erbsen, *Hoppe Seyler's Z. Physiol. Chem.*, 326, 200, 1961.
59. **Chew, L. H., Carper, R., and Issaq, H. J.,** Effects of metal substitution in pig liver monoamine oxidase, *Biochem. Biophys. Res. Commun.*, 66, 217, 1975.
60. **Crabbe, J. C. and Bardsley, W. G.,** The inhibition of human placental diamine oxidase by substrate analogues, *Biochem. J.*, 139, 183, 1974.
61. **Mondovì, B., Rotilio, G., Costa, M. T., and Finazzi-Agrò, A.,** Diamine oxidase (pig kidney), in *Methods in Enzymology*, Vol. 17B, Tabor, H. and Tabor, C. W., Eds., Academic Press, New York, 1971, 735.
62. **Hill, J. M.,** Diamine oxidase (pea seedling), in *Methods in Enzymology*, Vol. 17B, Tabor, H. and Tabor, C. W., Eds., Academic Press, New York, 1971, 730.
63. **Yamada, H. and Adachi, O.,** Amine oxidase *(Aspergillus niger), in Methods in Enzymology*, Vol. 17B, Tabor, H. and Tabor, C. W., Eds., Academic Press, New York, 1971, 705.
64. **Seiler, N.,** Polyamine metabolism and function in brain, *Neurochem. Inter.*, 3, 95, 1981.
65. **DeMello, F. G., Bachrach, U., and Nirenberg, M.,** Ornithine and glutamic acid decarboxylase activities in the developing chick retina, *J. Neurochem.*, 27, 847, 1976.
66. **Bachrach, U.,** Metabolism of amines. III. The degradation and synthesis of γ-aminobutyric acid by *Pseudomonas aeruginosa*, *Biochem. J.*, 77, 417, 1960.
67. **Razin, S., Bachrach, U., and Gery, I.,** Formation of β-alanine from spermine and spermidine by *Pseudomonas aeruginosa, Nature (London)*, 181, 700, 1958.
68. **Kakimoto, Y., Nakajima, T., Kumon, A., Matsuoka, Y., Imaoka, N., Sano, I., and Kanazawa, A.,** Putreanine, *N*-(4-aminobutyl)-3-aminopropionic acid. An amino acid occurring uniquely in the mammalian central nervous system, *J. Biol. Chem.*, 244, 6003, 1969.
69. **Imaoka, N. and Matsuoka, Y.,** Isolation and identification of spermic acid, *N,N'-bis*(2-carboxyethyl)-1,4-diaminobutane, from bovine brain, *J. Neurochem.*, 22, 859, 1974.
70. **Seiler, N., Bolkenius, F. N., Knödgen, B., and Mamont, P.,** Polyamine oxidase in rat tissues, *Biochim. Biophys. Acta*, 615, 480, 1980.
71. **Hirsch, J. G.,** Spermine oxidase: an enzyme with specificity for spermine and spermidine, *J. Exp. Med.*, 97, 345, 1953.
72. **Tabor, C. W., Tabor, H., and Bachrach, U.,** Identification of the aminoaldehydes produced by the oxidation of spermine and spermidine with purified plasma amine oxidase, *J. Biol. Chem.*, 239, 2194, 1964.
73. **Bachrach, U. and Rosenkranz, H. S.,** Inhibition of bacterial macromolecular syntheses by polyamine POX-3, *Biochem. Pharmacol.*, 18, 243, 1969.
74. **Bachrach, U.,** Oxidized polyamines, *Ann. N.Y. Acad. Sci.*, 171, 939, 1970.
75. **Kimes, B. W. and Morris, D. R.,** Preparation and stability of oxidized polyamines, *Biochim. Biophys. Acta*, 228, 223, 1971.
76. **Alarcon, R. A.,** Acrolein. IV. Evidence for the formation of the cytotoxic aldehyde acrolein from enzymatically oxidized spermine or spermidine, *Arch. Biochem. Biophys.*, 137, 365, 1970.
77. **Bachrach, U., Don, S., and Wiener, H.,** Antivirus action of acrolein, glutaraldehyde and oxidized spermine, *J. Gen. Virol.*, 13, 415, 1971.
78. **Eilon, G. and Bachrach, U.,** Interaction of oxidized polyamines with DNA. III. Association with nucleosides, mono- and polynucleotides, *Biochim. Biophys. Acta*, 179, 464, 1969.
79. **Bachrach, U. and Eilon, G.,** Interaction of oxidized polyamines with DNA. I. Evidence for the formation of cross-links, *Biochim. Biophys. Acta*, 145, 418, 1967.
80. **Persky, S., Ephrati-Elizur, E., and Bachrach, U.,** Interaction of oxidized polyamines with DNA. II. Inactivation of transforming DNA, *Biochim. Biophys. Acta*, 149, 459, 1967.
81. **Bachrach, U. and Leibovici, J.,** Inhibitory effect of oxidized spermine on the multiplication of coliphage T5, *J. Mol. Biol.*, 19, 120, 1966.

82. **Bachrach, U. and Eilon, G.,** The effect of spermine and oxidized spermine on the enzymic degradation of DNA, *Biochim. Biophys. Acta*, 179, 494, 1969.
83. **Hirsch, J. G. and Dubos, R.,** The effect of spermine on tubercle bacilli, *J. Exp. Med.*, 95, 191, 1952.
84. **Bachrach, U. and Persky, S.,** Antibacterial action of oxidized spermine, *J. Gen. Microbiol.*, 37, 195, 1964.
85. **Tabor, C. W. and Rosenthal, S. M.,** Pharmacology of spermine and spermidine. Some effects on animals and bacteria, *J. Pharmacol. Exp. Ther.*, 116, 139, 1956.
86. **Katz, E., Goldblum, T., Bachrach, U., and Goldblum, N.,** Antiviral action of oxidized spermine: inactivation of certain animal viruses, *Isr. J. Med. Sci.*, 3, 575, 1967.
87. **Bachrach, U., Rabina, S., Loebenstein, G., and Eilon, G.,** Antiviral action of oxidized spermine-inactivation of plant viruses, *Nature (London)*, 208, 1095, 1965.
88. **Bachrach, U., Abzug, S., and Bekierkunst, A.,** Cytotoxic effect of oxidized spermine on Ehrlich arcites cells, *Biochim. Biophys. Acta*, 134, 174, 1967.
89. **Byrd, W. J., Jacobs, D. M., and Amoss, M. S.,** Synthetic polyamines added to cultures containing bovine sera, reversibly inhibit *in vitro* parameters of immunity, *Nature (London)*, 267, 621, 1977.
90. **Rytomaa, R. and Kiviniemi, K.,** Control of DNA duplication in rat chloroleukemia by means of the granulocytic chalone, *Eur. J. Cancer*, 4, 595, 1968.
91. **Barfod, N. M.,** Association of chalones with polyamines: evidence for a chalone from JB-1 ascites tumour cells as spermine complex, in *Polyamines in Biomedical Research*, Gaugas, J. M., Ed., John Wiley & Sons, New York, 1980, 321.
92. **Gonzalez, R. and Verly, W. G.,** Isolation of an inhibitor of DNA synthesis specific for normal and malignant mammary cells, *Proc. Natl. Acad. Sci. U.S.A.*, 73, 2196, 1976.
93. **Gaugas, J. M.,** Biogenic diamines and polyamines in support and in inhibition of lymphocytic proliferation, in *Polyamines in Biomedical Research*, Ganges, J. M., Ed., John Wiley & Sons, New York, 1980, 343.
94. **Gaugas, J. M. and Curzen, P.,** Polyamine interaction with pregnancy serum in suppression of lymphocyte transformation, *Lancet*, 1, 18, 1978.
95. **Morgan, D. M. L.,** Polyamine oxidation and human pregnancy, in *Advances in Polyamine Research*, Vol. 4, Bachrach, U., Kaye, A., and Chayen, R., Eds., Raven Press, New York, 1983, 155.
96. **Ishizuka, M., Takeuchi, T., and Umezawa, H.,** Studies of the effect of spermidine in enhancing cellular immunity, *Proc. Jpn. Acad.*, 58, 327, 1982.
97. **Perin, A., Sessa, A., and Desiderio, M. A.,** Induction of diamine oxidase activity in some processes of growth, in *Advances in Polyamine Research*, Vol. 4, Bachrach, U., Kaye, A., and Chayen, R., Eds., Raven Press, New York, 1983, 175.
98. **Mondovì, B., Riccio, P., Riccio, A., and Marcozzi, G. S.,** Amine oxidase activity in malignant human brain tumors, in *Advances in Polyamine Research*, Vol. 4, Bachrach, U., Kaye, A., and Chayen, R., Eds., Raven Press, New York, *1983*, 183.
99. **Kobayashi, Y.,** Plasma diamine oxidase titres of normal and pregnant rats, *Nature (London)*, 203, 146, 1967.
100. **Tryding, N. and Willert, B.,** Determination of plasma diamine oxidase (histaminase) in clinical practice. A comparison between biological method and a radiochemical micromethod, *Scand. J. Clin. Lab. Invest.*, 22, 29, 1968.
101. **Quash, G., Calogero, H., Fossar, N., Ferdinand, A., and Taylor, D.,** Modification of diamine oxidase activity *in vitro* by metabolites of asparagine and differences asparagine decarboxylation in normal and virus-transformed baby hamster kidney cells, *Biochem. J.*, 157, 599, 1976.
102. **Bachrach, U.,** Metabolism of polyamines by cultured glioma cells. Effect of asparagine on γ-aminobutyric acid concentrations, *Biochem. J.*, 188, 387, 1980.
103. **Desiderio, M. A., Sessa, A., and Perin, A.,** Induction of diamine oxidase activity in rat kidney during compensatory hypertrophy, *Biochim. Biophys. Acta*, 714, 243, 1982.
104. **Perin, A., Sessa, A., and Desiderio, A.,** Polyamine levels and diamine oxidase activity in hypertrophic heart of spontaneously hypertensive rats and of rats treated with isoproterenol, *Biochim. Biophys. Acta*, 755, 344, 1983.
105. **Mondovì, B., Gerosa, P., and Cavaliere, R.,** Studies on the effect of polyamines and their products on Ehrlich ascites tumours, *Agents Actions*, 12, 450, 1982.
106. **Leete, E.,** The biogenesis of nicotine. V. New precursors of the pyrrolidine ring, *J. Am. Chem. Soc.*, 80, 2162, 1958.
107. **Cromwell, B. T.,** The role of putrescine in the synthesis of hyoscyamine, *Biochem. J.*, 37, 722, 1943.
108. **Diaper, D. G. M., Kirkwood, S., and Marion, L.,** The biogenesis of alkaloids. III. A study of hyoscyamine biosynthesis using isotopic putrescine, *Can. J. Chem.*, 29, 964, 1951.
109. **Tuppy, H. and Faltaous, M. S.,** Eine enzymatische Synthese der Alkaloide Hygrin, Cuskhygrin, Isopelletierin und Methylisopelletierin, *Monatsh. Chem.*, 91, 167, 1960.
110. **Schütte, H. R.,** Die Biosynthese von Alkaloiden. III. Uber die gegenseitige Umwandelbarkeit von Lupinin und Spartein sowie uber deren Biogeneseort in *Lupinus lutens*, *Atompraxis*, 7, 91, 1961.

111. **Schütte, H. R., Bohlmann, F., and Reusche, W.,** Die Biosynthese des Sparteins, *Arch. Pharm.*, 294, 610, 1961.
112. **Schütte, H. R.,** Zur Biosynthese der Lupinenalkaloide, *Arch. Pharm.*, 293, 1006, 1960.
113. **Schütte, H. R., Nowacki, E., and Schafer, C.,** Die Bildung von Lupanin und Hydroxylupanin aus Cedaverin (1,5-^{14}C)-dihydrochloride in *Lupinus angustifolius*, *Arch. Pharm.*, 295, 20, 1962.
114. **Hasse, K. and Berg, P.,** Bildung von Anabasin aus Cadaverin in Gegenwart von Planzenextrakten, *Biochem. J.*, 331, 349, 1959.
115. **Traub, W. and Piez, K. A.,** The chemistry and structure of collagen, *Adv. Prot. Chem.*, 25, 243, 1971.
116. **Chan, W. Y., Seale, T. W., Shukla, J. B., and Rennert, O. M.,** Polyamine conjugates and fetal polyamine concentrations in human amniotic fluid, *Clin. Chim. Acta*, 91, 233, 1979.
117. **Scalabrino, G. and Ferioli, M. E.,** Polyamines in mammalian tumors. II, *Adv. Cancer Res.*, 36, 1, 1982.
118. **Baylin, S. B., Beaven, M. A., Engelman, K., and Sjoerdsma, A.,** Elevated histaminase activity in medullary carcinoma of the thyroid gland, *N. Engl. J. Med.*, 283, 1239, 1970.
119. **Keiser, H. R., Beaven, M. A., Doppman, J., Wells, S., and Buja, L. M.,** Sipple's Syndrome: medullary thyroid carcinoma, pheochromocytoma, and parathyroid disease, *Ann. Intern. Med.*, 78, 561, 1973.
120. **Mendelsohn, G., Eggleston, J. C., Weisburger, W. R., Gann, D. S., and Baylin, S. B.,** Calcitonin and histaminase in C-cell hyperplasia and medullary thyroid carcinoma, *Am. J. Pathol.*, 92, 35, 1978.
121. **Lin, C. W., Orcutt, M. L., and Stolbach, L. L.,** Elevation of histaminase and its concurrence with Regan isoenzyme in ovarian cancer, *Cancer Res.*, 35, 2762, 1975.
122. **Lin, C. W., Inglis, N. R., Rule, A. H., Turksoy, R. N., Chapman, C. M., Kirley, S. D., and Stolbach, L. L.,** Histaminase and other tumor markers in malignant effusion fluids, *Cancer Res.*, 39, 4894, 1979.
123. **Küsche, J., Bieganski, T., Hesterberg, R., Stahlkrecht, C-D., Feussner, K-D., Stahlenberg, I., and Lorenz, W.,** The influence of carcinoma growth on diamine oxidase activity in human gastrointestinal tract, *Agents Actions*, 10, 110, 1980.
124. **Perin, A., Sessa, A., Baizini, M., and Desiderio, M. A.,** Diamine oxidase activity in regenerating liver and in hepatomas of the rat, *Ital. J. Biochem.*, 28, 385, 1979.
125. **Okuyama, T. and Kobayashi, Y.,** Determination of diamine oxidase activity by liquid scintillation counting, *Arch. Biochem. Biophys.*, 95, 242, 1961.
126. **Morgan, R. E. and Morgan, J. M.,** Plasma levels of aromatic amines in renal failure, *Metabolism*, 15, 479, 1966.
127. **Tam, C. F., Kopple, J. D., Wang, M., and Swendseid, M. E.,** Alterations of monoamine oxidase activity in uremia, *Kidney Int.*, 7, 328, 1975.
128. **Wang, M., Tam, C-F., Swendseid, M. E., and Kopple, J. D.,** Monoamine and diamine oxidase activities in uremic rats, *Life Sci.*, 17, 653, 1975.
129. **Giordano, C., Bloom, J., and Merrill, J. P.,** Effects of urea on physiologic systems. I. Studies on monoamine oxidase activity, *J. Lab. Clin. Med.*, 59, 396, 1962.
130. **Campbell, R. A., Grettie, D. P., Bartos, D., and Bartos, F.,** Clinical polyamine analysis: problems and promise, in *Advances in Polyamine Research*, Vol. 3, Calderera, C. M., Zappia, V., and Bachrach, U., Eds., Raven Press, New York, 1981, 409.
131. **Kopin, I. J.,** Storage and metabolism of catecholamines: the role of monoamine oxidase, *Pharmacol. Rev.*, 16, 179, 1964.
132. **Wyatt, R. J., Potkin, S. G., and Murphy, D. L.,** Platelet monoamine oxidase activity in schizophrenia: a review of the data, *Am. J. Psychiatr.*, 136, 377, 1979.
133. **Coursey, R. D., Buchsbaum, M. S., and Murphy, D. L.,** Platelet MAO activity evoked potentials in the identification of subjects biologically at risk for psychiatric disorders, *Br. J. Psychiatr.*, 136, 372, 1979.
134. **Sostek, A. J., Sostek, A. M., Murphy, D. L., Martin, E. B., and Born, W. S.,** Cord blood amine oxidase activities relate to arousal and motor functioning in human newborns, *Life Sci.*, 28, 2561, 1981.
135. **Weingold, A. B.,** Enzymatic indices of fetal environment, *Clin. Obstet. Gynecol.*, 11, 1081, 1968.
136. **Kobayashi, Y.,** Blood levels of diamine oxidase as a test for pregnancy, *J. Lab. Clin. Med.*, 62, 699, 1963.
137. **Elmfors, B., Tryding, N., and Tufvesson, G.,** The diagnosis of ruptured fetal membranes by measurement of the diamine oxidase (DAO) activity in vaginal fluid, *J. Obstet. Gynecol.*, 81, 361, 1974.
138. **Henningsson, A. Ch., Henningsson, S., and Nilsson, O.,** Diamines and polyamines in mammalian reproduction, in *Advances in Polyamine Research*, Vol. 4, Bachrach, U., Kaye, A., and Chayen, R., Eds., Raven Press, New York, 1983, 193.
139. **Sakamoto, S. and Samejima, K.,** Preparation of specific antibody to 2,3-trimethylene-4-quinazolone for immunoassay of Δ^1-pyrroline, *Chem. Pharm. Bull.*, 28, 916, 1980.
140. **Unemoto, T., Ikeda, K., Hayashi, M., and Miyaki, K.,** Studies on polyamines. I. A new fluorometric determination of spermine and spermidine, *Chem. Pharm. Bull.*, 11, 148, 1963.
141. **Bachrach, U. and Reches, B.,** Enzymatic assay for spermine and spermidine, *Anal. Biochem.*, 17, 38, 1966.
142. **Neufeld, E. and Chayen, R.,** An evaluation of three methods for the measurement of diamine oxidase (DAO) activity in amniotic fluid, *Anal. Biochem.*, 96, 411, 1979.

Chapter 3

ACETYLPOLYAMINES AS SUBSTRATES OF AMINE OXIDASES

Nikolaus Seiler

TABLE OF CONTENTS

I. INTRODUCTION

The structures of the compounds which are the topic of this review* are shown in Figure 1.

Acetylputrescine (*N*-acetyl-1,4-diaminobutane) was first found in chilled cultures of *E. coli*.[2] It is a normal constituent of various animal tissues, including brain,[3,4] and of human and rat urine.[5-7] Acetylcadaverine (*N*-acetyl-1,5-diaminopentane) and propionylcadaverine are found in human blood[8] and urine,[9] and also in low concentration in mouse brain.[8]

N^1-Acetylspermidine and N^8-acetylspermidine are the major excretory forms of spermidine.[5-7,10] Tissue levels of these compounds are usually below the detection level of the existing methods. Only for hamster epididymis have high N^1-acetylspermidine levels been reported.[11] Considerable amounts of acetylspermidine accumulate in chilled cultures of *E. coli* in the presence of spermidine in the culture medium,[2] or after inhibition of RNA biosynthesis.[12] Intoxication of rats or mice with carbon tetrachloride or thioacetamide, among others, induces a cytosolic acetylase in liver,[13,14] and as a consequence the levels of N^1-acetylspermidine and N^1-acetylspermine are considerably increased in this organ.[15-17] Similarly, treatment with secretin induces accumulation of N^1-acetylspermidine in rat pancreas.[18] Acetylpolyamine concentrations are increased in the liver as a physiological response to starvation.[19]

While these treatments influence N^1-acetylspermidine and N^1-acetylspermine concentrations, acetylputrescine and N^8-acetylspermidine concentrations remain unaltered, because they are the products of a nuclear acetyltransferase,[20,21] not of the above-mentioned inducible cytoplasmic enzyme.

While all the monoacetyl derivatives of the polyamines are natural products, none of the expected diacetyl derivatives have been detected. Nevertheless, we shall include diacetylspermine in our considerations, because it has been shown to be a specific substrate of tissue polyamine oxidase (PAO).[22]

One may conclude from these observations that:

1. Monoacetylation of the di- and polyamines is a significant physiological process in many organisms
2. A likely role for polyamine acetylation is the inactivation of the physiologically active polyamines

However, the fact that the two isomeric acetylspermidines are the products of two different enzymes, which are localized in two different cellular compartments, cytoplasm and nucleus, respectively, indicates an additional, more sophisticated function. One may assume, therefore, that excretion is not the only physiologically important process in which the acetylpolyamines are involved. A detailed knowledge of their metabolic fate is one of the prerequisites for the understanding of their functional role.

The studies of oxidative deamination of acetylated polyamines were provoked by the following observation: putrescine injected into mice was mainly degraded by diamine oxidase (DAO).[23] However, in an organ such as the mammalian brain, which does not contain significant activities of DAO, formation of 4-aminobutyrate from putrescine nevertheless occurred at a considerable rate.[24,25] The explanation of this finding was found by showing that tissues contained an acetyltransferase, capable of forming acetylputrescine, with acetyl-CoA as acetyl donor,[20] and by demonstration that acetylputrescine was a substrate of mono-

* For reasons of simplicity, the term "polyamine" will be used for the diamines putrescine and cadaverine and for spermidine and spermine. For the designation of the acetylpolyamines the nomenclature suggested by Tabor et al. will be used.

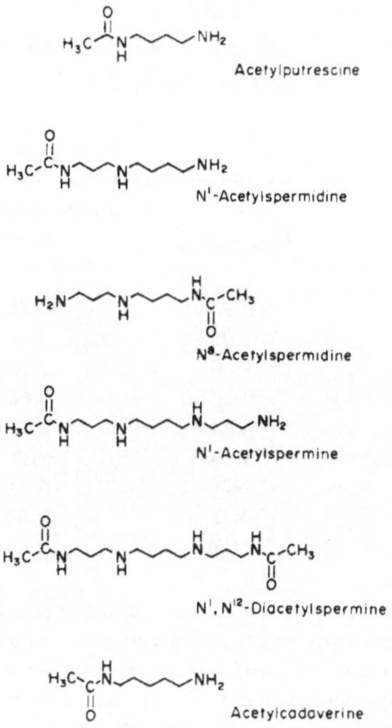

FIGURE 1. Chemical structure of natural acetylpolyamines.

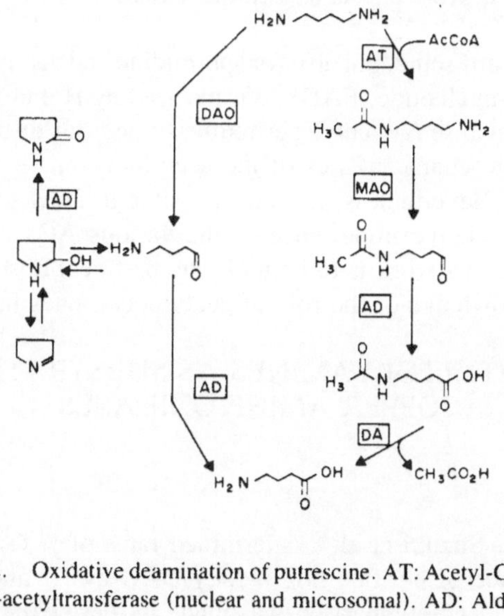

FIGURE 2. Oxidative deamination of putrescine. AT: Acetyl-CoA: Spermidine N^8-acetyltransferase (nuclear and microsomal). AD: Aldehyde dehydrogenase (or another enzyme capable of transforming an aldehyde to a carboxylic acid derivative). DA: Deacetylase (hydrolase).

amine oxidase (MAO). The reaction with MAO and an aldehyde dehydrogenase produced in vitro *N*-acetyl-4-aminobutyric acid. In vivo 4-aminobutyrate was formed by deacetylation of the latter compound.[26] Figure 2 depicts these reactions.

The well-known oxidative deamination of spermidine and spermine to the corresponding aldehydes by serum amine oxidase (SAO),[27] a copper AO,[28] is closely analogous to the oxidative deamination of putrescine by DAO. It was expected that by extension of this analogy a reaction sequence which is initiated by acetylation might exist not only for the

Table 1
RELATIVE RATES OF SUBSTRATE OXIDATION BY DIAMINE OXIDASES

	Enzyme source		
Substrate	**Hog kidney**	**Human pregnancy serum**	**Fetal calf serum**
Acetylputrescine	0.94	0.86	1.39
Acetylcadaverine	1.05	ND[a]	ND
N^1-Acetylspermidine	1.00	1.00	1.00
N^8-Acetylspermidine	0.20	0.64	0.61
N^1-Acetylspermine	0.44	ND	ND
N^1,N^{12}-Diacetylspermine	BDL[b]	ND	ND
Putrescine	ND	0.71	1.77
Spermidine	ND	0.64	0.87
Spermine	ND	0.65	1.07

Note: The data of the experiments with purified hog kidney DAO were obtained at 10 m*M*;[30] the data with partially purified enzymes of human pregnancy serum[31] and fetal calf serum[32] at 1 m*M* substrate concentrations (37°C) (mean values of two or more determinations). The original data of Suzuki et al.[30] and of Gahl and Pitot[31,32] were recalculated and are expressed as rates of H_2O_2 formation relative to N^1-acetylspermidine as substrate.

[a] ND: Not done.
[b] BDL: Below detection level of the analytical method.

oxidative degradation of putrescine, but also for spermidine and spermine. Polyamine oxidase (PAO), the flavin-adenin-nucleotide (FAD) AO, purified by Hölttä from rat liver,[29] seemed an appropriate enzyme because N^1-acetylspermidine turned out to be a substrate.

Until now, the substrate characteristics of the acetylpolyamines for AO have only been incompletely established. Nevertheless, it seems possible to draw some preliminary conclusions with regard to the relative importance of the various AOs in the catabolism of polyamines. The purpose of this work is to compile the pertinent observations, and to draw a picture of our present knowledge of the role of acetylpolyamines in polyamine catabolism.

II. ACETYLPOLYAMINES AS SUBSTRATES OF COPPER AMINE OXIDASES

A. Diamine Oxidase

1. In Vitro Studies

In a preliminary work, Suzuki et al.[30] determined rates of H_2O_2 formation from acetylpolyamines in the presence of purified hog kidney DAO at 0.1 and 10 m*M* substrate concentrations. Unfortunately, no comparison was made in these experiments with putrescine, so that it is difficult to evaluate the substrate characteristics from these experiments. At 0.1 m*M*, oxidation rates were slow with all substrates, whereas at 10-m*M* concentrations all acetyl derivatives, with the exception of diacetylspermine, were oxidatively deaminated by hog kidney DAO. Table 1 summarizes the results, showing that acetylcadaverine, acetylputrescine, and N^1-acetylspermidine were about equally effective, whereas N^8-acetylspermidine showed the lowest rate of oxidative deamination. Using two other preparations of DAO, Gahl and Pitot[31,32] obtained comparable results (Table 1). Again, N^8-acetylspermidine was oxidized at a considerably slower rate than N^1-acetylspermidine. In the case of human pregnancy serum DAO, N^1-acetylspermidine was an even better substrate than putrescine. Reaction products of the acetylated polyamines were not characterized in these studies.

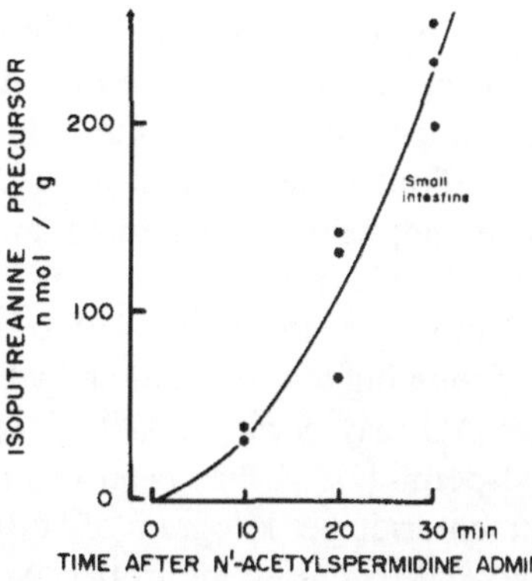

FIGURE 3. Accumulation of *N*-(3-acetamidopropyl)pyrrolidin-2-one in the small intestine of mice after i.p. administration of 1 mmol of N^1-acetylspermidine. 2 HCl per kilogram of body weight. The reaction product of N^1-acetylspermidine was determined by HPLC[33] after its hydrolysis to isoputreanine.

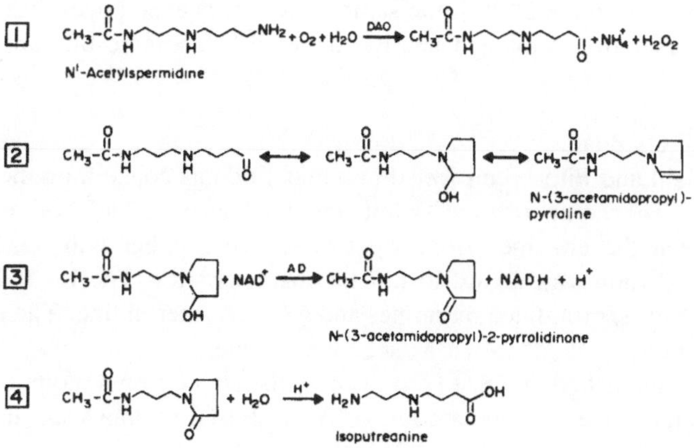

FIGURE 4. Reactions involved in the formation of *N*-(3-acetamidopropyl) pyrrolidin-2-one and isoputreanine from N^1-acetylspermidine. AD: Aldehyde dehydrogenase or another enzyme capable of transforming an aldehyde to a carboxylic acid derivative.

2. *Evidence for the In Vivo Oxidation of N^1-Acetylspermidine by a DAO-Like Enzyme*

It has recently been shown[33] that the i.p. injection of spermidine in mice is followed by a time-dependent increase of putreanine content in various tissues and of *N*-(3-aminopropyl)pyrrolidin-2-one, the lactam of isoputreanine, which can be hydrolyzed to isoputreanine. After i.p. administration of 1 mmol of N^1-acetylspermidine per kilogram of body weight, neither putreanine nor *N*-(3-aminopropyl)pyrrolidin-2-one was formed. However, the reaction product could be hydrolyzed to isoputreanine and thus be determined indirectly. The accumulation in the small intestine of about 200 nmol/g of the metabolite within 30 min (Figure 3) demonstrates the considerable capacity of this organ to metabolize N^1-acetylspermidine.

The fact that the reaction product was hydrolyzed to isoputreanine suggests the formation of *N*-(3-acetamidopropyl)pyrrolidin-2-one by analogy to the formation of *N*-(3-aminopropyl)pyrrolidin-2-one from spermidine along the reaction sequence which is formulated in Figure 4. The analogy between this reaction sequence and that of putrescine with DAO

(Figure 2) is obvious. It was shown by Lundgren et al.[34] that incubation of putrescine with slices of liver (but not of brain) forms pyrrolidin-2-one in addition to 4-aminobutyrate.

That DAO or a closely related copper AO is involved in this transformation is suggested by the high DAO activity in the small intestine of rodents,[35] and by the fact that pretreatment of the mice with 0.2 mmol of aminoguanidine per kilogram of body weight completely prevented the formation of *N*-(3-acetamidopropyl)pyrrolidin-2-one.[33]

While the above experiments demonstrate the capacity of rodents to oxidatively metabolize N^1-acetylspermidine in vivo, the following observations show that the oxidative deamination of N^1-acetylspermidine by copper AO may occur physiologically, or at least in situations of enhanced formation of N^1-acetylspermidine. It has been mentioned that treatment of mice and rats with 2 mℓ of carbon tetrachloride per kilogram of body weight causes the enhanced formation of N^1-acetylspermidine and of putrescine in the liver.[16,17] In agreement with the changes in liver, an augmented excretion of these compounds was found in urine. Spermidine and N^8-acetylspermidine excretion was not significantly altered by this treatment.

Administration of 0.8 mmol of aminoguanidine per kilogram of body weight to control animals mainly enhanced putrescine excretion, with little effect on N^1-acetylspermidine. However, if aminoguanidine-treated rats received, in addition, 2 mℓ of carbon tetrachloride per kilogram of body weight a very significant increase of putrescine and N^1-acetylspermidine excretion was observed in the 24-hr urine samples, whereas even under these conditions N^8-acetylspermidine excretion remained nearly unaltered.[36] These results are summarized in Figure 5.

B. Serum Amine Oxidase

The work of Gahl and Pitot[32] confirmed previous findings concerning the substrate specificity of SAO.[27] The comparison of adult and fetal enzyme did not reveal significant differences between the enzymes from the two sources neither with regard to substrate specificity (Table 2) nor with regard to kinetic characteristics. The K_m values were about the same (40 μ*M*) for spermidine, spermine, and N^8-acetylspermidine. The reaction product of N^8-acetylspermidine was identified as acetylputrescine.

Since it is well established that SAO attacks only the aminopropyl moiety of spermidine,[37] it has to be assumed that the preparation of N^1-acetylspermidine used in this work was contaminated by spermidine or N^8-acetylspermidine. In our hands,[38] N^1-acetylspermidine was completely stable in newborn calf serum whereas N^1-acetylspermine could be quantitatively transformed into N^1-acetylspermidine if incubated for extended periods. The formation of acetylputrescine from N^8-acetylspermidine, and of N^1-acetylspermidine from N^1-acetylspermine is formulated in Figure 6. Oxidative deamination to the corresponding aldehyde is followed by nonenzymatic elimination of acrolein, as has previously been established for spermidine and spermine.[37,39]

III. ACETYLPOLYAMINES AS SUBSTRATES OF FLAVIN-DEPENDENT AMINE OXIDASES

A. Monoamine Oxidase

Acetylputrescine, acetylcadaverine, propionylcadaverine, and N^8-acetylspermidine are substrates of mitochondrial MAO, whereas N^1-acetylspermidine, N^1-acetylspermine, and N^1-,N^{12}-diacetylspermine are not metabolized by this enzyme at significant rates.[30]

For acetylcadaverine and propionylcadaverine, K_m values of 40.5 and 25.6 m*M* have been reported.[40] These compounds are substrates of both A- and B-MAO.[40] This has to be assumed by analogy also for acetylputrescine.

There is no presently available evidence for the degradation in vivo of N^8-acetylspermidine by MAO. This reaction should yield N^1-acetyl-N^2-(2-carboxyethyl)putrescine (acetylputreanine) as reaction product.

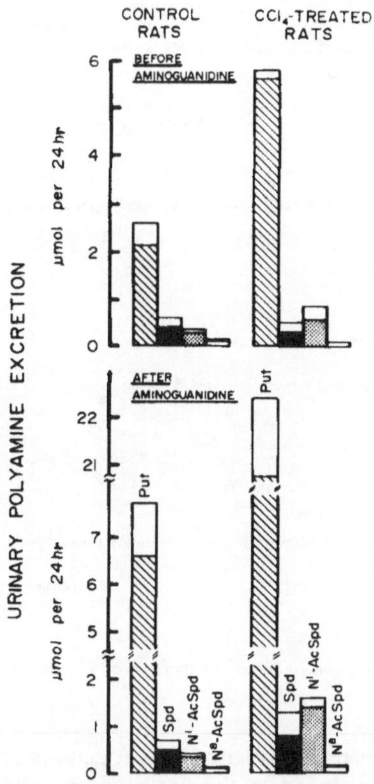

FIGURE 5. Polyamine excretion by carbon tetrachloride-intoxicated rats. The effect of treatment with aminoguanidine on the urinary polyamine pattern. Groups of three rats received either a single dose of 8 mℓ/kg maize oil (vehicle) or 2 mℓ/kg carbon tetrachloride dissolved in maize oil. Other groups were pretreated with 100 mg/kg aminoguanidine 6 hr before they were treated with vehicle or carbon tetrachloride. Polyamines were determined in the 24-hr urine.[36]

Table 2
RELATIVE RATES OF SUBSTRATE OXIDATION BY ADULT AND FETAL BOVINE SERUM AMINE OXIDASE

	Enzyme	
Substrate	**Adult bovine SAO**	**Fetal bovine SAO**
Spermidine	1.00	1.00
Spermine	0.81	0.76
Putrescine	0.05	0.06
N^1-Acetylspermidine	0.20	0.14
N^8-Acetylspermidine	0.66	0.65
Acetylputrescine	0.05	0.02

Note: The data are from Gahl and Pitot,[32] and are expressed as activities relative to spermidine as substrate (substrate conc. 1 m*M*, incubation at 37°C).

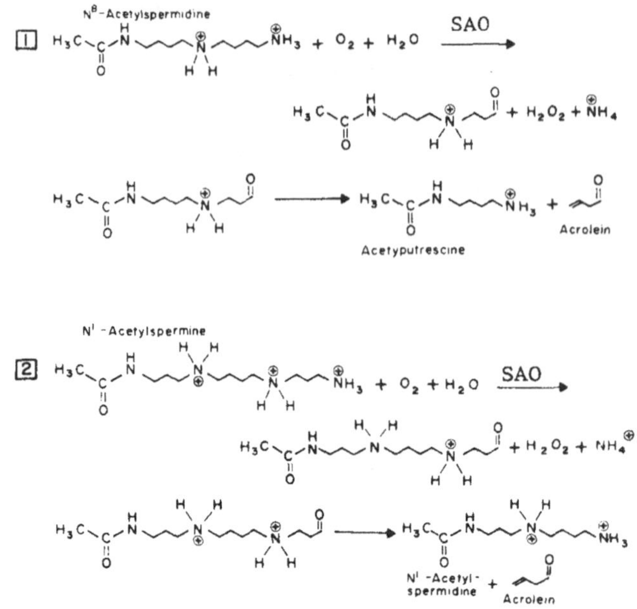

FIGURE 6. Reaction of N^8-acetylspermidine (1) and N^1-acetylspermine (2) with serum amine oxidase.

It has been shown that the transformation of exogenous acetylputrescine into 4-aminobutyric acid is significantly reduced in the brain of pargyline pretreated mice.[26] Administration of daily doses of 25 mg of pargyline or nialamide per kilogram of body weight on 7 consecutive days produced an increase of acetylcadaverine and propionylcadaverine in mouse brain of several hundred percent.[8] This shows that catabolism of acetylpolyamines by MAO is likely to be a physiological event.

B. Polyamine Oxidase

1. In Vitro Studies

The FAD-AO, which was purified and characterized by Hölttä[29] from rat liver, is an ubiquitous enzyme.[22] Although it is closely related to MAO, it is not located in mitochondria, but has been found in peroxisomes and cytosol.[29]

Spermine, and at a considerably slower rate, spermidine, is oxidized by this enzyme.[29] However, N^1-acetylspermidine, N^1-acetylspermine, and N^1,N^{12}-diacetylspermine are better substrates than the unsubstituted polyamines. Acetylputrescine and N^8-acetylspermidine are not substrates of PAO.[41]

In Figure 7 the reactions of the acetylpolyamines are shown. Although it has not been studied in detail, one has to assume that the reaction mechanism of PAO is analogous to that of MAO.[42] In the case of PAO, however, the methylene group of the acetamidopropyl moiety, which is adjacent to the secondary amino group, is preferentially oxidized, to form an iminium ion. The iminium ion is hydrolized to the aldehyde. Concomitant with the oxidation of the amine, the flavin is reduced.

Table 3 summarizes the kinetic data which were obtained with a partially purified preparation of rat liver PAO. It appears from these data that N^1-acetylspermine has the lowest K_m and the highest V_{max}. N^1,N^{12}-Diacetylspermine is also a good substrate. Since this compound, in contrast to the other acetylderivatives, is not a substrate of any other known tissue oxidase, it allows one to determine PAO activities in tissues with a high grade of specificity. For this purpose, H_2O_2 formation can be measured.[41] A more sensitive method

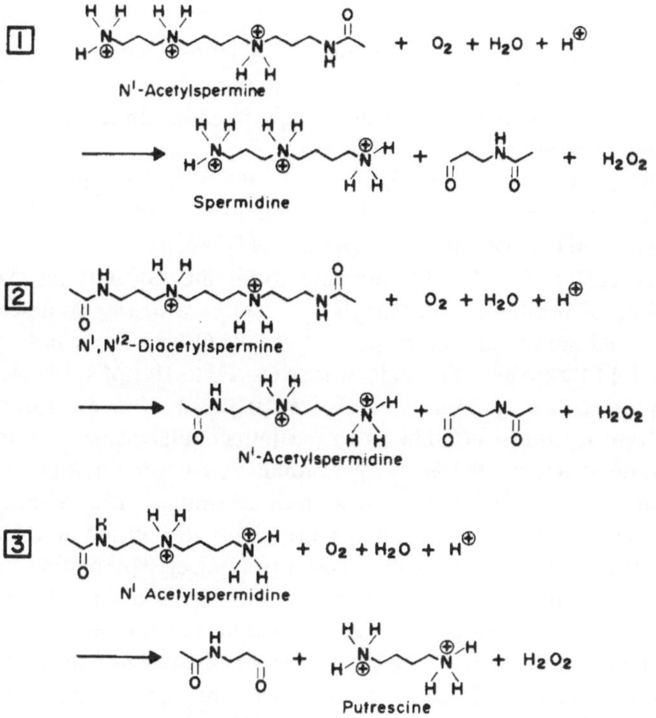

FIGURE 7. Substrates and reaction products of polyamine oxidase.

Table 3
SUBSTRATE CHARACTERISTICS OF POLYAMINE OXIDASE FROM RAT LIVER[41]

Substrate	Reaction product	K_m (μM)	V_{max} (nmol/min/mg protein)
Spermine	Spermidine	40	0.1
N^1-Acetylspermine	Spermidine	0.6	2.8
N^1,N^{12}-Diacetylspermine	N^1-Acetylspermidine	5.0	4.0
N^1-Acetylspermidine	Putrescine	14	2.5

is the determination of the reaction product, N^1-acetylspermidine, either by reaction with dansylchloride and thin-layer chromatographic separation of the reaction products, as was described originally,[22] or by direct determination of N^1-acetylspermidine by HPLC.[43]

The formation of the acetylpolyamines in tissues and their high affinity for PAO suggested that they are the natural substrates of this enzyme.[44] Of course, it is not possible to exclude spermine and spermidine as natural substrates of PAO on the grounds of their kinetic parameters. Both compounds are, as has been shown, effectively competing with the acetylpolyamines for the active site of the enzyme.[41] It will depend on the relative concentration of the free and acetylated compounds, which will be preferably degraded. In this regard, one should keep in mind that polyamines are presumed to be bound by electrostatic forces to macromolecular anionic sites, such as the phosphate groups of nucleic acids and membranes and carboxylic groups of proteins. The cellular concentration of free polyamines is presumably very low. Since binding to anionic sites is diminished by acetylation because of the diminution of the number of positive charges, even low concentrations of the acetylated compounds may effectively compete with the nonconjugated polyamines.[44]

2. In Vivo Studies

The degradation of [^{14}C] N^1-acetylspermidine to [^{14}C]putrescine has been shown to occur in mice.[45] Since, however, labeled spermidine is also transformed in vivo into putrescine,[46] it is not possible to draw conclusions from these experiments with regard to the significance of acetylpolyamines as natural substrates of PAO.

Evidence for the correctness of the suggestion[44] that N^1-acetylspermine and N^1-acetylspermidine are the actual substrates of PAO has been obtained from studies of acetylpolyamine metabolism in rat hepatoma tissue culture (HTC) cells.[38]

Exposure of HTC cells to 5 m*M* α-difluoromethylornithine (DFMO), an enzyme-activated irreversible inhibitor of ornithine decarboxylase,[47] causes a nearly complete depletion of cellular putrescine and spermidine contents, but the PAO activity is not affected by this treatment.[22] When DFMO-treated HTC cells were exposed to 10 μ*M* spermine, no formation of spermidine or putrescine was observed. However, if 10 μ*M* N^1-acetylspermine was added instead to the culture medium, in addition to the intracellular accumulation of N^1-acetylspermine, a very rapid accumulation of spermidine was observed, but putrescine levels remained low,[38] in accordance with the formation of spermidine from N^1-acetylspermine by PAO (Figure 7). On the other hand, cellular putrescine and spermidine contents increased in DFMO-treated HTC cells if they were exposed to N^1-acetylspermidine, whereas no putrescine formation was observed during the first 2 hr after addition of spermidine to the culture medium.[38] That both putrescine and spermidine are formed in these cells in the presence of N^1-acetylspermidine is due to the fact that cells treated with DFMO accumulate decarboxylated *S*-adenosylmethionine,[48] and are therefore rapidly transforming putrescine into spermidine, according to the reaction scheme in Figure 8.

These experiments clearly demonstrate that even though polyamine-deficient HTC cells contain high PAO activities, they are unable to form spermidine from spermine and putrescine from spermidine until the cytosolic acetyltransferase is induced by the accumulating polyamine. At 10 μ*M* spermidine in the culture medium, putrescine formation is observed after about 2 hr, when cellular spermidine levels are supernormal.[38] These cells, however, rapidly produce putrescine and spermidine from the corresponding monoacetyl derivatives.

Since N^8-acetylspermidine is not a substrate of PAO, as has been mentioned, HTC cells are not capable of forming putrescine from this compound.[49]

IV. PHYSIOLOGICAL SIGNIFICANCE OF OXIDATIVE DEAMINATION OF ACETYLPOLYAMINES

A. Copper Amine Oxidases

N^1-Acetylspermidine is the only acetylpolyamine for which in vivo evidence is available showing that an aminoguanidine-sensitive copper AO, probably DAO, is catalyzing its oxidative deamination. The aldehyde formed by this reaction is further oxidized to *N*-acetylisoputreanine lactam (*N*-(3-acetamidopropyl)pyrrolidin-2-one). This compound has not been directly determined, but conjugates of isoputreanine lactam are found in normal urine.[50] It can be assumed that the acetylderivative is among them. Enhanced formation of N^1-acetylspermidine seems to be accompanied by its enhanced catabolism by a copper AO.[36] Since N^1-acetylspermidine is an excretory product per se,[5,6,10] it is not evident what advantage its transformation to *N*-acetylisoputreanine lactam has.

DAO activity is high only in a few organs, among which the small intestine is most prominent.[35] Lactam formation from N^1-acetylspermidine, and also from spermidine, is therefore assumed to be a detoxification mechanism specific for that organ.

Since certain tumors have enhanced DAO activities,[51] it is not unlikely that lactam formation from spermidine and N^1-acetylspermidine will be enhanced, and could perhaps serve as a diagnostic tool. A single case report of enhanced isoputreanine excretion has already been published.[52]

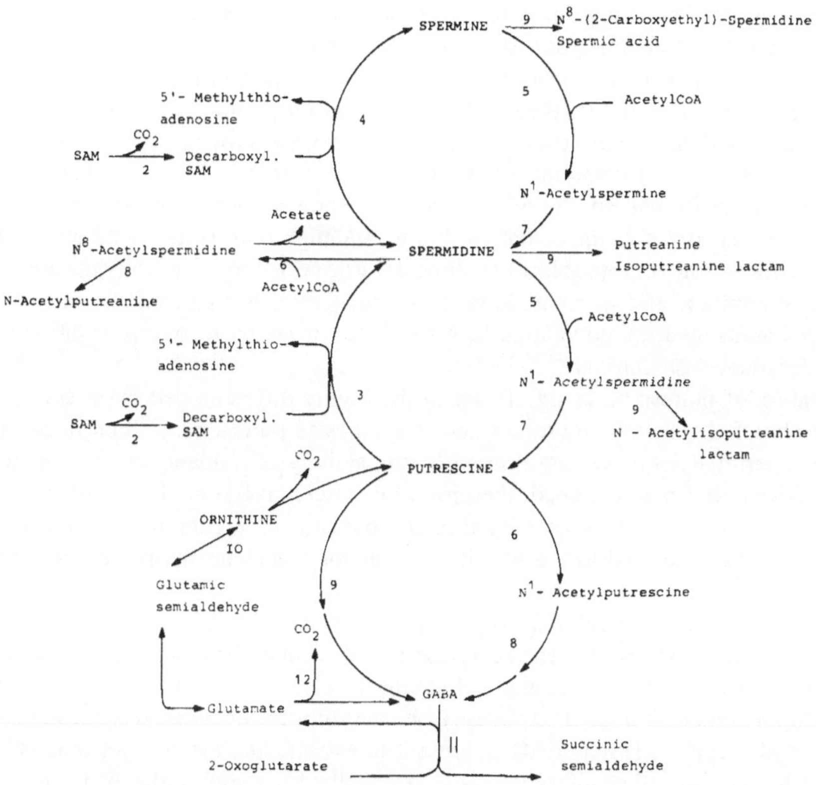

FIGURE 8. Metabolic scheme of the polyamines. (1) ornithine decarboxylase; (2) *S*-adenosylmethionine decarboxylase; (3) spermidine synthase; (4) spermine synthase; (5) acetyl-CoA: spermidine N^1-acetyltransferase (cytoplasmic); (6) acetyl-CoA: spermidine N^8-acetyltransferase (nuclear); (7) polyamine oxidase (PAO); (8) monoamine oxidase (MAO); (9) diamine oxidase (DAO); aldehyde dehydrogenase; (10) ornithine: 2-oxoacid aminotransferase; (11) 4-aminobutyrate: 2-oxoacid aminotransferase; (12) glutamate decarboxylase.

B. FAD-Dependent Amine Oxidases

1. MAO

It was mentioned in the Introduction that MAO is assumed to play a role in the removal of excessive putrescine from cells which do not contain significant DAO activities. In view of the ubiquitous occurrence of MAO, and the mentioned uneven distribution of DAO, the reaction sequence, which is initiated by acetylation, and followed by reaction with MAO (Figure 2), could in principle be of importance in several organs, but evidence for its physiological significance is only available for mouse brain.[26]

Enhanced cellular 4-aminobutyrate levels influence polyamine metabolism, and vice versa.[53] It remains to be clarified whether the MAO-catalyzed putrescine degradation is an essential link in the regulatory interrelations between 4-aminobutyrate and the polyamines.

2. PAO

Evidence has been presented above for the role of N^1-acetylspermidine and N^1-acetylspermine as natural substrates of PAO. The scheme in Figure 8 depicts the metabolic function of polyamine acetylation and oxidative degradation by PAO. As a result of this reaction sequence, spermine is degraded to spermidine, and spermidine to putrescine in a formal inversion of the synthetic reactions. All compounds involved in this reaction sequence can undergo degradative reactions, as is shown in Figure 8.

N[1]-Acetylspermine has up to now not been detected in urine. However, *N*[1]-acetylspermidine is, in addition to being a substrate of PAO, a normal excretory product, as has been mentioned repeatedly. It is not clear at present which factors determine, what proportion of *N*[1]-acetylspermidine is oxidatively degraded, and what proportion is excreted.

It can be expected that at least part of the putrescine which is formed by the PAO-catalyzed reaction is reutilized for spermidine synthesis, as is indicated in Figure 8. The same holds for the analogously formed spermidine. It has been shown that the proportion of reutilized putrescine is especially high in vertebrate brain.[54] A high reutilization rate allows the cell to maintain a high turnover rate at low ornithine decarboxylase activity. Only the aminopropyl moieties of spermidine and spermine have to be substituted. A high turnover rate is a means for the rapid adjustment of polyamine levels to local needs by relatively small changes of ornithine decarboxylase activity.[54,55]

Reutilization of putrescine is significant in the highly differentiated brain cells. Rapidly growing embryonal cells[56,57] and tumor cells[58] synthesize putrescine at high rates and spermidine and spermine syntheses are limited by the activity of *S*-adenosyl-methionine decarboxylase activity. It can be assumed, therefore, that interconversion of the polyamines along the PAO-catalyzed pathway is an advanced mechanism for regulating cellular polyamine concentrations, as compared to the primitive regulatory mechanisms present in embryonal and tumor cells.

It is well established that cellular polyamines can be regulated by regulating the activity of the biosynthetic enzymes.[59,60] The acetylase-initiated catabolic reactions are another complementary means to regulate cellular polyamine levels.[61,62] In this reaction sequence the acetyl-CoA:polyamine *N*[1]-acetyltransferase is the rate-limiting enzyme, which can be induced by a variety of means, whereas PAO is present in excess, and usually not inducible by an increase or decrease of cell proliferation rates. All available examples of enhanced acetylation rates — intoxication with thioacetamide,[13,15] carbon tetrachloride,[14,16,17] administration of spermidine,[63] and other compounds,[64] or starvation[19] can be interpreted as a means of the cell to get rid of excessive polyamines, whereby "excessive" indicates not necessarily a supernormal concentration, but only an amount above that which is physiologically needed under the specific circumstances.

V. CONCLUSION

Some acetylpolyamines are substrates of copper AO; others are substrates of flavine-dependent enzymes. *N*[1]-Acetylspermidine is a substrate of both types of enzymes. A physiologically highly significant role could only be established for *N*[1]-acetylspermine and *N*[1]-acetylspermidine, which are intermediates of the polyamine interconversion pathway. It is presently difficult to produce data which would allow one to estimate the quantitative significance of the transformation of spermine into spermidine, and of spermidine into putrescine in the various organs in vivo. Likewise, it is difficult to determine the role of polyamine reutilization, i.e., the proportion of putrescine and spermidine which is formed by monoacetylation and oxidative deamination by PAO and is utilized again for spermidine and spermine synthesis, respectively. Inhibitors of PAO will allow us to obtain the data, which will show the quantitative significance of acetylpolyamine oxidation by this enzyme under normal and pathophysiological conditions. They will also demonstrate the importance of the interconversion reaction, as compared with acetylpolyamine excretion and oxidative deaminations of polyamines by copper AO.

REFERENCES

1. **Tabor, H., Tabor, C. W., and DeMeis, L.,** Chemical synthesis of *N*-acetyl-1,4-diaminobutane, N^1-acetylspermidine and N^8-acetylspermidine, *Meth. Enzymol.*, 17b, 829, 1971.
2. **Tabor, C. W. and Dobbs, L. G.,** Metabolism of 1,4-diaminobutane and spermidine in *Escherichia coli:* the effects of low temperature during storage and harvesting of culture, *J. Biol. Chem.*, 245, 2086, 1970.
3. **Seiler, N., Al-Therib, M. J., and Knödgen, B.,** Occurrence of monoacetylputrescine in vertebrate tissue, *Hoppe-Seyler's Z. Physiol. Chem.*, 354, 589, 1973.
4. **Perry, T.L., Hansen, S., and MacDougall, L.,** Amines of human whole brain, *J. Neurochem.*, 14, 775, 1967.
5. **Abdel-Monem, M. M. and Ohno, K.,** Polyamine metabolism. II. N^1-(Monoaminoalkyl)- and *N*-(polyaminoalkyl)acetamides in human urine, *J. Pharm. Sci.*, 66, 1195, 1977.
6. **Seiler, N. and Knödgen, B.,** Determination of the naturally occurring monoacetyl derivatives of di- and polyamines, *J. Chromatogr.*, 164, 155, 1979.
7. **Noto, T., Tanaka, T., and Nakajima, T.,** Urinary metabolites of polyamines in rats, *J. Biochem.*, 83, 543, 1978.
8. **Dolezalova, H., Stepita-Klauco, M., Kucera, J., Uchimura, H., and Hirano, M.,** Blood concentrations of monoacylcadaverines in schizophrenia, in *Mass Spectrometry in Drug Metabolism*, Frigerio, A. and Ghisalberti, E. L., Eds., Plenum Press, New York, 1977, 201.
9. **Perry, T. L., Hansen, S., and MacDougall, L.,** Identity and significance of some pink spots in schizophrenia and other conditions, *Nature (London)*, 214, 484, 1967.
10. **Seiler, N., Koch-Weser, J., Knödgen, B., Richards, W., Tardif, C., Bolkenius, F. N., Schechter, P., Tell, G., Mamont, P., Fozard, J., Bachrach, U., and Grosshans, E.,** The significance of acetylation in the urinary excretion of polyamines, in *Advances in Polyamine Research*, Vol. 3, Caldarera, C. M., Zappia, V., and Bachrach, U., Eds., Raven Press, New York, 1981, 197.
11. **Matsuzaki, S., Hamana, K., Imai, K., and Matsuura, K.,** Occurrence in high concentrations of N^1-acetylspermidine and sym.-homospermidine in the hamster epididymis, *Biochem. Biophys. Res. Commun.*, 107, 307, 1982.
12. **Cohen, S. S.,** *Introduction to the Polyamines*, Prentice-Hall, Englewood Cliffs, N.J., 1971, 88.
13. **Matsui, I. and Pegg, A. E.,** Effect of thioacetamide, growth hormone or partial hepatectomy on spermidine acetylase activity of rat liver cytosol, *Biochim. Biophys. Acta*, 633, 87, 1980.
14. **Matsui, I. and Pegg, A. E.,** Increase in acetylation of spermidine in rat liver extracts brought about by treatment with carbon tetrachloride, *Biochem. Biophys. Res. Commun.*, 92, 1009, 1980.
15. **Seiler, N., Bolkenius, F. N., and Knödgen, B.,** Acetylation of spermidine in polyamine catabolism, *Biochim. Biophys. Acta*, 633, 181, 1980.
16. **Seiler, N., Bolkenius, F. N., Knödgen, B., and Haegele, K.,** The determination of N^1-acetylspermine in mouse liver, *Biochim. Biophys. Acta*, 676, 1, 1981.
17. **Abdel-Monem, M. M. and Merdinck, J.,** Formation of N^1-acetylspermidine in rat liver after treatment with carbon tetrachloride, *Life Sci.*, 28, 2017, 1981.
18. **Danzin, C., Bolkenius, F. N., Claverie, N., Wagner, J., and Grove, J.,** Secretin-induced accumulation of N^1-acetylspermidine and putrescine in the rat pancreas, *Biochem. Biophys. Res. Commun.*, 109, 1234, 1982.
19. **Seiler, N., Bolkenius, F. N., and Sarhan, S.,** Formation of acetylpolyamines in the liver of fasting animals, *Int. J. Biochem.*, 13, 1205, 1981.
20. **Seiler, N. and Al-Therib, M. J.,** AcetylCoA: 1,4-diaminobutane *N*-acetyltransferase. Occurrence in vertebrate organs and subcellular distribution, *Biochim. Biophys. Acta*, 354, 206, 1974.
21. **Blankenship, J. and Walle, T.,** In vitro studies of enzymatic synthesis and metabolism of *N*-acetylated polyamines, in *Advances in Polyamine Research*, Vol. 2, Campbell, R. A., Morris, D. R., Bartos, D., Daves, G. D., and Bartos, F., Eds., Raven Press, New York, 1978, 97.
22. **Seiler, N., Bolkenius, F. N., Knödgen, B., and Mamont, P.,** Polyamine oxidase in rat tissues, *Biochim. Biophys. Acta*, 615, 480, 1980.
23. **Seiler, N. and Eichentopf, B.,** 4-Aminobutyrate in mammalian putrescine catabolism, *Biochem. J.*, 152, 201, 1975.
24. **Seiler, N., Wiechmann, M., Fischer, H. A., and Werner, G.,** The incorporation of putrescine carbon into γ-aminobutyric acid in rat liver and brain in vivo, *Brain Res.*, 28, 317, 1971.
25. **Seiler, N. and Knödgen, B.,** Die Umwandlung von Glutaminsäure, Putrescin und Ornithin in die γ-Aminobuttersäure im Gehirn, *Hoppe-Seyler's Z. Physiol. Chem.*, 352, 97, 1971.
26. **Seiler, N. and Al-Therib, M. J.,** Putrescine catabolism in mammalian brain, *Biochem. J.*, 144, 29, 1974.
27. **Bachrach, U.,** Oxidation of polyamines and diamines, in *Polyamines in Biology and Medicine*, Morris, D. R. and Marton, L. J., Eds., Marcel Dekker, New York, 1981, 151.

28. **Mondovì, B. and Finazzi Agrò, A.,** Structure and function of amine oxidases, in *Structure and Function Relationships in Biochemical Systems*, Bossa Chiancone Finazzi Agrò, A. and Storm, Eds., Plenum Press, New York, 1982, 141.
29. **Hölttä, E.,** Oxidation of spermidine and spermine in rat liver: purification and properties of polyamine oxidase, *Biochemistry*, 16, 91, 1977.
30. **Suzuki, O., Matsumoto, T., Oya, M., and Katsumata, Y.,** Metabolism of acetylpolyamines by monoamine oxidase, diamine oxidase and polyamine oxidase, *Biochim. Biophys. Acta*, 677, 190, 1981.
31. **Gahl, W. A. and Pitot, H. C.,** Acetylated polyamines as substrates for human pregnancy serum diamine oxidase, *Life Sci.*, 29, 2177, 1981.
32. **Gahl, W. A. and Pitot, H. C.,** Polyamine degradation in foetal and adult bovine serum, *Biochem. J.*, 202, 603, 1982.
33. **Seiler, N., Knödgen, B., and Haegele, K.,** *N* -(3-Aminopropyl)pyrrolidin-2-one, a product of spermidine catabolism in vivo, *Biochem. J.*, 208, 189, 1982.
34. **Lundgren, D. W., Lloyd, H. A., and Hankins, J.,** Δ^1-Pyrroline: an intermediary metabolite in the conversion of putrescine to 2-pyrrolidone.
35. **Shaff, R. E. and Beaven, M. A.,** Turnover and synthesis of diamine oxidase (DAO) in rat tissues. Studies with heparin and cycloheximide, *Biochem. Pharmacol.*, 25, 1057, 1976.
36. **Seiler, N., Knödgen, B., Bink, G., Sarhan, S., and Bolkenius, F.,** Diamine oxidase and polyamine catabolism, in *Advances in Polyamine Research*, Vol. 4, Bachrach, U., Kaye, A., and Chayen, R., Eds., Raven Press, New York, 1983, 135.
37. **Tabor, H., Tabor, C. W., and Bachrach, U.,** Identification of the aminoaldehydes produced by the oxidation of spermine and spermidine in purified plasma amine oxidase, *J. Biol. Chem.*, 239, 2194, 1964.
38. **Mamont, P. S., Seiler, N., Siat, M., Joder-Ohlenbusch, A.-M., and Knödgen, B.,** Metabolism of acetyl derivatives of polyamines in cultured polyamine-deficient rat hepatoma cells, *Med. Biol.*, 59, 347, 1981.
39. **Kimes, B. W. and Morris, D. R.,** Preparation and stability of oxidized polyamines, *Biochim. Biophys. Acta*, 228, 223, 1971.
40. **Suzuki, O., Matsumoto, T., Oya, M., Matsumata, Y., and Stepita-Klauco, M.,** Monoacylcadaverines as substrates for both monoamine oxidase and diamine oxidase; low rates of activity, *Experientia*, 36, 535, 1980.
41. **Bolkenius, F. N. and Seiler, N.,** Acetylderivatives as intermediates in polyamine catabolism, *Int. J. Biochem.*, 13, 287, 1981.
42. **Silverman, R. B., Hoffmann, S. J., and Catus, W. B., III,** A mechanism for mitochondrial monoamine oxidase catalyzed amine oxidation, *J. Am. Chem. Soc.*, 102, 7126, 1980.
43. **Seiler, N. and Knödgen, B.,** High performance liquid chromatographic procedure for the simultaneous determination of the natural polyamines and their monoacetyl derivatives, *J. Chromatogr.*, 221, 227, 1980.
44. **Seiler, N.,** Amide-bond-forming reactions of polyamines, in *Polyamines in Biology and Medicine*, Morris, D. R. and Marton, L. J., Eds., Marcel Dekker, New York, 1981, 127.
45. **Seiler, N. and Bolkenius, F. N.,** Unpublished observations.
46. **Siimes, M.,** Studies on the metabolism of 1,4-^{14}C spermidine and 1,4-^{14}C spermine in the rat, *Acta Physiol. Scand.*, 298, 1, 1967.
47. **Metcalf, B. W., Bey, P., Danzin, C., Jung, M., Casara, P., and Vevert, J. P.,** Catalytic irreversible inhibition of mammalian ornithine decarboxylase (E.C.4.1.1.17) by substrate and product analogues, *J. Am. Chem. Soc.*, 100, 2551, 1978.
48. **Mamont, P. S., Danzin, C., Wagner, J., Siat, M., Joder-Ohlenbusch, A. M., and Claverie, N.,** Accumulation of decarboxylated *S*-adenosyl-L-methionine in mammalian cells as consequence of the inhibition of putrescine biosynthesis, *Eur. J. Biochem.*, 123, 499, 1982.
49. **Mamont, P. S., Joder-Ohlenbusch, A. M., and Siat, M.,** Metabolism of N^8-monoacetylspermidine in rat hepatoma cells. Investigation of its effect on the activity of L-ornithine decarboxylase, *Eur. J. Biochem.*, 133, 613, 1983.
50. **Seiler, N. and Knödgen, B.,** *N* -(3-Aminopropyl)pyrrolidin-2-one: a physiological excretory product deriving from spermidine, *Int. J. Biochem.*, 15, 907, 1983.
51. **Baylin, S. B. and Luk, G. D.,** Diamine oxidase (histaminase) activity in human disease, in *Polyamines in Biology and Medicine*, Morris, D. R. and Marton, L. J., Eds., Marcel Dekker, New York, 1981, 377.
52. **Muskiet, F. A., Stratingh, C. M., Fernouw-Ottevangers, D. C., and Halie, M. R.,** Mass spectrometric identification of isoputreanine, a metabolite of spermidine and/or spermine in human urine, *J. Chromatogr.*, 230, 142, 1982.
53. **Seiler, N.,** On the role of GABA in vertebrate polyamine metabolism, *Physiol. Chem. Phys.*, 12, 411, 1980.
54. **Antrup, H. and Seiler, N.,** On the turnover of polyamines spermidine and spermine in mouse brain and other organs, *Neurochem. Res.*, 5, 123, 1980.

55. **Seiler, N.,** Turnover of polyamines, in *Polyamines in Biology and Medicine,* Morris, D. R. and Marton, L. J., Eds., Marcel Dekker, New York, 1981, 169.
56. **Anderson, T. R. and Schanberg, S. M.,** Ornithine decarboxylase activity in developing rat brain, *J. Neurochem.,* 19, 1471, 1972.
57. **Schmidt, A. L. and Cantoni, G. L.,** Adenosylmethionine decarboxylase in developing rat brain, *J. Neurochem.,* 20, 1378, 1973.
58. **Jänne, J., Pösö, H., and Raina, A.,** Polyamines in rapid growth and cancer, *Biochim. Biophys. Acta,* 473, 241, 1978.
59. **Pegg, A. E. and Williams-Ashman, H. G.,** Biosynthesis of putrescine, in *Polyamines in Biology and Medicine,* Morris, D. R. and Marton, L. H., Eds., Marcel Dekker, New York, 1981, 3.
60. **Williams-Ashman, H. G. and Pegg, A. E.,** Aminopropyl group transfer in polyamine biosynthesis, in *Polyamines in Biology and Medicine,* Morris, D. R. and Marton, L. J., Eds., Marcel Dekker, New York, 1981, 43.
61. **Pegg, A. E., Hibasami, H., Matsui, I., and Bethell, D. R.,** Formation and interconversion of putrescine and spermidine in mammalian cells, *Adv. Enzyme Regul.,* 19, 427, 1980.
62. **Seiler, N., Bolkenius, F. N., and Rennert, O. M.,** Interconversion, catabolism and elimination of the polyamines, *Med. Biol.,* 59, 334, 1981.
63. **Matsui, I., Pösö, M., and Pegg, A. E.,** Conversion of exogenous spermidine into putrescine after administration to rats, *Biochim. Biophys. Acta,* 719, 199, 1982.
64. **Pegg, A. E., Seely, Y. E., Pösö, H., DellaRagione, F., and Zagon, I. S.,** Polyamine biosynthesis and interconversion in rodent tissues, *Fed. Proc. Fed. Am. Soc. Exp. Biol.,* 41, 3065, 1982.

Chapter 4

BACTERIAL AND FUNGAL AMINE OXIDASES

Hidehiko Kumagai and Hideaki Yamada

TABLE OF CONTENTS

I. INTRODUCTION

In this chapter, the purification, characterization, and reaction mechanism of microbial amine oxidases (AOs) summarized in Table 1 are described.

II. AMINE OXIDASE FROM *ASPERGILLUS NIGER*

A. Formation and Purification

An AO was inducibly formed in the mycelia of fungi when a monoamine or a diamine was the sole nitrogen source.[1] Ammonia, nitrate, urea, and amino acids were inactive in enzyme formation. Maximum formation of the enzyme was found when *A. niger* was cultivated with the feeding of a substrate *n*-butylamine.[2] Crystalline preparation was obtained from the mycelia by 220-fold purification in the yield of 83%. Crystals were needles or rods and showed a pink color.

B. Properties

The enzyme has a molecular weight of 252,000 and is composed of two identical subunits with a molecular weight of 125,000. It contains two atoms of Cu^{2+} per mole of enzyme.[3] The enzyme oxidized preferentially aliphatic monoamines of C_3 to C_6. Benzylamine, phenethylamine, histamine, and agmatine were oxidized as well. Aliphatic diamines of C_4 to C_6 were oxidized but at rather low rates. The enzyme inhibits absorption maxima at 280 and 480 to 500 nm and a shoulder at 410 nm. The extinction coefficient at 280 nm ($E_{1\,cm}^{1\%}$) of the enzyme is 11.8. The absorption in the visible region was partially bleached by the addition of excess sodium ditionite or by amine substrates.[4] These spectrophotometric properties resemble those of the AOs from bovine serum or from pig kidney.

Recently, the similarity of the chromophore isolated from copper-containing AOs to pyrroloquinoline quinone, a novel coenzyme, was indicated spectrophotometrically by Ameyama et al.[5]

C. Reaction Mechanism

Titration of the enzyme with phenylhydrazine showed that there are two carbonyl groups per mole of the enzyme and that these carbonyl groups are directly related to the activity. Anaerobic titration of the enzyme with butylamine and successive addition of oxygen gave results indicating that the enzyme reaction is composed of two steps of reaction: the transamination, which takes place anaerobically, and the oxidative deamination, which takes place in the presence of molecular oxygen.[6]

$$E(CHO)_2 + 2RCH_2NH_2 = E(CH_2NH_2)_2 + 2RCHO$$

$$E(CH_2NH_2)_2 + 2O_2 + 2H_2O = E(CHO)_2 + 2H_2O_2 + 2NH_3$$

where $E(CHO)_2$ and $E(CH_2NH_2)_2$ are two forms of enzyme carbonyl form and amine form, and RCH_2NH_2 and RCHO are an amine as substrate and the corresponding aldehyde produced.

β-Br-Ethylamine is both a substrate and an irreversible inhibitor of AO from *A. niger*. The enzyme catalyzes the nonoxidative elimination of HBr from β-Br-ethylamine to form acetaldehyde. Investigation of the reaction of the enzyme with this haloamine provided results indicating a reaction mechanism as shown in Figure 1.

Data on the overall reaction rate and that obtained by the "stopped" and "continuous flow" methods were analyzed, and the following scheme was proposed as the mechanism of the AO reaction.

Table 1
BACTERIAL AND FUNGAL AMINE OXIDASES

Coenzyme	Enzyme name (oxidase)	Source	Substrate	Ref.
Cu^{2+}	Amine	*A. niger*	Aliphatic amines (C_3 to C_6) diamines (C_4 to C_6), bez, phn	1—8
	Agmatine	*P. chrysogenum*	Agmatine, histamine, put	9,10
FAD	Putrescine	*M. rubens*	put, cadaverine, spd	13—16
	Polyamine	*P. chrysogenum*	spm, spd	17,18
	Polyamine	*A. terreus*	spm, spd	17,19
	Tyramine	*S. lutea*	Tyramine, dopamine	11,12

Note: bez, benzylamine; phn, phenethylamine; put, putrescine; spd, spermidine; spm, spermine. Differential determination procedure for put, spd, and spm is described in References 20 to 22.

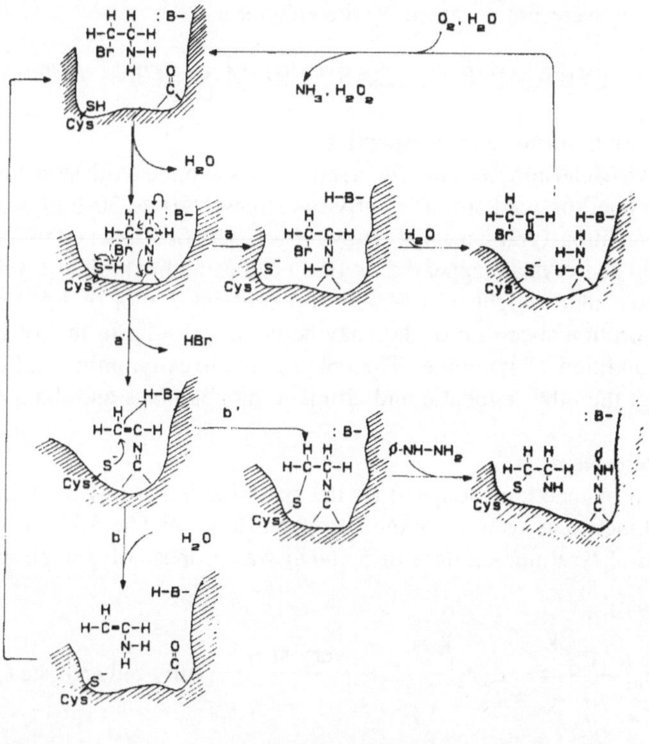

FIGURE 1. Schematic presentation of reaction of *A. niger* amine oxidase with β-Br-ethylamine. :B- and H-B- represent a basic residue in the enzyme protein in its deprotonated and protonated form. ϕ-NH-NH$_2$, phenyl hydrazine.

$$E_L + S \overset{K}{\rightleftarrows} X \overset{k_{+2}}{\rightarrow} E_M + P$$

$$E_M + O_2 \overset{k_{+3}}{\rightarrow} Y \overset{k_{+4}}{\rightarrow} E_L + H_2O_2 + NH_3$$

where E_L and E_M are aldehyde and amine forms of the enzyme, X and Y are intermediates, S is the substrate, and P is the product.

III. AGMATINE OXIDASE FROM *PENICILLIUM CHRYSOGENUM*

A. Formation and Purification

Agmatine oxidase catalyzes oxidative deamination of agmatine to form γ-guanidinobutyraldehyde, NH_3, and H_2O_2. Agmatine oxidase was found in mycelia of fungi belonging to the genera *Aspergillus, Penicillium, Absidia*, etc. when they were grown in agmatine-containing medium. The maximal formation of the enzyme of *P. chrysogenum* was observed in the early stationary phase of growth. The enzyme was purified as hexagonal pink board-crystals with an overall yield of 30%.

B. Properties

The enzyme has a molecular weight of 160,000 and is composed of two identical subunits. The prosthetic group was identified as copper and its content was determined to be two atoms per mole of the enzyme. The enzyme rapidly oxidized agmatine and slowly oxidized histamine, putrescine, 1,3-diaminopropane, and cadaverine. Monoamines, polyamines, and guanidyl derivatives were not oxidized by the enzyme.

IV. TYRAMINE OXIDASE FROM *SARCINA LUTEA*

A. Formation, Purification, and Properties

A Gram-positive bacterium, *Sarcina lutea*, contains an amine oxidase specific for tyramine when it grows on bouillon medium. The activity increased 20- to 30-fold when it was grown on a medium containing tyramine. The enzyme was purified as crystals from the cells in the yield of 25%. The crystals appeared as fine needles and exhibited a yellow color. The enzyme has a molecular weight of 129,000 and contains 2 mol of FAD per mole of the enzyme. The absorption spectrum of the enzyme is very similar to that of free FAD and is bleached by the addition of tyramine. The enzyme oxidizes tyramine and dopamine at the same rate, but not the other aliphatic and aromatic monoamines and diamines.

B. Reaction Mechanism

The mechanism of reaction catalyzed by the tyramine oxidase was studied based on the overall reaction kinetics as well as by the "stopped flow".[12] The following scheme for the acton mechanism of tyramine oxidase of *S. lutea* was proposed from the data:

$$E_{ox} + S \underset{k_{-1}}{\overset{k_{+1}}{\rightleftarrows}} E_{ox}S \xrightarrow{k_{+2}} X \xrightarrow{\text{very slow}} \text{fully reduced state}$$

$$X + O_2 \xrightarrow{k_{ox}} E_{ox} + P + H_2O_2 + NH_3$$

where E_{ox} represents the oxidized form of the enzyme, $E_{ox}S$ the enzyme(oxidized form)-substrate complex, X the intermediary compound formed in the reaction with the substrate, S the substrate(tyramine), and P the product(*p*-hydroxyphenylacetoaldehyde).

V. PUTRESCINE OXIDASE FROM *MICROCOCCUS RUBENS*

Putrescine oxidase occurred in cells of *M. rubens* when it was grown on an ordinary medium. The enzyme content in the cell was increased by two to three times when the bacterium was cultured on a medium containing 0.01 to 0.02% putrescine. The enzyme was

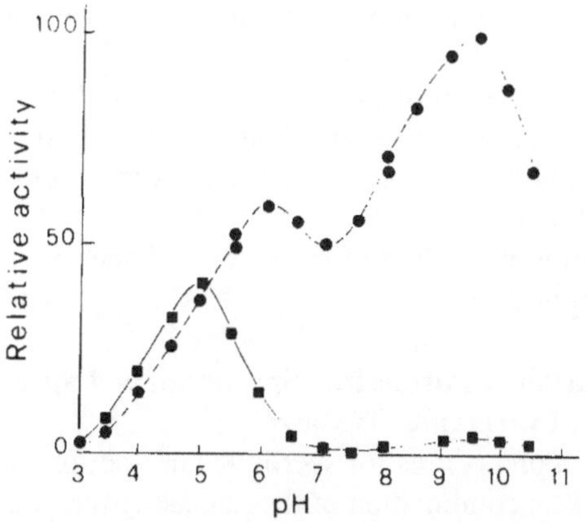

FIGURE 2. pH Dependence of PAO from *P. chrysogenum*.

purified to the homogeneous state in the yield of 11%.[13] The enzyme has a molecular weight of 82,000[15] and contains 1 mol of FAD per mole of enzyme. The absorption spectrum exhibits maxima at 380 and 458 nm, and these absorptions are reduced by the addition of a substrate. Treatment of the purified enzyme with acid ammonium sulfate results both in liberation of FAD and in complete loss of the activity.[14] The original activity is fully restored when the apoenzyme is incubated with added FAD. Beside putrescine, the enzyme oxidatively degrades cadaverine and spermidine.

DeSa investigated the reaction mechanism of the enzyme.[15] Okada et al. reported purification using affinity chromatography.[16]

VI. POLYAMINE OXIDASE FROM *PENICILLIUM CHRYSOGENUM*

Polyamine oxidase (PAO) was found in mycelium of fungi belonging to the genera *Aspergillus, Mucor, Penicillium, Rhizopus*, etc. when they were grown in a medium containing spermine or spermidine as the sole source of nitrogen.[17] PAO of *P. chrysogenum* was purified and crystallized in the yield of 35% from the mycelia. The enzyme has a molecular weight of 160,000 and is composed of two identical subunits. The enzyme contains 2 mol of FAD per mole of enzyme. The enzyme oxidized spermine and spermidine, but not diamines and monoamines. The oxidation products of spermine and spermidine by the enzyme were identified as putrescine, 3-amino propionaldehyde, and H_2O_2. Oxidation of spermine and spermidine showed a quite different pH dependency, as shown in Figure 2.[18]

VII. POLYAMINE OXIDASE FROM *ASPERGILLUS TERREUS*

Polyamine oxidase (PAO) of *A. terreus* was purified and crystallized in the yield of 40%. PAO has a molecular weight of 130,000 and is composed of two identical subunits. It contains 2 mol of FAD per mole of enzyme. It only catalyzes oxidation of spermine and spermidine.[19] The ratio of oxidation of spermine vs. spermidine is largely changed depending on pH as shown by the enzyme from *P. chrysogenum*.

VIII. DETERMINATION OF POLYAMINES WITH AMINE OXIDASES

A. Differential Determination of Spermine and Spermidine with Fungal Polyamine Oxidases

Kinetic properties of PAO from *P. chrysogenum* and *A. terreus* were investigated and differential determination of spermine and spermidine was carried out using these enzymes. The initial rate was assayed with *Penicillium* enzyme and the end point was measured after addition of *Aspergillus* enzyme. Spermine (25 to 200 nmol) and spermidine (25 to 250 nmol) were assayed by solving two equations obtained simultaneously from the rate assay method and the end point assay method. The calculated values were in close agreement with those obtained by an amino acid analyzer.[20]

A differential determination procedure of spermine and spermidine with beef plasma AO was also developed and reported.[21]

B. Differential Determination of Putrescine, Spermine, and Spermidine with Polyamine Oxidases from Fungi and Putrescine Oxidase[22]

Differential determination procedures for spermine and spermidine using beef plasma AO or fungal PAOs were based on combination of the rate assay method and the end point assay method. In the rate assay method, absorbance change obtained was very little and intensive care was required to estimate accurate enzyme activity. Moreover, putrescine could not be assayed by those methods.

A differential determination procedure for spermine, spermidine, and putrescine using end point assay method alone was developed using fungal PAOs and putrescine oxidase from *M. rubens*.

Method 1:
1. Spermidine and spermine were first oxidized with PAO.
2. To the reaction mixture, putrescine oxidase was added to oxidize putrescine.
3. Putrescine and spermidine in another reaction mixture were oxidized with putrescine oxidase.

Method 2:
1. Putrescine and spermidine were first oxidized with putrescine oxidase.
2. To the reaction mixture, PAO was added to oxidize spermine.
3. Spermidine and spermine in another reaction mixture were oxidized with PAO.

The amounts of putrescine, spermidine, and spermine were determined from the absorbance values at each step. Polyamines (5 to 40 nmol) were able to be assayed by either method (Method 1 or 2).

REFERENCES

1. **Yamada, H., Adachi, O., and Ogata, K.,** Amine oxidase of microorganisms. I. Formation of amine oxidase by fungi, *Agric. Biol. Chem.*, 29, 117, 1965.
2. **Adachi, O. and Yamada, H.,** Amine oxidase of microorganisms. VII. An improved purification procedure and further properties of amine oxidase of *Aspergillus niger*, *Agric. Biol. Chem.*, 33, 1707, 1969.
3. **Kumagai, H., Kishimoto, N., and Yamada, H.,** Subunit structure of amine oxidase from *Aspergillus niger*, *Agric. Biol. Chem.*, 42, 893, 1978.
4. **Yamada, H., Adachi, O., and Ogata, K.,** Amine oxidases of microorganisms. III. Properties of amine oxidase of *Aspergillus niger*, *Agric. Biol. Chem.*, 29, 864, 1965.
5. **Ameyama, M., Hayashi, M., Matsushita, K., Shinagawa, E., and Adachi, O.,** Microbial production of pyrroloquinoline quinone, *Agric. Biol. Chem.*, 48, 561, 1984.
6. **Suzuki, H., Ogura, Y., and Yamada, H.,** Stoichiometry of the reaction by amine oxidase from *Aspergillus niger*, *J. Biochem.*, 69, 1065, 1971.
7. **Kumagai, H., Uchida, H., and Yamada, H.,** Reaction of fungal amine oxidase with β-bromoethylamine, *J. Biol. Chem.*, 254, 10913, 1979.

8. **Suzuki, H., Ogura, Y., and Yamada, H.,** Kinetic studies on the amine oxidase reaction, *J. Biochem.*, 72, 703, 1972.
9. **Isobe, K., Tani, Y., and Yamada, H.,** A new enzyme, agmatine oxidase, from fungi, *Agric. Biol. Chem.*, 46, 1345, 1982.
10. **Isobe, K., Tani, Y., and Yamada, H.,** Crystallization and characterization of agamatine oxidase from *Penicillium chrysogenum, Agric. Biol., Chem.*, 46, 1353, 1982.
11. **Kumagai, H., Matsui, H., Ogata, K., and Yamada, H.,** Properties of crystalline tyramine oxidase from *Sarcina lutea, Biochim. Biophys. Acta*, 171, 1, 1968.
12. **Kumagai, H., Yamada, H., Suzuki, H., and Ogura, Y.,** Action mechanism of tyramine oxidase from *Sarcina lutea, J. Biochem.*, 69, 137, 1971.
13. **Adachi, O., Yamada, H., and Ogata, K.,** Purification and properties of putrescine oxidase of *Micrococcus rubens, Agric. Biol. Chem.*, 30, 1202, 1966.
14. **Yamada, H., Adachi, O., and Ogata, K.,** Putrescine oxidase, a diamine oxidase requiring flavin adenine dinucleotide, *Agric. Biol. Chem.*, 29, 1148, 1965.
15. **DeSa, R. J.,** Putrescine oxidase from *Micrococcus rubens.* Purification and properties of the enzyme, *J. Biol. Chem.*, 247, 5527, 1972.
16. **Okada, M., Kawashima, S., and Imahori, K.,** Affinity chromatography of putrescine oxidase from *Micrococcus rubens* and spermidine dehydrogenase from *Serratia marcescens, J. Biochem.*, 85, 1225, 1979.
17. **Yamada, H., Isobe, K., and Tani, Y.,** Oxidation of polyamines by fungal enzymes, *Agric. Biol. Chem.*, 44, 2469, 1980.
18. **Isobe, K., Tani, Y., and Yamada, H.,** Crystallization and characterization of polyamine oxidase from *Penicillium chrysogenum, Agric. Biol. Chem.*, 44, 2651, 1980.
19. **Isobe, K., Tani, Y., and Yamada, H.,** Crystallization and characterization of polyamine oxidase from *Aspergillus terreus, Agric. Biol. Chem.*, 44, 2749, 1980.
20. **Isobe, K., Tani, Y., and Yamada, H.,** Kinetic properties of fungal polyamine oxidases and their application to differential determination of spermine and spermidine, *Agric. Biol. Chem.*, 44, 2955, 1980.
21. **Yamada, H., Isobe, K., Tani, Y., and Hiromi, K.,** A differential determination of procedures for spermine and spermidine with beef plasma amine oxidase, *Agric. Biol. Chem.*, 43, 2487, 1979.
22. **Isobe, K., Tani, Y., and Yamada, H.,** Differential determination procedure for putrescine, spermidine and spermine with polyamine oxidase from fungi and putrescine oxidase, *Agric. Biol. Chem.*, 45, 727, 1981.

Chapter 5

SPERMIDINE DEHYDROGENASE FROM *SERRATIA MARCESCENS*: A HEME-FLAVIN ENZYME

Celia White Tabor

TABLE OF CONTENTS

I. INTRODUCTION

The oxidation of spermidine by *Serratia marcescens* was first described 25 years ago by Razin et al.[1] Subsequently, the enzyme has been purified to homogeneity[2-5] and the reaction well characterized.[6,7] The pure protein differs from most other enzymes that oxidize amines in that it is a dehydrogenase with an absolute requirement for an electron carrier. It contains two prosthetic groups: flavin adenine dinucleotide (FAD) and a heme.[2,3]

II. REACTION

The reaction carried out by the *S. marcescens* spermidine dehydrogenase is shown in Equation 1:*[6]

$$NH_2CH_2CH_2CH_2NHCH_2CH_2CH_2CH_2NH_2 \rightarrow NH_2CH_2CH_2CH_2NH_2 + CHOCH_2CH_2CH_2NH_2 \quad (1)$$

1 2 3 4 5 6 7 8

spermidine — 1,3-diaminopropane — γ-aminobutyraldehyde

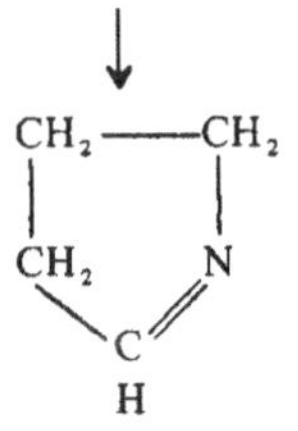

Δ^1-pyrroline

The enzyme cleaved spermidine specifically between the secondary amino group at position 4 and C-5, yielding diaminopropane and the unstable γ-aminobutyraldehyde, which cyclizes to form Δ^1-pyrroline. The crude extract utilized oxygen, but a tenfold purified preparation was unable to utilize oxygen directly. By providing an artificial electron carrier, such as phenazine methosulfate, the ability to oxidize spermidine was restored.[7] This reaction provides a biosynthetic path for 1,3-diaminopropane, a diamine which is occasionally found in natural materials. This enzyme differs from several others that oxidize spermidine at the secondary nitrogen. For example, the enzyme from *Micrococcus rubens*[9] (see Chapter 4) cleaves at the same site as the *Serratia* enzyme, but forms H_2O_2 as a product. Cleavage of spermidine between C-3 and the secondary amino nitrogen, N-4, to form putrescine and 3-aminopropionaldehyde has been shown both with a crude *Pseudomonas* enzyme[10] and with a purified liver enzyme.[11]

III. ASSAY METHODS

The enzyme can be assayed in a crude extract by oxygen utilization or by the reaction of the product Δ^1-pyrroline with *o*-aminobenzaldehyde.[7]

The reaction rate with the purified enzyme is most conveniently assayed by a spectrophotometric method based on the reduction of the electron acceptors, potassium ferricyanide or dichlorophenolindophenol.[2] Other spectrophotometric assays have depended upon the formation of the colored derivative of the product Δ^1-pyrroline with *o*-aminobenzaldehyde, using phenazine methosulfate as the electron carrier.[2] The enzyme can also be assayed after

* The numbering system used here is the modified aza-nomenclature described in Reference 8.

Table 1
PURIFICATION OF SPERMIDINE DEHYDROGENASE FROM *SERRATIA MARCESCENS*

Step	Volume (mℓ)	Total protein (mg)	Total units[a]	Recovery (%)	Specific activity (units/mg protein)
Extract	6725	48,100	4400	100	0.09
DEAE I chromatography	3900	4960	1850	42	0.37
Calcium phosphate cellulose adsorption	500	725	1330	30	1.84
DEAE chromatography	8	76	680	15	9.0
Hydroxylapatite chromatography	10	6.4	450	10	70.0
Acrylamide gel electrophoresis	1.6	0.6	250	6	420.0

Note: This represents a typical purification from about 700 g of wet cell paste.

[a] A unit is the amount of enzyme that oxidizes 1 μmol of spermidine in 1 min, using assay A in Reference 2.

From Tabor, C. W. and Kellogg, P. D., *J. Biol. Chem.*, 245, 54, 24, 1970. With permission.

Table 2
AFFINITY-CHROMOTOGRAPHY PURIFICATION OF SPERMIDINE DEHYDROGENASE FROM *SERRATIA MARCESCENS*

Step	Total protein (mg)	Total activity (units)	Specific activity (unit/mg)	Purification	Recovery (%)
Cell extracts[a]	165,000	981	0.006	1.0	100
DEAE-Cellulose	1256	460	0.366	61	46.9
Affinity column[b]	0.91	397	436	72,700	40.5

[a] The starting material was 300 g of cells of *S. marcescens*.
[b] A 1,8-diaminooctane-Sepharose 4B column was used.

From Okada, M., Kawashima, S., and Imahori, K., *J. Biochem.*, 85, 1225, 1979. With permission.

electrophoresis on a nondenaturing acrylamide gel by adding a reaction mixture containing spermidine, phenazine methosulfate, and *p*-nitroblue tetrazolium to the gel.[2]

IV. PURIFICATION

Spermidine dehydrogenase is present in *S. marcescens* (ATCC 25179) in low concentration. By growing the cells at 30°C in a salts-citrate medium without added carbohydrate, but with added *N*-(3-aminopropyl)-1,3-diaminopropane or spermidine, and with vigorous aeration, the enzyme yield can be increased up to 20-fold over that in cells grown in rich medium or with glucose.[2]

The enzyme appeared to be particulate in a crude extract prepared by sonication in 0.2 *M* phosphate buffer, pH 7.0, 0.01% mercaptoethanol, and sedimented with the ribosomal fraction. Solubilization of the activity was markedly increased if the cell pellet had been suspended in water without buffer before sonication.[2]

The enzyme was purified about 4000-fold to homogeneity by the procedure given in Table 1. A later procedure with another strain of *S. marcescens*, based on affinity chromatography, resulted in much better recovery (Table 2).[4] The apparent large total increase in the purification ratio in this procedure can be explained by the very low specific activity of the starting extracts.

V. PROPERTIES

The enzyme is a monomer, of approximately 76,000 mol wt. The pure spermidine dehydrogenase shows a major absorption peak at 414 nm. This absorption was due to the presence of the heme moiety, iron-protoporphyrin IX[2], since the spectrum of the pyridine hemochromogen derivative of the pure enzyme was identical with that of iron-protoporphyrin IX. Flavin adenine dinucleotide (FAD) was also found in a partially purified enzyme as well as in the 4000-fold purified protein.[2] The pure enzyme was partially inactivated by acid ammonium sulfate precipitation; the spectrum of the soluble cofactor in the supernatant was similar to that of FAD. In addition, the inactive protein was partially reactivated by incubation with FAD.[2] The ratio of protein to heme to FAD was 1:0.8:0.9.[2]

The enzyme reaction was linear with respect to substrate and time. The K_m for spermidine was too low to determine ($<10^{-8}$ *M*). The pH optimum was at 7.2. The K_m for spermine, which also served as a substrate, was 0.02 m*M*; the pH optimum was at 8.8. The enzyme was inhibited 95% by *p*-hydroxymercuribenzoate and was partially inhibited by *N*-ethylmaleimide. Carbonyl-binding reagents did not inhibit the enzyme. As discussed above, the enzyme could not utilize oxygen directly, but required an electron carrier. The natural acceptor may be cytochrome *c*, since this compound can serve as an electron carrier for the purified enzyme.[2] Thus, this enzyme differs from the usual amine oxidases (AOs) which can utilize oxygen directly, and should be classified as a dehydrogenase. At this time, this is the only AO or dehydrogenase which has been shown to contain both a heme and a flavin cofactor. Two equivalents of ferricyanide were reduced per mole of spermidine or spermine oxidized; thus, spermine was oxidized at only one of the secondary amine groups. The enzyme also oxidized *N*-(3-aminopropyl)-1,3-propanediamine and *N,N'-bis*(3-aminopropyl)-1,3-propane-diamine, but at a slower rate than spermidine. N^1-Acetylspermidine* (monoacetylspermidine B)[2] was a substrate, but N^8-acetylspermidine* (monoacetylspermidine A)[2] was not oxidized.

From an extensive study of a number of amine substrates, Okada et al. concluded that many triamines and tetramines served as substrates.[12] Those amines that contained a 4-aminobutylimino moiety, $NH_2(CH_2)_4NH$, were more active as substrates than those not containing this moiety; each of these amines, upon oxidation, formed Δ^1-pyrroline stoichiometrically. Tetramines containing a 3-aminopropylimino group were oxidized to form 1,3-diaminopropane. If an amine contained both the 4-aminobutylimino and 3-aminopropylimino group, the 4-aminobutylimino group was preferentially oxidized by the enzyme. Diamines inhibited the enzyme competitively, with long-chain diamines being the more potent inhibitors. From these studies, Okada et al.[12] proposed a model for the active site that required a substrate with a free amino group at the terminus and an imino group in the middle of the chain.

Serratia spermidine dehydrogenase has been utilized for an enzymatic assay for spermidine.[13] However, because of the wide variety of amines that serve as substrates, this assay is useful only if other amines are not present.

* The numbering system used here is the modified aza-nomenclature described in Reference 8.

REFERENCES

1. **Razin, S., Gery, I., and Bachrach, U.**, The degradation of natural polyamines and diamines by bacteria, *Biochem. J.*, 71, 551, 1959.
2. **Tabor, C. W. and Kellogg, P. D.**, Identification of flavin adenine dinucleotide and heme in a homogeneous spermidine dehydrogenase from *Serratia marcescens*, *J. Biol. Chem.*, 245, 5424, 1970.
3. **Tabor, C. W. and Kellogg, P. D.**, Spermidine dehydrogenase *(Serratia marcescens)*, *Meth. Enzymol.*, 17B, 746, 1971.
4. **Okada, M., Kawashima, S., and Imahori, K.**, Affinity chromatography of putrescine oxidase from *Micococcus rubens* and spermidine dehydrogenase from *Serratia marcescens*, *J. Biochem.*, 85, 1225, 1979.
5. **Okada, M., Kawashima, S., and Imahori, K.**, Purification of spermidine dehydrogenase from *Serratia marcescens* by affinity chromatography, *Meth. Enzymol.*, 94, 303, 1983.
6. **Bachrach, U.**, Spermidine oxidase from *Serratia marcescens*, *J. Biol. Chem.*, 237, 3443, 1962.
7. **Campello, A. P., Tabor, C. W., and Tabor, H.**, Resolution of spermidine dehydrogenase from *Serratia marcescens:* requirements for flavin adenine dinucleotide and an additional electron carrier, *Biochem. Biophys. Res. Commun.*, 19, 6, 1965.
8. **Tabor, H., Tabor, C. W., and de Meis, L.**, Chemical synthesis of *N*-acetyl-1,4-diaminobutane, N^1-acetylspermidine, and N^8-acetylspermidine, *Meth. Enzymol.*, 17B, 829, 1971.
9. **Adachi, O., Yamada, H., and Ogata, K.**, Purification and properties of putrescine oxidase of *Micrococcus rubens*, *Agric. Biol. Chem.*, 30, 1202, 1966.
10. **Padmanabhan, R. and Kim, K.**, Oxidation of spermidine by a *Pseudomonas*, *Biochem. Biophys. Res. Commun.*, 19, 1, 1965.
11. **Hölltä, E.**, Oxidation of spermidine and spermine in rat liver: purification and properties of polyamine oxidase, *Biochemistry*, 16, 91, 1977.
12. **Okada, M., Kawashima, S., and Imahori, K.**, Substrate binding characteristics of the active site of spermidine dehydrogenase from *Serratia marcescens*, *J. Biochem.*, 85, 1235, 1979.
13. **Bachrach, U. and Oser, I. S.**, Enzymatic assay for spermidine, *J. Biol. Chem.*, 238, 2098, 1963.

Chapter 6

PLANT AMINE OXIDASES

Augusto Rinaldi, Giovanni Floris, and Anna Giartosio

TABLE OF CONTENTS

I. INTRODUCTION

Since Werle and co-workers[1,2] first demonstrated the presence of amine oxidase (AO) activity in plant extracts, plant AOs from different species of several families have been partially purified and characterized.[3] The investigations on these enzymes have been concerned with organ distribution, onset and regulation of synthesis, and molecular and catalytic properties.

The classification used for animal enzymes (Cu-AO and FAD-AO) cannot be applied to the plant enzymes since very little is at present known about FAD-containing plant AOs. Oxidase activity towards monoamines has been found[3] in several plants, but the enzymes responsible have not been isolated and characterized. This chapter will therefore deal only with enzymes oxidizing diamines (Cu-AOs) and polyamines (PAOs).

Cu-AOs have been extracted from various sources: pea *(Pisum sativum)*,[4-11] lupine *(Lupinus luteus)*,[12] tobacco (*Nicotiana tabacum*),[13] soybean *(Glycine max)*,[14] chick-pea *(Lathyrus sativus)*,[15] broad bean *(Vicia faba)*,[16-17] lentil (*Lens esculenta)*,[18] and a perennial Mediterranean shrub *(Euphorbia characias)*.[19]

Only the enzymes from the epicotyls of pea (PSAO)[9,11] and lentil (LSAO),[18] from the leaves of *Vicia faba* (VLAO),[16] and from the latex of *Euphorbia characias* (ELAO)[19] have been purified to homogeneity and well characterized. It is noteworthy that all these enzymes oxidize mainly diamines; in some cases they also act on mono- or polyamines.

In the plant kingdom, PAOs have been found in the Gramineae family.[3] The only enzymes isolated and partially characterized are those from oats *(Avena sativa)*,[20] barley *(Hordeum vulgare)*,[20] and maize (*Zea mais)* seedlings.[21,22]

II. Cu-Amine Oxidases

In this section the molecular and functional properties of the four best known Cu-AOs (E.C.1.4.3.6.) will be described and compared. When possible, reference will be made to other less-studied enzymes.

A. Molecular Features

Tables 1 and 2 summarize some of the molecular properties of plant Cu-AOs. All the enzymes examined are dimers, formed by apparently identical subunits, and all contain two copper (Cu II) atoms per mole. The carbohydrate content seems to be fairly constant (12 to 14%). Comparison of the amino acid compositions of the four enzymes shows no similarity.

In the *Euphorbia* enzyme two –SH groups per dimer are titratable either in the presence of aminic substrate or in denaturing conditions.[23] No other example of reactive thiols has been reported.

Spectral features of Cu-AOs (see next section) suggest the presence in these enzymes of an organic prosthetic group. The nature of this cofactor has not yet been discovered and is the subject of current intense investigation. There is a general agreement that the cofactor is tightly bound to the enzymatic protein, probably in covalent linkage, although its removal from the pea enzyme has been reported.[24] A charge transfer interaction between the organic cofactor and Cu(II) has been suggested.[24]

B. Spectroscopic Features

All the purified enzymes are pink, with an absorption band in the visible region centered around 500 nm (Table 3). The color is stable in the pH range 5 to 9.[6,16] The 500-nm absorption band is only slightly affected by the removal of copper (Table 3), indicating the presence of a second prosthetic group. This prosthetic group has more than one electronic transition, since the CD spectra of the lentil enzyme shows a large positive band centered at about 520 nm, clearly resulting from more than one component (Figure 1).[25]

Table 1
SOME PROPERTIES OF PLANT Cu-AOs

Properties	ELAO[19]	LSAO[11,18]	PSAO[8,11]	VLAO[16]
Molecular weight				
Gel chromatography	140,000	—	180,000	126,000
$s_{20,w}(M_r)$	6.0(186,000)[a]	7.3(187,000)	7.7(184,000)	—
SDS-PAGE	72,000	78,000	85,000	74,000
Cu content				
%	0.09	0.082	0.085	0.107
g·atom/mol	2	2	2	2
Isoelectric point	5.2	7.4	7.4	—
Partial specific volume	0.736	0.732	—	—
Carbohydrate content (neutral sugar %)	12	14	13	—

[a] M_r and $s_{20,w}$ values reported in Reference 19 were erroneous.[35]

Table 2
AMINO ACID COMPOSITION OF Cu-AOs FROM VARIOUS SOURCES

Amino acid	ELAO[19]	LSAO[18]	PSAO[11]	VLAO[16]
Asp	120	121	63	107
Thr	98	67	37	62
Ser	100	74	37	65
Glu	110	104	52	82
Pro	72	80	13	16
Gly	70	77	34	67
Ala	54	48	28	56
Val	76	87	55	70
Met	12	23	1	1
Ile	96	66	48	72
Leu	76	76	42	73
Tyr	46	43	21	19
Phe	60	50	29	53
His	40	33	22	31
Lys	72	56	33	51
Arg	42	41	23	33
½ Cys	12	14	4	10
Trp	26	14	ND	ND
Total	1182	1077	542	868

Note: ND = Not determined. The values under the columns ELAO and LSAO represent the nearest integer per mole of dimer; the values under PSAO and VLAO represent the nearest integer value per mole of methionine.

The aerobic addition of amine substrate to LSAO is followed by a rapid disappearance of the 500-nm absorption, which slowly reappears after exhaustion of substrate.[26] Also, the 500-nm band of the copper-free enzyme is rapidly bleached by addition of substrate, but in this case the absorption band is not restored even after dialysis.[7,26] In the presence of aminic substrate and in the absence of oxygen the native enzyme 500-nm absorption band disappears and an intermediate with characteristic spectral properties (a band at 350 nm and two sharp bands centered at 430 and 460 nm) is stabilized, conferring a yellow color to the solution (Figure 2). The yellow peaks gradually disappear upon admission of air. These absorption

Table 3
OPTICAL PROPERTIES AND EPR PARAMETERS OF PLANT Cu-AOs FROM VARIOUS SOURCES

	ELAO[13]	LSAO[18]	PSAO[6,7,11]
Spectral maxima			
Native	490($\epsilon_M = 2 \times 10^3$)	495($\epsilon_M = 2.6 \times 10^3$)	500
	278($\epsilon_M = 1.75 \times 10^5$)	278($\epsilon_M = 2 \times 10^5$)	280
Copper free		480($\epsilon_M = 2.6 \times 10^3$)	480
		278($\epsilon_M = 2 \times 10^5$)	280
EPR spectrum (g_m)	2.07	2.07	—

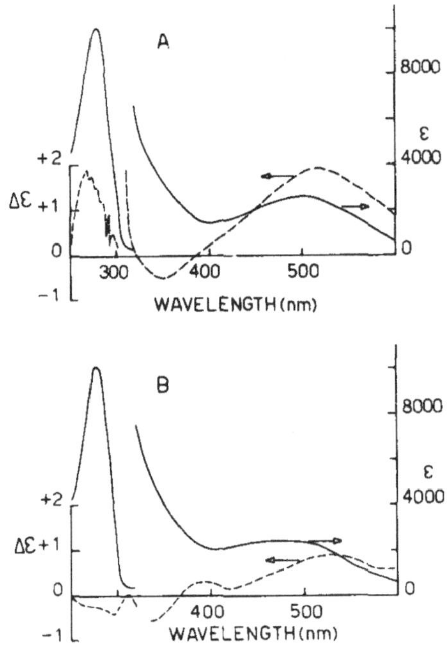

FIGURE 1. Electronic and CD spectra of LSAO. (A) Native LSAO in 0.1 *M* phosphate buffer, pH 7; (B) copper-free LSAO, same buffer as A.

bands are not observed with copper-free enzymes.[7,26] They are related to the enzyme rather than to the substrate, since their shape and position are the same with different substrates. However, the relative intensities of the yellow bands depend upon the nature of the substrate. In the case of LSAO their intensity is increased by CN^- addition.[25]

The two copper atoms are undistinguishable on the basis of EPR spectra, and the relative spectroscopic parameters, consistent with the presence of Cu(II) in a tetragonal field, are very similar in all plants and animal enzymes (see Chapter 12). The EPR spectra of LSAO native, copper-free, and reconstituted with copper are shown in Figure 3.[25]

C. Substrate Specificity

Specificity varies widely among the different enzymes. The best substrates for all of them are putrescine and cadaverine. A wide specificity is shown by PSAO,[5,7,27-29] VLAO,[16] chick pea AO,[15] and soybean AO,[14] which act on mono-, di-, and polyamines. A restricted specificity is shown instead by ELAO, which oxidizes only aliphatic diamines of critical dimensions[19] and by LSAO, which acts on the same compounds and also on spermidine.[18]

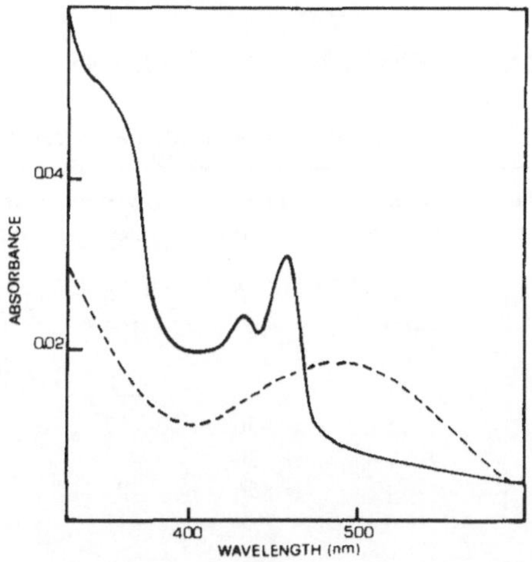

FIGURE 2. Reaction of LSAO with putrescine. Absorption spectra of 7.2×10^{-6} *M* LSAO in 0.1 *M* phosphate buffer, pH 7, in anaerobic conditions. (----), no addition; (—) after addition of putrescine (final concentration 1.7×10^{-2} *M*).

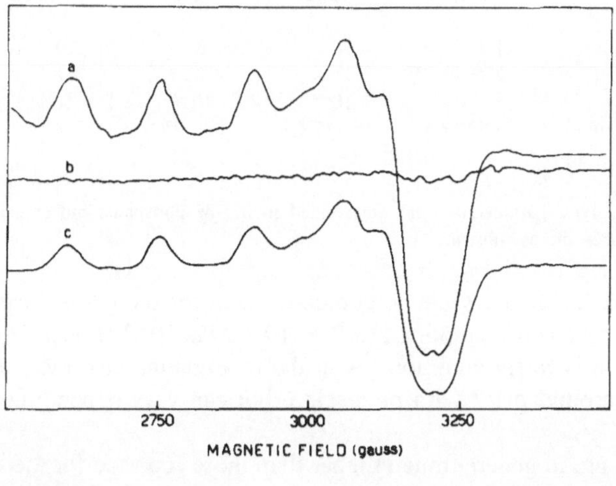

FIGURE 3. EPR spectra of native, copper-free, and reconstituted LS-AO. (a) Native LSAO (8.4×10^{-5} *M*) in 0.1 *M* phosphate buffer, pH 7; (b) copper-free LSAO (1.4×10^{-4} *M*), same buffer as (a); (c) same as (b) 48 hr after the addition of 5×10^{-4} *M* $CuCl_2$ and dialysis against 0.1 *M* phosphate buffer, pH 7, 10^{-4} *M* EDTA. EPR conditions: microwave frequency, 9.15 GHz; microwave power, 15 mW; field modulation amplitude, 10 gauss; temperature, 123K.

As an example, the relative reaction rates with the most significant substrates of several enzymes are shown in Table 4.

A general observation concerning substrate specificity has been made upon comparison of plant and animal enzymes: plant AOs seem to be directed more to putrescine and cadaverine oxidation, while the mammalian ones seem to be directed more towards histamine oxidation.[30-32]

D. Catalytic Activity Parameters

Apparent K_m values for putrescine seem to be of the same order of magnitude for all the

Table 4
SUBSTRATE SPECIFICITY OF PLANT CU-AOs

	Relative reaction rate (% of putrescine)[a]						
Substrate	*Euphorbia characias*[19]	*Lens esculenta*[18]	*Pisum sativum*[27]	*Glycine max*[14]	*Lathyrus sativus*[15]	*Lupinus luteus*[12]	*Vicia faba*[17]
1,3-Diaminopropane	0	0	0	0	0	0	2
Putrescine	100	100	100	100	100	100	100
Cadaverine	100	91	111	183	81	58	90
1,6-Diaminohexane	56	31	44	—	21	21	—
Agmatine	43	15	56	52	19	26	64
Histamine	0	0	30	4	—	5	9
Spermidine	0	4	56	81	31	—	52
Spermine	0	0	8	15	18	—	8

[a] All reported values were obtained at pH 7.0 in phosphate buffer.

Table 5
CATALYTIC PROPERTIES OF PLANT CU-AOs

	ELAO[19]	LSAO[18]	PSAO[8,11]	VLAO[16]
pH Optimum	7.1	7.1	7.0	7.0
K_m (*M*)	2×10^{-4}	2.3×10^{-4}	8.2×10^{-5}	3.7×10^{-5}
Specific activity (μmol/min mg)	27(37°C)	63(37°C)	70(37°C)	22(30°C)

Note: Catalytic parameters were determined in 0.1 *M* phosphate buffer using putrescine as substrate.

enzymes investigated (Table 5), while no comparisons of the oxygen K_m can be made, since only one apparent value has been reported (8.2×10^{-5} *M* at 10^{-3} *M* tryptamine for PSAO).[33]

Substrate inhibition is fairly common, as in the mammalian enzymes, and pH optimum is constantly found around pH 7 with putrescine, but can vary depending on the nature of the substrate used.

Specific activities are in general much higher than those reported for the animal enzymes.

E. Inhibitors

Metal chelators inhibit all plant Cu-AOs noncompetitively.[6,12,14,19] The different enzymes have quite different sensitivities. Diethyldithiocarbamate, which is generally used to obtain copper-free enzymes, is reported to be a powerful inhibitor of the *Vicia* enzyme,[16,17] of the lentil enzyme ($K_i = 2.5 \times 10^{-5}$ *M*),[18] and of the *Euphorbia* enzyme ($K_i = 4.3 \times 10^{-4}$ *M*),[19] but it suppresses the enzymic activity of the *Lathyrus* enzyme only at concentrations higher than 10^{-2} *M*.[15]

The prosthetic group probably contains a carbonyl function since carbonyl group reagents form irreversible adducts with all Cu-AOs with concomitant loss of catalytic activity. The most effective agents are hydrazines and hydrazides. The reaction with phenylhydrazine is characterized by high affinity, no oxygen consumption,[34] and the formation of an adduct showing a strong absorption band with a maximum at 415 to 430 nm ($\epsilon = 44$ to 48×10^3 M^{-1} cm^{-1}).[7,35] The band is stable, is present also in the copper-free enzyme, and is displaced at lower wavelength, but not abolished by enzymatic proteolysis.[34] In the pea enzyme, however, it has been reported that the 415-nm absorption fades after some time, being replaced by a new absorption band at 342 nm.[33]

Different from the other plant enzymes, ELAO has been found to be sensitive to thiol reagents in the presence of aminic substrate. Half-site reactivity is found towards *N*-ethylmaleimide[23] and iodoacetate;[36] both reagents produce 50% inactivation and modification of only one thiol per dimer; 100% inactivation and modification of both reactive thiols has been obtained with 4,4′-dithiodipyridine.

A series of reversible inhibitors of PSAO has been described. Carnosine, anserine, and impromidine appear to behave as noncompetitive inhibitors, while other imidazolic derivatives do not affect the catalytic activity.[30,32] Methylglyoxalbis(guanylhydrazone) (MGBG) behaves instead as a competitive inhibitor ($K_i = 2.6 \times 10^{-4}$ *M*) if it is not preincubated with the enzyme. After 1-hr incubation, the kinetics seem to reflect a noncompetitive inhibition ($K_i = 1.7 \times 10^{-5}$ *M*). The affinity of MGBG for the pea enzyme has permitted the use of an affinity chromatography technique which uses MGBG-Sepharose to purify PSAO.[11]

Hydrogen peroxide, which is one of the products of the enzymatic reaction, is an inactivating agent for all Cu-AOs if not rapidly removed by the action of catalase. In the pea enzyme, protection by catalase is abolished in the presence of added cupric ions.[37] Hydrogen peroxide inactivation of ELAO is concomitant with the modification of the reactive thiols;[23] nothing else is known about the mechanism of inhibition in this and other Cu-AOs.

F. Stoichiometry of the Catalyzed Reaction

When substrate interacts with native Cu-AO in the absence of oxygen, or with copper-free enzyme in air, aldehyde and ammonia production should be stoichiometric with the active site concentration since the enzyme cannot be reoxidized. Stoichiometry of amine consumption and aldehyde production has been determined for the native and copper-free lentil enzyme.[26] In both cases, 2 mol of aminic substrate were found to be oxidized and 2 mol of aldehyde produced per mole of enzyme.*

Since phenylhydrazine binds irreversibly to the enzyme, the reaction stoichiometry may be determined by following the adduct spectral features. In all four purified AOs, 1 mol of phenylhydrazine was found to react per mol of enzyme, almost abolishing the catalytic activity (residual activity is about 1% in LSAO). Since the binding of phenylhydrazine is prevented by the presence of amine substrate,[27,35] it might be proposed that the same binding site is involved in both interactions.

Discrepancy exists, therefore, between the two sets of results obtained with LSAO, i.e., titration with substrate (two sites per dimer) and with phenylhydrazine (one site per dimer). The presence on the dimeric molecule of two binding sites showing an anticooperative behavior might be an attractive hypothesis. Other explanations can, however, be proposed.

G. Catalytic Sequence and Oxygen Binding

Preliminary results obtained by stopped-flow experiments with LSAO indicate that, upon mixing the enzyme with substrate under anaerobic conditions, the bleaching of the visible absorption precedes the formation of the yellow intermediate; on the other hand, upon mixing with oxygen this intermediate disappears with a rate equal to the reappearance of the 490-nm absorption band. The kinetically determined affinity of the reduced enzyme for oxygen appears to be fairly high ($K_D \leq 10^{-5}$ *M*) and the observed rate constant seems O_2 concentration dependent.[38]

There are indications that the binding of oxygen to AOs triggers a conformational variation. Ammonia produced by reaction of substrates with the lentil enzyme is not released in the absence of oxygen and in some cases (i.e., in the reaction of LSAO with putrescine) the

* Ammonia is not released in the absence of oxygen and could therefore be determined only after reaction of copper-free LSAO with substrate: 2 mol of ammonia was produced per mole of dimer.

Table 6
SOME PROPERTIES OF OATS, BARLEY, AND MAIZE SEEDLING PAO

Properties	Oats[20,43]	Barley[20]	Maize[22]
Molecular weight			
Gel chromatography	85,000	85,000	65,000
$s_{20,w}(M_r)$	7.26(130,000)	—	5.95(65,000)
SDS-PAGE	—	—	65,000
Activation energy (kj/mol)	19.7	31.0	—
Absorption maxima (visible range)	350—450	—	380—460
Cofactors	ND	ND	FAD

Note: ND = Not determined.

same appears to be true for the aldehydic product.[26] Product release after oxygen admission is observed also with copper-free LSAO, suggesting that oxygen binds to the enzyme also in the absence of copper.

III. POLYAMINE OXIDASES

A. Introduction

Polyamine oxidases (PAOs) (E.C.1.5.3.3.) provide the only specific pathway of spermine and spermidine degradation in higher plants. The enzyme present in extracts of barley leaves[39] catalyzes the oxidation of spermidine and spermine at the secondary amino groups to give 1-pyrroline and 1-(3-aminopropyl)-pyrroline, respectively. 1,3-Diaminopropane and hydrogen peroxide are coproducts in both cases. Also, several Cu-AOs (e.g., PSAO) can oxidize polyamines, but these enzymes attack only primary amino groups, with the formation of aldehyde, ammonia, and hydrogen peroxide.[40] In general, PAOs from plants resemble those from bacteria, in that they attack the secondary amino groups of the polyamines to produce diamine. A number of different plant species were investigated for PAO activity. The enzymatic activity was found in oats *(Avena sativa)*, maize (*Zea mays*), barley *(Hordeum vulgare)*, wheat (*Triticum vulgare*), and rye (*Secale cereale)* in order of diminishing activity.[28] Only the enzymes from oat seedlings,[20] maize shoots,[21,22] and barley shoots[39] have been purified and partially characterized. They will be described in the following section.

B. Molecular and Catalytic Properties

Very little is known about the molecular features of plant PAOs. Some properties of the enzymes purified from oats, barley, and maize seedlings are listed in Table 6.

Maize PAO appears to be a monomer. The molecular weights of the other two enzymes were obtained only under nondenaturing conditions, so that it is not possible to establish the presence or absence of a quaternary structure.

The absorption spectrum of oats PAO shows only weak shoulders at 350 and 450 nm,[20] while the enzyme from maize shows two maxima at 380 and 460 nm.[22] Under aerobic conditions the visible absorption bands of the maize enzyme are decreased by sodium dithionite or spermidine addition.[22]

The flavoprotein nature of maize PAO had been inferred by inhibition studies performed with acridine compounds such as quinacrine, rivanol, and acriflavine.[41] Recently, the yellow chromophore of the purified enzyme was extracted and identified as FAD by paper chromatography.[22]

The barley enzyme is inhibited by sodium sulfite: this inhibition is consistent with the possibility that this enzyme also contains FAD as the prosthetic group.

Table 7
OPTIMUM pH AND RELATED K_m FOR OATS, BARLEY, AND MAIZE PAO

Enzyme	Substrate	Optimum	K_m	Ref.
Oats	Spermidine	6.5	8×10^{-6} *M*	43
	Spermine	6.6	2×10^{-6} *M*	
Barley	Spermidine	8.0	2×10^{-5} *M*	44
	Spermine	4.8	3×10^{-5} *M*	
Maize	Spermidine	6.0	1.2×10^{-5} *M*	22
	Spermine	5.5	2.3×10^{-5} *M*	

Under aerobic conditions the maize enzyme can reduce *p*-benzoquinone but not other electron carriers such as 2,2-dichlorophenol-indophenol, phenazine methosulfate, and ferricyanide. At variance with other flavoprotein oxidases, maize PAO is not inhibited but stimulated in the presence of sodium sulfite.[21]

PAO from maize, oats, and barley show the same substrate specificity. All these enzymes oxidize spermidine and spermine, and their catalytic activity is influenced by pH (Table 7). Putrescine, cadaverine, histamine, and other amines which are substrates for Cu-AOs are not oxidized. Analogs of spermidine and spermine, in which the carbon chain is shortened or lengthened, cause competitive inhibition of oats[42] and barley[43] enzymes. Spermidine analogs having 8 and 10 methylene groups are effective as inhibitors of barley PAO (K_i = 10^{-5} *M* and 5×10^{-6} *M*, respectively).

Several metal binding agents, thiol and carbonyl reagents, were without effect on PAO from different sources. Also, parnate (*trans*-2-phenylcyclopropylamine), which is a potent inhibitor of animal MAOs,[44] does not inhibit the maize PAO. Maize enzymic activity is inhibited by $CuSO_4$ but is unaffected by $AgNO_3$ or $CdSO_4$, suggesting that Cu^{++} might react with the flavin component of the enzyme.[21] The oats enzyme is activated by the chlorides of Li^+, Na^+, K^+, and Rb^+ in order of increasing effectiveness,[45] while Mg^{++} and Ca^{++} ions have an inhibitory effect.

IV. PHYSIOLOGICAL ASPECTS

Very little is known at present about the physiological role and the regulation mechanism of the enzymes catalyzing the degradation of amines and polyamines in the plant kingdom, although a great bulk of literature is to be found on the distribution and multiplicity of function of polyamines.[46]

A role of AOs in the production of GABA,[47] β-alanine,[48] and hydrogen peroxide[27] has also been proposed.

In recent years some data concerning onset of synthesis and cellular localization, mainly of PSAO and barley PAO, have been reported and will be briefly reviewed here.

PSAO appears to be absent from ungerminated seeds, but activity increases during the early periods of germination.[27,49] The enzyme activity is greatest in the growing regions and decreases with the maturity and senescence of the plant.[50] In pea seedlings the cotyledon enzyme differs from the embryo enzyme in respect to electrophoretic mobility, thermal stability, and substrate affinity. Moreover, only the cotyledon enzyme is induced by its substrates[51] and by phytic acid.[52]

PSAO activity appears to be regulated by a feedback mechanism through auxins and is influenced by various compounds affecting the metabolism of these hormones.[51-55]

Natural inhibitors of Cu-AOs have been detected in germinating seeds. Two distinct inhibitors have been reported to be present in germinated seeds of *Sorghum vulgare*. They are not inactivated by TCA: one is heat labile and inhibits PSAO in vitro; the second is heat stable and inhibits pea seeds germination.[56] A lipid-protein complex has been found to be responsible for the apparent absence of AO activity in groundnut *(Arachis ipogea)* cotyledons: while the embryo enzyme increase may be followed during germination, the cotyledons enzyme is not detected if Triton X-100 is not added to the extracts.[57]

Preliminary work with oats, barley, and maize has suggested the association of PAO with a particulate fraction.[20,39,41] A more recent study demonstrates that PAO normally occurs in the cell wall[58] and that it is particularly active in vascular tissue.[58,59] Nothing is known at present about PAO activity regulation in vivo.

ACKNOWLEDGMENT

The original studies reported here were supported in part by the Italian National Research Council, Special Project on the "Chimica Fine", Contract No. 83.00310.95.

REFERENCES

1. **Werle, E. and Raub, A.,** Über Vorkommen, Bildung und Abbau biogener Amine bei Pflanzen unter besonderer Berucksichtigung des Histamins, *Biochem. Z.*, 318, 538, 1948.
2. **Werle, E. and Zabel, A.,** Über die Verbreitung der Histaminase im Pflanzenreich, *Biochem. Z.*, 318, 554, 1948.
3. **Smith, T. A.,** Plant amines, in *Encyclopedia of Plant Physiology*, Vol. 8, Bell, E. A. and Charlwood, B. V., Eds., Springer-Verlag, Berlin, 1980, 433.
4. **Werle, E., Trautschold, I., and Aures, D.,** Reinigung und Charakterisierung der Diamineoxidase aus Erbsen, *Z. Physiol. Chem.*, 326, 200, 1961.
5. **Mann, P. J. G.,** Purification and properties of the amine oxidase of pea seedlings, *Biochem. J.*, 59, 609, 1959.
6. **Mann, P. J. G.,** Further purification and properties of the amine oxidase of pea seedlings, *Biochem. J.*, 79, 623, 1961.
7. **Hill, J. M. and Mann, P. J. G.,** Further properties of the diamine oxidase of pea seedlings, *Biochem. J.*, 91, 171, 1964.
8. **McGowan, R. E. and Muir, R. M.,** Purification and properties of amine oxidase from epicotyls of *Pisum sativum, Plant Physiol.*, 47, 644, 1971.
9. **Srivastava, S. K. and Prakash, V.,** Purification and properties of pea cotyledon and embryo diamine oxidase, *Phytochemistry*, 16, 189, 1977.
10. **Macholan, L. and Haubrova, J.,** Isolation and some characteristics of diamine oxidase from etiolated pea seedlings, *Collection Czechoslov. Chem. Commun.*, 41, 2987, 1976.
11. **Yanagisawa, H., Hirasawa, E., and Suzuki, Y.,** Purification and properties of diamine oxidase from pea epicotyls, *Phytochemistry*, 20, 2105, 1981.
12. **Schutte, H. R., Knofel, D., and Heyer, O.,** Biosynthese von Lupinenalkaloiden 19. Anreicherung und Charakterisierung der Diaminoxydase aus Keimlingen von *Lupinus luteus, Z. Pflanzenphysiol.*, 55, 110, 1966.
13. **Mizusaki, S., Tanabe, Y., Noguchi, M., and Tamaki, E.,** *N*-Methylputrescine oxidase from tobacco roots, *Phytochemistry*, 11, 899, 1972.
14. **Suzuki, Y.,** Some properties of the amine oxidase of soybean seedlings, *Plant Cell Physiol.*, 14, 413, 1973.
15. **Suresh, M. R., Ramakrishna, S., and Adiga, P. R.,** Diamine oxidase of *Lathyrus sativus* seedling, *Phytochemistry*, 15, 483, 1976.
16. **Matsuda, H. and Suzuki, Y.,** Purification and properties of the diamine oxidase from *Vicia faba* leaves, *Plant Cell Physiol.*, 22, 737, 1981.
17. **Matsuda, H. and Suzuki, Y.,** Some properties of the amine oxidase in *Vicia faba* seedlings, *Plant Cell Physiol.*, 18, 1131, 1977.
18. **Floris, G., Giartosio, A., and Rinaldi, A.,** Diamine oxidase from *Lens esculenta* seedlings: purification and properties, *Phytochemistry*, 22, 1871, 1983.

19. **Rinaldi, A., Floris, G., and Finazzi-Agrò, A.,** Purification and properties of diamine oxidase from *Euphorbia latex, Eur. J. Biochem.*, 127, 417, 1982.
20. **Smith, T. A.,** Polyamine oxidase from barley and oats, *Phytochemistry*, 15, 633, 1976.
21. **Hirasawa, E. and Suzuki, Y.,** Purification and properties of a polyamine oxidase from *Zea mays, Phytochemistry*, 14, 99, 1975.
22. **Suzuki, Y. and Yanagisawa, H.,** Purification and properties of maize polyamine oxidase: a flavoprotein, *Plant Cell Physiol.*, 21, 1085, 1980.
23. **Floris, G., Giartosio, A., and Rinaldi, A.,** Essential sulfhydryl groups in diamine oxidase from *Euphorbia characias* latex, *Arch. Biochem. Biophys.*, 220, 623, 1983.
24. **Kluetz, M. D., Adamson, K., and Flynn, J. E., Jr.,** Cryoenzymology and spectrophotometry of pea seedling diamine oxidase, *Biochemistry*, 19, 1617, 1980.
25. **Finazzi-Agrò, A., Rinaldi, A., and Rotilio, G.,** Unpublished results.
26. **Rinaldi, A., Giartosio, A., Floris, G., Medda, R., and Finazzi-Agrò, A.,** Lentil seedlings amine oxidase: preparation and properties of the copper-free enzyme, *Biochem. Biophys. Res. Commun.*, 120, 242, 1984.
27. **Kenten, R. H. and Mann, P. J. G.,** The oxidation of amines by extracts of pea seedlings, *Biochem. J.*, 50, 360, 1952.
28. **Werle, E. and Pechmann, E.,** Über die Diamineoxydase der Pflanzen und ihre adaptative Bildung durch Bakterien, *Ann. Chem.*, 562, 44, 1949.
29. **Werle, E. and Hartung, G.,** Zur Kenntnis der Diamineoxydase mit besonderer Berucksichtigung des Enzymes aus Rotklee, *Biochem. Z.*, 328, 228, 1956.
30. **Bieganski, T., Kushe, J., Feussner, K. D., Hesterberg, R., Richter, H., and Lorenz, W.,** Human intestinal diamine oxidase: substrate specificity and comparative inhibitor study, *Agents Actions*, 10, 108, 1980.
31. **Bieganski, T., Kushe, J., Feussner, K. D., Hesterberg, R., Richter, H., and Lorenz, W.,** The importance of human intestinal diamine oxidase in the oxidation of histamine and/or putrescine, *Arch. Immunol. Ther. Exp.*, 28, 901, 1980.
32. **Bieganski, T., Osinska, Z., and Maslinski, C.,** Inhibition of plant and mammalian diamine oxidases by hydrazine and guanidine compounds, *Int. J. Biochem.*, 14, 949, 1982.
33. **Yamasaki, E. F., Swindell, R., and Reed, D. J.,** Some aspects of catalysis by the amine oxidase of pea seedlings, *Biochemistry*, 9, 1206, 1970.
34. **Rinaldi, A., Floris, G., Vecchini, P., and Giartosio, A.,** Unpublished results.
35. **Rinaldi, A., Floris, G., Sabatini, S., Finazzi-Agrò, A., Giartosio, A., Rotilio, G., and Mondovì, B.,** Reaction of beef plasma and lentil seedlings Cu-amine oxidases with phenylhydrazine, *Biochem. Biophys. Res. Comm.*, 115, 841, 1983.
36. **Rinaldi, A., Floris, G., and Giartosio, A.,** Unpublished results.
37. **Hill, B. J. and Mann, P. J. G.,** The inhibition of pea-seedling diamine oxidase by chelating agents, *Biochem. J.*, 85, 198, 1962.
38. **Brunori, M., Belleli, A., Floris, G., Giartosio, A., Rinaldi, A., and Finazzi-Agrò, A.,** Manuscript in preparation.
39. **Smith, T. A.,** Purification and properties of the polyamine oxidase of barley plants, *Phytochemistry*, 11, 899, 1972.
40. **Hasse, K. and Schüher, K.,** Derivatives of γ-aminobutyraldehydes and δ-aminovaleraldehyde, *Biochem. Z.*, 336, 20, 1962.
41. **Suzuki, Y. and Hirasawa, E.,** Polyamine oxidase from *Zea mays* shoots, *Phytochemistry*, 12, 2863, 1973.
42. **Smith, T. A.,** Further properties of the polyamine oxidase from oat seedlings, *Phytochemistry*, 16, 1647, 1977.
43. **Smith, T. A.,** Polyamine oxidation by enzymes from *Hordeum vulgare* and *Pisum sativum* seedlings, *Phytochemistry*, 13, 1075, 1974.
44. **Zeller, E. A. and Sarkar, S.,** Amine oxidases. XIX. Inhibition of monoamine oxidase by phenylcyclopropylamine and iproniazid, *J. Biol. Chem.*, 237, 2333, 1962.
45. **Smith, T. A.,** Polyamine oxidase (oat seedlings), in *Methods in Enzymology*, Vol. 94, Tabor, H. and Tabor, G. W., Eds., Academic Press, New York, 1983, 311.
46. **Morris, D. R. and Marton, L. J.,** Polyamines in biology and medicine, in *The Biochemistry of Diseases*, Vol. 8, Marcel Dekker, New York, 1981.
47. **Terano, S. and Suzuki, Y.,** Biosynthesis of γ-aminobutyric acid from spermine in maize seedlings, *Phytochemistry*, 17, 550, 1978.
48. **Terano, S. and Suzuki, Y.,** Formation of β-alanine from spermine and spermidine in maize shoots, *Phytochemistry*, 17, 148, 1978.
49. **Werle, E., Beaucamp, G., and Shirren, V.,** Amine oxidase of pea seedlings and its importance in germination, *Planta*, 53, 125, 1959.

50. **Srivastava, S. K., Raj, A. D., and Naik, B. I.,** Polyamine metabolism during aging and senescence of pea leaves, *Ind. J. Exp. Biol.*, 19, 437, 1981.
51. **Srivastava, S. K., Prakash, V., and Naik, B. I.,** Regulation of diamine oxidase activity in germinating pea seeds, *Phytochemistry*, 16, 185, 1977.
52. **Prakash, V. and Sindhu, R. K.,** Effect of phytic acid on diamine oxidase activity in germinating pea seedlings, *Experientia*, 35, 855, 1979.
53. **Srivastava, S. K., Naik, B. I., and Raj, A. D.,** Effect of phenolics on plant amine oxidase, *Phytochemistry*, 21, 1519, 1982.
54. **Srivastava, S. K., Prakash, V., and Naik, B. I.,** Effect of ethrel and chloroethanol on pea diamine oxidase activity, *Phytochemistry*, 16, 1297, 1977.
55. **Sindhu, R. K. and Desai, H. V.,** Partial purification and characterization of diamine oxidase from groundnut embryo, *Ind. J. Biochem. Biophys.*, 17, 194, 1980.
56. **Joshi, B. H. and Prakash, V.,** Naturally occurring pea diamine oxidase inhibitors obtained from *Sorghum vulgare* during germination, *Experientia*, 38, 315, 1982.
57. **Sindhu, R. K. and Desai, H. V.,** Inhibitor of diamine oxidase in cotyledons of groundnut seedlings, *Phytochemistry*, 19, 317, 1980.
58. **Smith, T. A.,** Polyamine oxidase in higher plants, *Biochem. Biophys. Res. Commun.*, 41, 1452, 1980.
59. **Kaur-Shawheey, R., Flores, H. E., and Galston, A. W.,** Polyamine oxidase in oat leaves: a cell wall-localized enzyme, *Plant Physiol.*, 68, 494, 1981.

Chapter 7

ANIMAL INTRACELLULAR AMINE OXIDASES

Bruno Mondovì and Pierluigi Riccio

TABLE OF CONTENTS

I. INTRODUCTION

Amine oxidases (AO) have been found in a large number of tissues of practically all animal species. The major portion of this chapter is devoted to the copper-dependent enzymes, though some FAD-dependent amine oxidases will also be discussed. Special emphasis is placed on the swine kidney and human placental enzymes.

A good review of nonrecent papers has been reported by Kapeller-Adler.[1]

II. SWINE KIDNEY DIAMINE OXIDASE

The structure and catalytic mechanism common to all copper amine oxidases, including the swine kidney enzyme (SKDAO), are discussed in Chapters 10 to 12.

SKDAO (diamine:oxygen oxidoreductase, EC 1.4.3.6) is a Cu^{2+}-containing enzyme which oxidatively deaminates several diamines. Good substrates of SKDAO are those primary diamines which have a hydrocarbon chain length ranging from four to seven carbon atoms.[2,3] Monoamines, such as propylamine and butylamine, are also oxidized by SKDAO, although much less readily than diamines,[4] indicating the importance of the second amino group contribution to the enzyme-substrate affinity. Histamine is also a good substrate of this enzyme.

DAO and histaminase activities have long been attributed to distinct enzymes, but experimental evidence indicates that SKDAO is identical to histaminase.[5-7] The latter enzyme was first described by Best in 1929,[8,9] but more than 3 decades had to pass before DAO and histaminase activity were demonstrated to be present in the same enzyme molecule:[6] a swine kidney cortex extract containing DAO and histaminase activities was fractionated by column chromatography on DEAE and CM cellulose and by column electrophoresis. The ratio of cadaverine oxidase activity to histaminase activity remained nearly constant in all stages of purification. It was found that the optimal conditions for the assay of cadaverine and histamine oxidation by this enzyme differed; this fact should explain the different results obtained by other authors.[10] In addition, histaminase prepared according to the method of Kapeller-Adler and MacFarlane[10] and SKDAO purified according to Mondovì et al.[6] were active towards both cadaverine and histamine as substrates.

We have recently observed that immobilized AOs show an altered affinity for some substrates as compared to the free enzymes;[109] this should be taken into account when considering DAO activity in vivo, there being a possible association between these enzymes and subcellular structures (e.g., membranes) (see Chapter 9).

The oxidation of substrates, other than diamines and histamine, is discussed in Chapter 3.

Some substrates at high concentration inhibit the enzyme activity:[11-14] the presence of a noncatalytic inhibitory substrate binding site has been claimed for SKDAO to explain this phenomenon,[11,12,15] though more recent kinetic studies on the swine serum enzyme suggest the occurrence of an "abortive" complex formation with substrate during the last step of the catalytic cycle, leading to the "resting" form of the enzyme[16-18] (see Chapter 10). NMR spectroscopy provides evidence for the reduction of water proton relaxation rates in the presence of high concentration of a substrate analog, 3-methyl-4-(β-dimethylammonium) ethyl imidazole.[19] This suggests that inhibition by excess substrate may occur via displacement of a copper-coordinated water molecule (see Chapter 12).

A discussion on the localization of SKDAO is reported in Chapter 9. This enzyme was obtained as an homogeneous protein by Mondovì et al.[6,20] and crystallized by Yamada et al.[21]

SKDAO, like other enzymes of its class, shows a pinkish-yellow color due to its absorption maximum at 480 nm, which diminishes upon the addition of substrate under anaerobic conditions. The absorption peak is restored by adding oxygen (see Chapter 11).

The molecular weight determination by sedimentation-diffusion and approach to equilibrium methods corresponded, respectively, to 135,000 and 119,500, but higher molecular weights were also reported. The true molecular weight appears to be approximately 180,000. Kluetz and Schmidt[22] calculated a molecular weight of 172,000 on the basis of sedimentation-diffusion and approach to equilibrium ultracentrifugation. The dependence of sedimentation coefficients upon the enzyme concentration accounts for these discrepancies and indicates that SKDAO undergoes an association-dissociation phenomena. Moreover, Pionetti[23] observed that SKDAO also undergoes an association-dissociation reaction that occurs under the influence of the oxygen concentration, which is one of the substrates. He reported a molecular weight of about 185,000 for the form obtained in an air-saturated solution (form I) and of about 725,000 for a form which is present in a practically oxygen-free medium (form IV).

In the presence of 4 *M* guanidine, the sedimentation coefficient dropped from 8.5 to about 5.[24] When 0.07 *M* mercaptoethanol was added to the guanidine-treated enzyme, an additional drop of the sedimentation coefficient to about 3 was obtained, indicating the presence of a disulfide bridge in the monomer.[24]

The enzyme molecule contains an average of 11.6 μmol of copper per milligram of protein, which corresponds to 2 copper atoms per enzyme molecule.[20] The above-mentioned association phenomena explain that the correspondence of the copper content to the enzyme molecule is not exact as referred to in previous reports.[6]

The DDC-treated SKDAO is catalytically inactive.[20] The activity is restored by adding suitable amounts of cupric copper; however, an excess of this metal inhibits the enzyme activity. Copper should be involved in the reoxidation step of the reduced enzyme. EPR spectral studies indicate that the copper signal is only slightly affected by substrate addition which does not bind directly to the metal ion.[16,20,25-27] A transient oxidoreduction mechanism has been proposed.[28] A detailed discussion on copper will be presented in Chapter 12.

SKDAO, like the other copper AOs, contains a second organic prosthetic group which was long thought to be pyridoxal phosphate (PLP).[5,24,29-32] However, a direct and unambiguous demonstration on the presence of PLP in the enzyme molecule was never available, as for all the Cu-AOs. One hypothesis that should be mentioned was that proposed by Hamilton[33] on the basis of the observation that FAD-dependent MAO, treated with oxidizing agents, changed its substrate specificity and thus became a diamine-oxidizing carbonyl reagent-sensitive enzyme (see Chapter 19). This author hypothesized an interesting model involving a modified flavine functioning as a carbonyl-carrying cofactor which also accounted for the role of copper in mediating the electron transfer from reduced cofactor to bound oxygen.[33]

In a recent report, the involvement of pyrroloquinoline quinone as the possible prosthetic group of copper AOs was suggested,[34] and this hypothesis was soon confirmed by the isolation of the novel coenzyme from a purified preparation of BSAO.[35] Due to the very recent time of the discovery, this new finding has not been considered in Chapter 10, which deals with the mechanism of action of the copper AOs. This, however, does not seem to prejudicate the effectiveness of the discussion, in that the presence of such a new coenzyme would not necessarily imply a mechanism of action different from that proposed in Chapter 10.

As concerns the mechanism of enzyme-substrate interaction, only some problems strictly related to SKDAO will be discussed in the present chapter; other information is available in Chapter 10.

Experiments, carried out with a series of compounds derived from chemical modification of substrates, proved very useful for the understanding of the kind of interaction that occurs between substrate and enzyme, as well as the structure of the active site of the latter. It is reasonable to believe that the active center of SKDAO contains binding sites for the amino

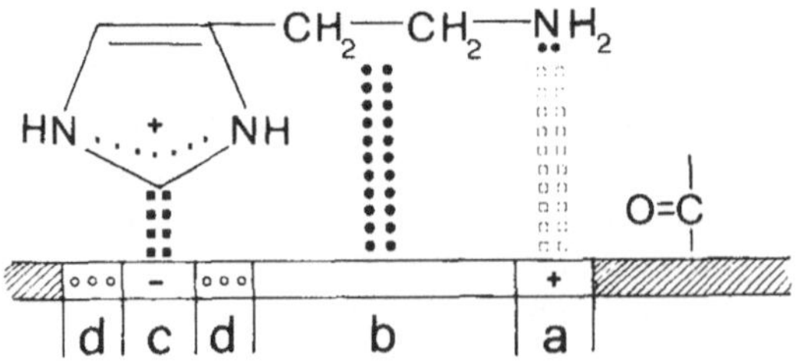

FIGURE 1. Schematic model for enzyme-substrate interaction at SKDAO active site. Histamine is represented. C=O: carbonyl function of the organic cofactor. (a) Electrophilic oxidizing binding site, which binds the unprotonated amino group to be removed (□□□: nucleophilic attraction). (b) Hydrophobic site which binds the hydrocarbon chain of substrate (●●●: Van der Waals attraction). (c) Negatively charged binding site, located at 6 to 9 Å distance from site a, which binds the second protonated amino group or the imidazole ring of histamine (■■■: electrostatic attraction). (d) Hydrophilic region surrounding site c.

group to be removed, for the aliphatic hydrocarbon chain of substrates, and for the other nitrogen-carrying function at the opposite and of the substrate. However, the interpretation of several experiments led to conflicting opinions on whether the protonated or unprotonated form of the substrate was required.

With respect to the pH dependence, it was at first assumed that with both cadaverine (and other diamines) and histamine, the real substrate was represented by the dicationic form of the diamine.[36] The affinity of the enzyme towards histamine is higher at a lower pH.[11] This fact confirmed the results obtained with less pure enzyme preparations.[2,4,37] It was also observed that a lowering of K_m values for histamine, as the pK of protonation of the imidazole nitrogen is approached and the incapability of the enzyme to oxidize lysine while oxidizing its methyl ester with a K_m value comparable to that of butylamine, could be explained by the presence of a negatively charged group at site c[11] (see Figure 1), as earlier suggested by Zeller[37] and Bardsley et al.[2] A specific interaction between this site and the imidazolyl nucleus of the substrate has been claimed to be the most effective determinant of the substrate specificity of SKDAO.[11]

On the other hand, experimental evidence was available which induced some authors to suggest that it was the monocationic form of diamines which was recognized by the enzyme's active site. Zeller and co-workers[4,38] observed, in fact, that hydroxy amines were good substrates of SKDAO, whereas monoquaternary derivatives were not; therefore, they believed the amino group being removed to be in the protonated form, and the other amino group (or the imidazole nitrogen of histamine) to be uncharged.

Bardsley et al.[15] excluded the possibility that the substrate could act as a nucleophile towards an electrophylic enzymic group, as described by Zeller et al., on the basis of an experiment demonstrating the ability of SKDAO to oxidize compounds containing one amino group and one onium group, being obviously impossible for the latter to act as a nucleophile. The only compounds incapable of being oxidized were those containing a trimethylammonium group, as previously observed by Zeller.[4,38] Thus they hypothesized a negatively charged binding site for the amino group to be removed where the cofactors are located and, at 6 to 9 Å from this oxidizing site, another negatively charged binding site for the second amino group surrounded by a hydrophilic region; between the two sites, a hydrophobic region should be present which interacts with the hydrocarbon chain of substrate.[15]

To date, the most reliable hypothesis appears to be that supported by some recent steady-state and transient-state kinetic studies which were carried out on the SSAO by Lindström et al.[39] According to the authors, it is probably the unprotonated form of the amino group to be removed that acts as a nucleophile towards a protonated group at the active site of the enzyme molecule (see Figure 1; for details, see also Chapter 10). There is no reason to believe that other copper AOs, including SKDAO, could have differently structured active sites. Actually, a Schiff-base formation occurs between a carbonyl function and an uncharged, nucleophylic amino group, and this type of enzyme-substrate association scheme is favorable to the covalent binding of substrate by the enzyme's carbonyl cofactor.

Concerning the other two sites (b and c, see Figure 1), the above-mentioned hypothesis[15] should be considered as the most reliable.

Most Cu-AOs, including SKDAO, are functional as dimeric enzymes composed of two apparently identical subunits. Despite this, phenylhydrazine titration of the enzymes from various sources indicate the presence of one carbonyl cofactor per dimer.[40-42] Moreover, the addition of substrate to the enzymes in anaerobiosis gave evidence of the release of 1[13] or 2[43-45] mol of aldehyde per mole of enzyme. The problem appears to be intriguing indeed.

Several hypotheses can be proposed: a single active site[46] located between the two subunits which participate in defining its structure;[47] a single carbonyl cofactor shared between two active sites[42] two independent active sites which would be functional one at a time (''flip flop'').[42] Further experimenting is needed to shed light upon this problem.

It has been ascertained that SKDAO operates by a ping-pong bi-ter (double displacement) mechanism.[13,16,48-50] The sequence of substrate binding and product release will be discussed in Chapter 10.

Presteady state kinetic studies on SKDAO, following optical density decrease at 500 nm, proved the presence of two distinct phases: a fast one followed by a plateau, then a second slower optical density decrease. The reduced enzyme reacts with oxygen with a velocity comparable to the fast phase of reduction. Substrate and oxygen concentration dependence of the various phases was observed.[28]

The mechanism of diamine oxidation by SKDAO and other kinetic problems encountered in its investigation will not be discussed in depth in the present chapter (for a detailed discussion see Chapters 10 to 12).

H_2O_2, produced during the reaction with substrate, exerts an inhibitory effect on the enzyme activity. The mechanism of enzyme inactivation by H_2O_2 is not clear; however, experimental evidence demonstrates that substrate is required to obtain a decrease of enzyme activity, as observed in both pea seedlings[51] and swine kidney[52] enzymes, indicating that the reduced form of the enzyme is affected. Inactivation occurs at neutral pH values; at alkaline pH, inactivation has not been observed. It is worthwhile recalling that the steady-state ratio $H_2O_2/{\cdot}O_2^-$ strongly favors the peroxide at neutral pH, whereas the contrary happens at alkaline pH.

The production of superoxide anion radicals by SKDAO during its catalytic cycle was suggested by Rotilio et al.[53] Recently, the employment of superoxide dismutase active copper chelates and native 2Cu, 2Zn-superoxide dismutase showed a significant decrease of SKDAO activity in the presence of the former, whereas the latter had no effect, suggesting that the production of ${\cdot}O_2^-$ occurs at a site scarcely accessible to such a large molecule as native superoxide dismutase.[54]

SKDAO shows a noteworthy resistance towards heat denaturation. Actually, all purification methods of this enzyme include a heat denaturation step. SKDAO is further protected against heat denaturation by the presence of some inhibitors, the most effective of which was arcaine. This inhibitor allows the enzyme to retain almost its full activity even after 30 min of incubation at 60°C.[55]

SKDAO does not appear to be immunologically related to the serum enzyme: cross reactions were not observed between serum and kidney enzymes, while bovine kidney and swine kidney DAOs share immunological determinants.[56]

III. HUMAN PLACENTAL AMINE OXIDASES

Among the numerous reports on the subject, it appears rather difficult to synthesize the main findings on human placental amine oxidases because of contrasting results reported by different authors.

As a result of several studies on human placental tissue, amine oxidizing activity has been detected and attributed to three distinct enzyme species: two of them are soluble amine oxidases[57,58] and the remaining is an insoluble, mitochondrial monoamine oxidase.[59]

Bardsley et al.[60] isolated the two soluble enzymes: one, a classical DAO having very similar properties and showing comparable substrate specificity to that of swine kidney; the other shows typical monoamine oxidases (MAO) properties except that it is soluble, cytoplasmic in origin, and oxidizes many diamines, although rather weakly in comparison to catecholamines.

MAO from human placental mitochondria was purified by Youdim and Sandler:[59] two isoenzymes appeared to be present in the human placental preparation.

A. Soluble Diamine Oxidase

Human placental diamine oxidase (HPDAO) is a Cu^{2+}-containing glycoprotein[59,61,62] which is active towards several diamines and histamine.[60] HPDAO is highly sensitive to aminoguanidine and, to a minor degree, to semicarbazide.[61] Crabbe et al.[62] suggested that HPDAO differed from other copper amine oxidases in terms of molecular weight of the active form of the enzyme, which would be a monomer weighing about 80,000. Higher molecular weight species were found to be multiple aggregates of the monomer.

The properties of HPDAO are very similar to those of the pregnancy serum DAO. This latter was purified by Baylin and Margolis,[63] who reported a molecular weight and substrate specificity similar to those of the placental enzyme. This would support the hypothesis that the pregnancy serum enzyme originates in the placenta, but we should cautiously consider that DAO activity can be also present in the serum of some nonpregnant women and sometimes in males.[60]

It has been established that HPDAO and the DAO present in the effusion of cancer patients and tumor extracts are biochemically and immunologically identical.[64,65] The development of a radioimmunoassay procedure employing a monospecific antiserum enabled Lin and Angellis to detect as little as 15 pg of DAO,[66] suggesting its use as a biochemical marker in human cancer (see Chapter 17).

B. Soluble and Mitochondrial Monoamine Oxidases

Very little is known about the soluble MAOs. Its activity can be easily discriminated from that of DAO because of its insensitivity to aminoguanidine and its resistance to purification procedures which destroy HPDAO.[60]

Little more information is available concerning the mitochondrial MAO from human placenta. It was suggested to be a single enzyme, different from MAO of other organs, which closely resembled some properties of type A MAO.[67,68] Other authors reported the probable existence of two isoenzymes in human placental preparations on the basis of a study employing four different substrates (tyramine, benzylamine, 5 HT, and PEA)[69] and polyacrilamide gel electrophoresis.[59]

Liver, brain, platelet, and placental MAOs are immunologically related and may also be identical.[70]

The possible functional significance of this enzyme system during pregnancy could be the protection of the organism against the high amounts of biogenic amines produced by the fetoplacental unit. AOs probably represent a focal control point in the metabolic activity of the fetoplacental unit during gestation, the regulation of its activity implementing not only biogenic amines metabolism, but also placental collagen biosynthesis and probably placental growth by influencing the rate of cell growth and division[62] (see Chapter 14).

Pregnant rats, injected with aminoguanidine, excreted five to ten times the amount of putrescine put out by the nontreated controls. Cadaverine excretion was also much higher. The increase in spermidine output by aminoguanidine-treated rats was less marked.[71] These facts confirm the suggestion that polyamine synthesis is greatly increased in pregnancy.

A spermidine oxidase activity was observed to be elevated in human pregnancy serum. Spermidine and diamine oxidase activities may be catalyzed by a single enzyme protein.[72]

IV. POLYAMINE OXIDASE

The existence in rat liver of an enzyme capable of forming putrescine from spermidine and spermidine from spermine, was demonstrated by Hölltä and was named "polyamine oxidase" (PAO).[73]

PAO utilizes O_2 as the electron acceptor; 3-aminopropionaldehyde and H_2O_2 are released as products of both spermidine and spermine oxidation. In addition to these polyamines, N^1-acetyl spermidine can be utilized by PAO as substrate (see Chapter 3).

The enzymes from rat liver appears to consist of a single polypeptide chain having a molecular weight of about 60,000.[73]

The enzyme is probably localized in peroxisomes (see Chapter 9), though PAO activity was also found in the cytosolic fraction; this may be due to a leakage of the enzyme protein out of the peroxisomes during the purification procedure or to the existence of an isoenzyme.[73] PAO appears to be a flavoprotein with FAD as the prosthetic group since the native enzyme could be resolved to an apoenzyme that is partially reactivated by FAD. Furthermore, the enzyme activity is abolished by inhibitors of FAD-dependent enzymes.[73]

PAO appears to contain sulfhydryl groups essential for the catalytic activity since sulfhydryl reagents inhibit the enzyme and exogenous thiols are needed for full activity.[73] The enzyme activity is also suppressed by carbonyl group reagents; however, pyridoxal phosphate is not likely to be the carbonyl-carrying function involved since compounds like isonicotinic acid hydrazine and canaline, which are typical inhibitors of pyridoxal enzymes, do not inhibit PAO.[73]

PAO may be of potential importance in regulating polyamine concentration in mammalian tissues. The enzyme activity should be associated with growth processes and malignant transformation[73] (see Chapter 17).

V. ANIMAL AMINE OXIDASES FROM OTHER SOURCES

Because of their role in controlling an essential metabolic pathway of cell metabolism, amine oxidizing enzymes are probably present in the cells of all tissues. However, due to the various kinds of metabolic reactions taking place in different organs (and in the different tissues composing each organ), which reflect their peculiar structural as well as functional differences, it is not surprising to find that distinct enzymes and different levels of activity for the same enzyme are present in each case.

Among the large number of reports on the matter, we have tried to summarize some information concerning the type of amine oxidase(s) present in various animal tissues.

A. Liver

Mouse,[74] rabbit,[75-77] swine,[78-81] and rat[59,73,82,83] liver AOs have been widely investigated. Mammalian liver contains at least three distinct oxidases which catalize the breakdown of mono-, di-, and polyamines and histamine.

Syatkin and Galaev[74] reported the presence of DAO activity in mouse liver homogenates; the enzyme deaminated putrescine, spermidine, and spermine.

Subcellular localization studies on rabbit liver gave evidence of the presence of DAO activity in the microsomal fraction:[75] subfractionation of microsomes indicated the presence of the enzyme in the smooth endoplasmic reticulum[76] (see Chapter 9).

Carper et al. reported the isolation of a copper-containing mitochondrial MAO from swine liver;[79,80] the presence of copper in this enzyme is questionable. In fact, Yasunobu et al.[43] observed a copper protein impurity present in the Triton X-100 extracts of beef liver mitochondria.

Swine liver microsomes contain a flavoenzyme which is able to oxidize *sec*- and *tert*-amines. This enzyme also catalyzed the *N*-oxidation of a large number of synthetic drugs and naturally occurring alkaloids.[81]

PAO from rat liver has been described above.

B. Intestine

Human intestinal DAO activity was investigated by Kusche et al.[84,85] and Bieganski et al.[86,87] Histamine, naturally methylated histamine, and putrescine are good substrates of this enzyme. DAO activity decreases from the distal duodenum towards the pylorus.[85] The enzyme was localized in the mucosa and was distributed primarily in the cytoplasm, the only exception being the guinea pig where it was located in the particulate fraction.[87] Aminoguanidine strongly inhibited human intestinal DAO.[88]

During the purification procedure a soluble monoamine oxidizing enzyme was separated from DAO. The most favored substrates of intestinal MAO were tyramine, tryptamine, and serotonin, but it was not possible to classify it as either type A or B MAO.[89]

Because of its primary role in histamine catabolism, the intestinal DAO is considered to play a protective role against histamine in diseases such as ischemic bowel syndrome, mesenteric infarction, and ulcerative colitis[87] (see Chapters 15 and 18).

Further information on the physiopathological role of intestinal DAO is available in Chapters 14, 15, and 17.

C. Blood Vessels

Blood vessels appear to contain different AOs. Lysyl oxidase, one of these enzymes and probably the best characterized, is discussed in depth in Chapter 8.

With respect to the carbonyl reagent sensitivity, two different amine oxidizing activities were described in rat aorta: one semicarbazide-sensitive, the other resistant to this inhibitor.[90] With tyramine as the substrate, a considerable part of rat aorta AO activity was inhibited by 10^{-4} *M* semicarbazide; the residual activity was only slightly affected even in the presence of 10^{-3} *M* inhibitor.[90]

The employment of MAO inhibitors revealed the presence of three different kinds of amine oxidases in rat heart and mesenteric artery.[91] Norepinephrine was deaminated by a MAO that was blocked by clorgyline and pargyline; PEA and benzylamine were both oxidized by two distinct enzymes: one sensitive and the other resistant to the MAO inhibitor pargyline. The pargyline-resistant enzyme is identical to the above-mentioned semicarbazide-sensitive AO: its activity was found to be enhanced in the mesenteric artery of spontaneously hypertensive rat,[91] which was considered to be a suitable model of essential hypertension. Oxidation of 5-HT was not affected by semicarbazide[90] (see Chapter 8).

The semicarbazide-resistant enzymes were identified with the mitochondrial MAO, located in the outer mitochondrial membrane.[90] The semicarbazide-sensitive AO was mainly localized in the microsomal fraction and was ascribed to plasma membrane.[90]

A carbonyl reagent-sensitive AO was partially purified from rabbit aorta.[92] This enzyme possessed a substrate specificity pattern very similar to that found in rabbit plasma.

Caramona[93] studied the localization of both type A and B MAO in blood vessels with respect to their innervation pattern: type A MAO has both intra- and extraneuronal locations in vascular structures, whereas type B MAO is predominantly, if not exclusively, extraneuronal.

A copper-containing AO was isolated from bovine brain microvessels.[94] It was shown that the enzyme content in microvessels is significantly higher as compared with large blood vessels such as the aorta.

D. Thymus

Rat thymus DAO was described by Beaven and De Jong,[95] who found a high DAO activity which slightly declined after thymus involution in old rats. Depletion of thymus lymphocytes by the administration of a single dose of dexamethazone produced a 40 to 60% decrease in thymus weight, but not a decrease of DAO activity per thymus, indicating that the enzyme was located in thymic tissue other than lymphocyte.[95,96] However, eosinophils were found to contain DAO activity which increased in acute inflammatory diseases such as active asthma, cold-induced urticaria, and parasitic infestation[97] (see Chapter 18).

E. Connective Tissue

Amine oxidase in connective tissue is supposed to participate in the cross-linking reaction of collagen and elastin.

Dual amine oxidase activity was demonstrated in bovine ligamentum nuchae[98] with both free lysine and aortic elastin as substrates.

Connective tissue AOs are dealt with in Chapter 8.

F. Fibroblasts

DAO activity was measured in homogenates of in vitro cultured human fibroblasts from normal and cystic fibrosis patients.[99] The enzyme activity per microgram of protein peaked during the period of rapid proliferation and reached its lowest level when fibroblasts grew in the static (confluent) phase. No difference was found between normal and cystic fibrosis fibroblasts regarding DAO specific activity, but these latter exhibited a slightly lower protein content during logarithmic growth as compared with normal cells.

G. Dental Pulp

An amine oxidase was partially purified from bovine dental pulp[100] which oxidized preferentially spermidine and spermine, benzylamine, and peptidyllysine, such as lysine-vasopressine and prolyllysylglycinamide, but not collagen and elastin. The enzyme contains copper and appears to be very similar to the serum enzyme, except for the affinity towards a peptidyllysyl substrate. The dental pulp AO oxidizes lysine-vasopressine about ten times faster than serum AO.

H. Salivary Glands

Salivary glands contain a DAO that is able to rapidly oxidize putrescine, cadaverine, histamine, and, to a lesser degree, hexamethylene diamine and benzylamine.[101]

MAO activity in rat salivary glands occurs mainly in parenchymal cells and is located in both mitochondrial and microsomal fractions.[102]

I. Human Seminal Fluid

Zeller was the first to describe the presence of high DAO concentrations in human seminal plasma.[103] The enzyme was able to oxidize spermine present in human semen.

Hölttä et al.[104] obtained the enzyme as an homogeneous protein. The enzyme had its highest affinity for diamines, but the polyamines spermidine and spermine were also degraded at concentrations that could be considered physiological in human semen. The enzyme showed very similar properties to other DAO and particularly to the placental enzyme. In both cases the substrate spectrum was much the same except that Bardsley et al.[60] reported that spermidine was not oxidized by their purified placental enzyme preparation.

Human spermatozoa, separated from the seminal plasma, appeared to contain spermine and a distinct DAO.[105] Spermine, but not DAO activity, could be removed from spermatozoa by washing the cells with buffers containing a high concentration of salts.

The physiological significance of the human seminal plasma DAO is not known. It was suggested to be involved in the formation (or removal) of compounds which could influence the motility and survival of spermatozoa.[104]

The anthelmintic drug and DAO inhibitor levamisole, like many other DAO inhibitors, have been demonstrated to inhibit spermatozoa motility.[106]

VI. CONCLUDING REMARKS

This chapter reports some information on the large number of different amine oxidizing activities to focus on a problem which has not yet been evidenced in full. The problem arises from the observation that when examining amine oxidizing enzymes we usually find, together with structural and functional similarities between each other, some constant differences concerning substrates and inhibitor specificity patterns, molecular weights, optimal environmental conditions, etc.

The question is whether many distinct genes codifying many different enzyme proteins exist, or rather a few genes for a limited number of AOs, whose differences must therefore be ascribed to post-transcriptional or post-translational modifications.

It has already been mentioned that immobilized AOs change their affinity towards substrates and that the same probably occurs in vivo. Moreover, association-dissociation reactions, which we have observed as a general rule in all the purified AOs, as a dependence of different factors (some of which probably still are not known), may further influence the catalytic activity of these enzymes.[107]

The environmental effect and the possible presence of endogenous inhibitors as regulators of AO activity should be also taken into account.[108] Further investigation might be directed towards the elucidation of physiological regulation mechanisms of AO activity.

ACKNOWLEDGMENT

The original studies reported here were supported in part by the Italian National Research Council, Special Project on the "Chimica Fine", Contract No. 83.00310.95.

REFERENCES

1. **Kapeller-Adler, R.,** *Amine Oxidases and Methods for their Study,* Wiley-Interscience, New York, 1970.
2. **Bardsley, W. G., Hill, C. M., and Lobley, R. W.,** A reinvestigation of the substrate specificity of pig kidney diamine oxidase, *Biochem. J.,* 117, 169, 1970.
3. **Hill, C. M. and Bardsley, W. G.,** Histamine and relaed compounds as substrates of diamine oxidase (histaminase), *Biochem. Pharmacol.,* 24, 253, 1975.
4. **Fouts, J. R., Blanksma, L. A., Carbon, J. A., and Zeller, E. A.,** Monoamines as substrates of diamine oxidase, *J. Biol. Chem.,* 225, 1025, 1957.
5. **Davison, A. N.,** Pyridoxal phosphate as coenzyme of diamine oxidase, *Biochem. J.,* 64, 546, 1956.
6. **Mondovì, B., Rotilio, G., Finazzi-Agrò, A., and Scioscia-Santoro, A.,** Purification of pig kidney diamine oxidase and its identity with histaminase, *Biochem. J.,* 91, 408, 1964.
7. **Zeller, E. A.,** Identity of histaminase and diamine oxidase, *Fed. Proc.,* 24, 766, 1965.
8. **Best, C. H.,** The disappearance of histamine from autolysing lung tissue, *J. Physiol. (London),* 67, 256, 1929.
9. **Best, C. H. and Mac Henry, E. W.,** The inactivation of histamine, *J. Physiol. (London),* 70, 349, 1930.

10. **Kapeller-Adler, R. and McFarlane, H.,** Purification and identification of hog kidney histaminase, *Biochim. Biophys. Acta*, 67, 542, 1963.
11. **Costa, M. T., Rotilio, G., Finazzi-Agrò, A., Vallogini, M. P., and Mondovì, B.,** On the active site of diamine oxidase: kinetic studies, *Arch. Biochem. Biophys.*, 147, 8, 1971.
12. **Wang, T. M., Achee, F. M., and Yasunobu, K. T.,** Amine oxidase. XI. Beef plasma amine oxidase, a sulfhydryl protein, *Arch. Biochem. Biophys.*, 128, 106, 1968.
13. **Oi, S., Inamasu, M., and Yasunobu, K. T.,** Mechanistic studies of beef plasma amine oxidase, *Biochemistry*, 9, 3378, 1970.
14. **Kelly, D., Knowles, P. F., Yadav, K. D. S., Bardsley, W. G., Lepp, P., and Wright, R. D.,** Steady-state kinetic studies on benzylamine oxidase from pig plasma, *Eur. J. Biochem.*, 114, 133, 1981.
15. **Bardsley, W. G., Ashford, J. S., and Hill, C. M.,** Synthesis and oxidation of aminoalkyl-onium compounds by pig kidney diamine oxidase, *Biochem. J.*, 122, 557, 1971.
16. **Lindström, A., Olsson, B., and Pettersson, G.,** Kinetics of the interaction between pig plasma benzylamine oxidase and substrate, *Eur. J. Biochem.*, 35, 70, 1973.
17. **Lindström, A., Olsson, B., and Pettersson, G.,** Transient kinetics of benzaldehyde formation during the catalytic action of pig plasma benzylamine oxidase, *Eur. J. Biochem.*, 42, 377, 1974.
18. **Olsson, B., Olsson, J., and Pettersson, G.,** The kinetics of reoxidation of reduced benzylamine oxidase, *Eur. J. Biochem.*, 74, 329, 1977.
19. **Kluetz, M. D. and Schmidt, P. G.,** Proton magnetic relaxation dispersion in solutions of the cuproprotein diamine oxidase, *Biophys. J.*, 29, 283, 1980.
20. **Mondovì, B., Rotilio, G., Costa, M. T., Finazzi-Agrò, A., Chiancone, E., Hansen, R. E., and Beinert, H.,** Diamine oxidase from pig kidney. Improved purification and properties, *J. Biol. Chem.*, 242, 1160, 1967.
21. **Yamada, H., Kumagai, H., Kawasaki, H., Matsui, H., and Ogata, K.,** Crystallization and properties of diamine oxidase from pig kidney, *Biochem. Biophys. Res. Commun.*, 29, 723, 1967.
22. **Kluetz, M. D. and Schmidt, P. G.,** Diamine oxidase, molecular weight and subunit analysis, *Biochem. Biophys. Res. Commun.*, 76, 40, 1977.
23. **Pionetti, J. M.,** Analytical-band centrifugation of the active form of pig kidney diamine oxidase, *Biochem. Biophys. Res. Commun.*, 58, 495, 1974.
24. **Mondovì, B., Costa, M. T., Finazzi-Agrò, A., and Rotilio, G.,** Pig kidney diamine oxidase: a copper-pyridoxal enzyme, in *Symposium on Pyridoxal Enzymes*, Maruzen Co., Tokio, Japan, 1968, 143.
25. **Yamada, H. and Yasunobu, K. T.,** Monoamine oxidase II: copper, one of the prosthetic groups of monoamine oxidase, *J. Biol. Chem.*, 237, 3077, 1962.
26. **Buffoni, F., DellaCorte, L., and Knowles, P. F.,** The nature of copper in pig plasma benzylamine oxidase, *Biochem. J.*, 106, 575, 1968.
27. **Grant, J., Kelly, I., Knowles, P. F., Olsson, J., and Pettersson, G.,** Changes in the copper centres of benzylamine oxidase from pig plasma during the catalytic cycle, *Biochem. Biophys. Res. Commun.*, 83, 1216, 1978.
28. **Mondovì, B., Rotilio, G., Finazzi-Agrò, A., and Antonini, E.,** Amine oxidases: a new class of copper oxidases, in *Magnetic Resonance in Biological Research*, Franconi, C., Ed., Gordon & Breach, New York, 1971, 233.
29. **Goryachenkova, B. V. and Ershova, E. A.,** Studies of substrate specificity and cofactors for diamine oxidase of hog kidney, *Biokhimiya*, 30, 165, 1965.
30. **Mondovì, B., Costa, M. T., Finazzi-Agrò, A., and Rotilio, G.,** Pyridoxal phosphate as a prosthetic group of pig kidney diamine oxidase, *Arch. Biochem. Biophys.*, 119, 373, 1967.
31. **Finazzi-Agrò, A., Guerrieri, P., Costa, M. T., and Mondovì, B.,** On the nature of cromophore in pig kidney diamine oxidase, *Eur. J. Biochem.*, 74, 435, 1977.
32. **Blaschko, H. and Buffoni, F.,** Pyridoxal phosphate as a constituent of the histaminase (benzylamine oxidase) of pig plasma, *Proc. R. Soc. (London) B*, 163, 45, 1965.
33. **Hamilton, G. A.,** Oxidases with monocopper reactive sites. Galactose oxidase and copper-containing amine oxidases, in *Copper Proteins*, Spiro, T. G., Ed., John Wiley & Sons, New York, 1981, chap. 5.
34. **Ameyama, M., Hayashi, M., Matsushita, K., Shinagawa, E., and Adachi, O.,** Microbial production of pyrroloquinoline quinone, *Agric. Biol. Chem.*, 48, 561, 1984.
35. **Lobenstein-Verbeek, C. L., Jongejan, J. A., Frank, J., and Duine, J. A.,** Bovine serum amine oxidase: a mammalian enzyme having covalently bound PQQ as prosthetic group, *FEBS Lett.*, 170, 305, 1984.
36. **Blaschko, H., Friedman, P. J., Hawes, R., and Nilsson, K.,** The amine oxidases of mammalian plasma, *J. Physiol.*, 145, 384, 1959.
37. **Zeller, E. A.,** Diamine oxidase, in *The Enzymes*, Vol. 2, Summer, J. B. and Myrbäck, K., Eds., Academic Press, New York, 1951, 536.
38. **Zeller, E. A., Fouts, J. R., Carbon, J. A., Lazanas, J. C., and Voegtli, W.,** Uber die Substratspezifität der Diamin-oxydase, 12, Mitteilung über Amin-oxydasen, *Helv. Chim. Acta*, 39, 1632, 1956.

39. **Lindström, A., Olsson, B., Pettersson, G., and Szymanska, J.,** Kinetics of the interaction between pig plasma benzylamine oxidase and various monoamines, *Eur. J. Biochem.*, 47, 99, 1974.
40. **Lindström, A. and Pettersson, G.,** Active-site titration of pig-plasma benzylamine oxidase, *Eur. J. Biochem.*, 34, 564, 1973.
41. **Rinaldi, A., Floris, G., Sabatini, S., Finazzi-Agrò, A., Giartosio, A., Rotilio, G., and Mondovì, B.,** Reaction of beef plasma and lentil seedlings Cu-amine oxidases with phenylhydrazine, *Biochem. Biophys. Res. Commun.*, 115, 841, 1983.
42. **Falk, M. C.,** Stoichiometry of phenylhydrazine inactivation of pig plasma amine oxidase, *Biochemistry*, 22, 3740, 1983.
43. **Yasunobu, K. T., Ishizaki, H., and Minamiura, N.,** The molecular, mechanistic and immunological properties of amine oxidases, *Mol. Cell Biochem.*, 13, 3, 1976.
44. **Berg, K. A. and Abeles, R. H.,** Mechanism of action of plasma amine oxidase: products released under anaerobic conditions, *Biochemistry*, 19, 3186, 1980.
45. **Rinaldi, A., Giartosio, A., Floris, G., Medda, R., and Finazzi-Agrò, A.,** Lentil seedlings amine oxidase: preparation and properties of the copper-free enzyme, *Biochem. Biophys. Res. Commun.*, 120, 242, 1984.
46. **Mondovì, B., Sabatini, S., and Befani, O.,** The active site of copper amine oxidases, *J. Mol. Catal.*, 23, 325, 1984.
47. **Knowles, P. F. and Yadav, K. D. S.,** Amine oxidases, in *Copper Proteins and Copper Enzymes*, Vol. II, Lontie, R., Ed., CRC Press, 1984, chap. 4.
48. **Finazzi-Agrò, A., Rotilio, G., Costa, M. T., and Mondovì, B.,** Evidence for a ping-pong mechanism in the diamine oxidase reaction, *FEBS Lett.*, 14, 31, 1969.
49. **Bardsley, W. G., Crabbe, M. G. C., and Shindler, J. S.,** Kinetics of the diamine oxidase reaction, *Biochem. J.*, 131, 459, 1973.
50. **Taylor, C. E., Taylor, R. S., Rasmussen, C., and Knowles, P. F.,** A catalytic mechanism for the enzyme benzylamine oxidase from pig plasma, *Biochem. J.*, 130, 713, 1972.
51. **Mann, P. G. J.,** Purification and properties of the amine oxidase of pea seedlings, *Biochem. J.*, 59, 609, 1955.
52. **Mondovì, B., Rotilio, G., Finazzi-Agrò, A., and Costa, M. T.,** Diamine oxidase inactivation by hydrogen peroxide, *Biochim. Biophys. Acta*, 132, 521, 1967.
53. **Rotilio, G., Calabrese, L., Finazzi-Agrò, A., and Mondovì, B.,** Indirect evidence for the production of superoxide anion radicals by pig kidney diamine oxidase, *Biochim. Biophys. Acta*, 198, 618, 1970.
54. **Younes, M. and Weser, U.,** Involvement of superoxide in the catalytic cycle of diamine oxidase, *Biochim. Biophys. Acta*, 526, 644, 1978.
55. **Floris, G., Rinaldi, A., and Finazzi-Agrò, A.,** Heat denaturation of pig kidney diamine oxidase: effect of arcaine, *Ital. J. Biochem.*, 30, 251, 1981.
56. **Oratore, A., Guerrieri, P., Ballini, A., Mondovì, B., and Finazzi-Agrò, A.,** Preparation and properties of antibody against pig kidney diamine oxidase, *FEBS Lett.*, 104, 154, 1979.
57. **Danforth, D. N. and Gorham, F.,** Placental histaminase, *Am. J. Physiol.*, 119, 294, 1937.
58. **Thompson, R. H. S. and Tickner, A.,** Observations on the monoamine oxidase activity of placenta and uterus, *Biochem. J.*, 45, 125, 1949.
59. **Youdim, M. B. H. and Sandler, M.,** Isoenzymes of soluble monoamine oxidase from human placental and rat liver mitochondria, *Biochem. J.*, 105, 43 P, 1967.
60. **Bardsley, W. G., Crabbe, J. C., and Scott, I. V.,** The amine oxidases of human placenta and pregnancy plasma, *Biochem. J.*, 139, 169, 1974.
61. **Paolucci, F., Cronenberger, L., Plan, R., and Pacheco, H.,** Purification et propriétés de la diamine: oxygène oxydoreductase du placenta humain, *Biochimie*, 53, 735, 1971.
62. **Crabbe, M. G. C., Waight, R. D., Bardsley, W. G., Barker, R. W., Kelly, I. D., and Knowles, P. F.,** Human placental diamine oxidase, *Biochem. J.*, 155, 679, 1976.
63. **Baylin, S. B. and Margolis, S.,** Purification of histaminase (diamine oxidase) from human pregnancy plasma by affinity cromatography, *Biochim. Biophys. Acta*, 397, 294, 1975.
64. **Lin, C. W. and Kirley, S. D.,** Histaminase as an oncoplacental enzyme: biochemical and immunological evidence, in *Oncodevelopmental Gene expression*, Fishamn, W. H. and Sell, S., Eds., Academic Press, New York, 1976, 469.
65. **Lin, C. W., Inglis, N. R., Rule, A. H., Turksoy, R. N., Chapman, L. M., Kirley, S. D., and Stolbach, L. L.,** Histaminase and other tumor markers in malignant effusion fluids, *Cancer Res.*, 39, 4894, 1979.
66. **Lin, C. W. and Angellis, D.,** Tumor and placental histaminase. II. A procedure for radioimmunoassay, *Oncodevelop. Biol. Med.*, 2, 281, 1981.
67. **Egashira, T.,** Studies on monoamine oxidase (Report SVIII). Enzymic properties of placental monoamine oxidase, *Jpn. J. Pharmacol.*, 26, 493, 1976.
68. **Kikuchi, R. and Kinemuchi, H.,** Evidence for existence of type A MAO in mitochondria from human placenta, *Folia pharmacol. japon*, 74, 763, 1978.

69. **Oguchi, K., Kobayashi, S., and Koide, R.,** Studies on human placental monoamine oxidase using mixed substrates, *Jpn. J. Pharmacol.*, 30, 157, 1980.
70. **Russel, S. M., Davey, J., and Mayer, R. J.,** Immunochemical characterization of monoamine oxidase from human liver, placenta, platelets and brain cortex, *Biochem. J.*, 181, 15, 1979.
71. **Rojansky, N., Neufeld, E., and Chayen, R.,** Excretion of polyamines by the pregnant rat following inhibition of diamine oxidase, *Biochim. Biophys. Acta*, 586, 1, 1979.
72. **Gahl, W. A., Vale, A. M., and Pitot, H. C.,** Spermidine oxidase in human pregnancy serum, *Biochem. J.*, 201, 161, 1982.
73. **Hölttä, E.,** Oxidation of spermidine and spermine in rat liver; purification and properties of polyamine oxidase, *Biochemistry*, 16, 91, 1977.
74. **Slatkin, S. P. and Galaev, Y. V.,** Oxidation of putrescine, spermidine and spermine by mouse liver diamine oxidase, *Biokhimiya*, 42, 1010, 1977.
75. **Argento-Cerù, M. P., Sartori, C., and Autuori, F.,** Diamine oxidase in rabbit liver. I. Assay conditions and subcellular localization, *Eur. J. Biochem.*, 34, 369, 1973.
76. **Argento-Cerù, M. P., Sartori, C., and Autuori, F.,** Diamine oxidase in rabbit liver. II. Sub microsomal distribution, *Eur. J. Biochem.*, 34, 375, 1973.
77. **Argento-Cerù, M. P., Sartori, C., and Autuori, F.,** Association of diamine oxidase with microsomal membranes in rabbit liver, *Life Sci.*, 15, 1909, 1974.
78. **Carper, W. R., Stoddard, D. D., and Martin, D. F.,** The isolation of pig liver monoamine oxidase, *Biochem. Biophys. Res. Commun.*, 54, 721, 1973.
79. **Carper, W. R., Stoddard, D. D., and Martin, D. F.,** Pig liver monoamine oxidase. I. Isolation and characterization, *Biochim. Biophys. Acta*, 334, 287, 1974.
80. **Chew, L. H., Carper, W. R., and Issaq, H. J.,** Effects of metal substitution on pig liver monoamine oxidase, *Biochem. Biophys. Res. Commun.*, 66, 217, 1975.
81. **Ziegler, D. M. and Mitchell, C. H.,** Microsomal oxidase. IV. Properties of a mixed-function amine oxidase isolated from pig liver microsomes, *Arch. Biochem. Biophys.*, 150, 116, 1972.
82. **Collins, G. G. S. and Youdim, M. B. H.,** The binding of ^{14}C phenethylhydrazine to rat liver monoamine oxidase, *Biochem. Pharmacol.*, 24, 703, 1975.
83. **Kimura, T., Kodama, M., and Nagata, C.,** Purification of a mixed-function amine oxidase from rat liver microsomes, *Biochem. Biophys. Res. Commun.*, 110, 640, 1983.
84. **Kusche, J., Lorenz, W., and Schmidt, J.,** Oxidative deamination of biogenic amines by intestinal amine oxidases: histamine is specifically inactivated by diamine oxidase, *Hoppe-Seyler's Z. Physiol. Chem.*, 356, 1485, 1975.
85. **Kusche, J., Lorenz, W., Stahlknecht, C.-D., Friedrich, A., Schmidt, A., Boo, K., and Reichert, G.,** Diamine oxidase activity in gastric and duodenal mucosa of man and other mammals with special reference to the pyloric junction, *Agents Actions*, 8(4), 366, 1978.
86. **Bieganski, T., Kusche, J., Feussner, K. D., Hesterberg, R., Richter, H., and Lorenz, W.,** The importance of human intestinal diamine oxidase in the oxidation of histamine and/or putrescine, *Arch. Immunol. Ther. Exp.*, 28, 901, 1980.
87. **Bieganski, T., Kusche, J., Lorenz, W., Hesterberg, R., Stahlknecht, C.-D., and Feussner, K.-D.,** Distribution and properties of human intestinal diamine oxidase and its relevance for the histamine catabolism, *Biochim. Biophys. Acta*, 756, 196, 1983.
88. **Bieganski, T., Kusche, J., Feussner, K.-D., Hesterberg, R., Richter, H., and Lorenz, W.,** Human intestinal diamine oxidase: substrate specificity and comparative inhibitor study, *Agents Actions*, 10, 108, 1980.
89. **Kusche, J., Feussner, K.-D., and Lorenz, W.,** Intestinal monoamine oxidase: does it have a role in histamine catabolism?, *Agents Actions*, 12, 53, 1982.
90. **Wibo, M., Duong, A. T., and Godfraind, T.,** Subcellular localization of semicarbazine-sensitive amine oxidase in rat aorta, *Eur. J. Biochem.*, 112, 87, 1980.
91. **Fuentes, J. A. and Urdin, M. C.,** Cardiovascular amine oxidase activity and hypertension, in *Dynamics of Neurotransmitter Function*, Hanin, I., Ed., Raven Press, New York, 1984, 57.
92. **Rucker, R. B. and Goettlich-Riemann, W.,** Properties of rabbit aorta amine oxidase, *Proc. Soc. Exp. Biol. Med.*, 139, 286, 1973.
93. **Caramona, M. M.,** Localization of monoamine oxidase of type A and B in blood vessels with different innervation patterns, *Naunyn-Schmiedeberg's Arch. Pharmacol.*, 324, 185, 1983.
94. **Mehrabian, Z. B. and Nalbandyan, R. M.,** Benzylamine oxidase from brain microvessels, *FEBS Lett.*, 164, 89, 1983.
95. **Beaven, M. A. and de Jong, W.,** Presence of histaminase and ornithine decarboxylase activities in rat thymus, *Biochem. Pharmacol.*, 22, 257, 1973.
96. **Atkins, F. L. and Beaven, M. A.,** Ornithine decarboxylase and histaminase (diamine oxidase) activities in rat thymus and their relationship to the thymus lymphocytes, *Biochem. Pharmacol.*, 24, 763, 1975.

97. **Herman, J. J.,** Eosinophil diamine oxidase activity in acute inflammation in humans, *Agents Actions*, 12, 46, 1982.
98. **Serafini-Fracassini, A., Ventrella, G., Field, M. J., Hinnie, J., Onyezili, N. I., and Griffiths, R.,** Characterization of a structural glycoprotein from bovine ligamentum nuchae exhibiting dual amine oxidase activity, *Biochemistry*, 20, 5424, 1981.
99. **Gahl, W. A. and Pitot, H. C.,** Diamine oxidase activity in fibroblasts from normal and cystic fibrosis patients, *Pediat. Res.*, 14, 118, 1980.
100. **Nakano, G., Harada, M., and Nagatsu, T.,** Purification and properties of an amine oxidase in bovine dental pulp and its comparison with serum amine oxidase, *Biochim. Biophys. Acta*, 341, 366, 1974.
101. **Kusche, J., Lorenz, W., Hahn, H., and Werle, E.,** Occurrence and properties of diamine oxidases in salivary glands and gastric mucosa of man and other mammals, *Naunyn-Schmiedebergs Arch. Pharmak.*, 264, 265, 1969.
102. **Nemoto, K., Hashimoto, S., Ikeno, T., and Kuzuya, H.,** Partial characterization of monoamine oxidase in the salivary gland of rats, *Jpn. J. Pharmacol.*, 30, 243, 1980.
103. **Zeller, E. A.,** Ueber das Vorkommen der Diamin-oxydase in menschlichen Sperma, *Helv. Chim. Acta*, 24, 117, 1941.
104. **Hölttä, E., Pulkkinen, P., Elfving, K., and Jänne, J.,** Oxidation of polyamines by diamine oxidase from human seminal plasma, *Biochem. J.*, 145, 373, 1975.
105. **Pulkkinen, P., Kanerva, S., Elfving, K., and Jänne, J.,** Association of spermine and diamine oxidase activity with human spermatozoa, *J. Reprod. Fert.*, 43, 49, 1975.
106. **Kavanagh, J. P., Brightwell, R., Bardsley, W. G., and Schnieden, H.,** Levamisole and other diamine oxidase inhibitors as inhibitors of sperm motility, *Arch. Androl.*, 7, 51, 1981.
107. **Guerrieri, P., Finazzi-Agrò, A., Costa, M. T., Rotilio, G., and Mondovì, B.,** Salt effect on diamine oxidase activity, *Ital. J. Biochem.*, 25, 160, 1976.
108. **Buffoni, F., Banchelli, G., Ignesti, G., Pirisino, R., and Raimondi, L.,** The presence of an inhibitor of benzylamine oxidase in human blood plasma, *Biochem. J.*, 211, 767, 1983.
109. **Mondovì, B. and Rigo, A.,** Unpublished results.

Chapter 8

LYSYL OXIDASE (CHARACTERIZATION AND CLINICAL IMPORTANCE)

Franca Buffoni

TABLE OF CONTENTS

I. INTRODUCTION

Many of the unique properties of connective tissue are due to the formation of covalent cross-linking of the two proteins, collagen and elastin. Certain lysyl residues are converted into peptide-bound α-aminoadipic-δ-semialdehyde.[1,2] This aldehyde may then react to form cross-links by either an aldol condensation with a second allysyl residue or by a Schiff base formation with the ε-amino group of lysyl or hydroxylysyl residues.

Lysyl oxidase (LAO) is the enzyme responsible for the formation of α-aminoadipic-δ-semialdehyde. This enzyme was discovered by Pinnell and Martin[3] in extracts of embryonic chick bone. The enzyme-dependent production of cross-linked collagen and the lysine-derived cross-linked precursor allysine[4] was soon demonstrated.

Various reports on this subject have been published.[5,6] The present report regards studies published before March 1983.

II. DISCOVERY OF LYSYL OXIDASE

Between 1961 and 1964 several study groups in the U.S. reported that a copper deficiency in chickens, turkeys, and pigs resulted in aortic ruptures and physical and chemical changes in elastin. Copper deficient elastin swelled more readily and contained a large amount of excess lysine.[7,8] These results suggested that during fibrogenesis some lysine sidechains on a soluble proelastin were oxidized under the influence of a copper enzyme. Furthermore, it has long been known that if lathyrogenic compounds are fed to young animals a marked reduction in the strength of collagenous tissues throughout the body follows. As a result of feeding β-aminopropionitrile, degenerative changes in the elastin of the aorta occur.

Desmosine and isodesmosine were isolated from elastin[9,10] and it was postulated that they could have arisen from lysine with the elimination of three ammonia residues. O'Dell et al.[11] and Miller et al.[12] showed that lathyrogen reduced the incorporation of lysine into the desmosine, which led to presume that lathyrogenic compounds might act as metabolic inhibitors in the desmosine biosynthesis process.

These observations indicated that a copper amine oxidase of the mammalian serum amine oxidase type could be responsible for the oxidation of the ε-amino lysine group of the elastic and collagenous fibers.

In 1965 it was shown that blood SAO was absent in copper-deficient pigs[13] and that the level of this enzyme could be restored by copper treatment.

The first observation that an enzyme was involved in the formation of cross-links in elastin and collagen was made by Pinnell and Martin.[3,14] These authors used a preparation of the tibia and femur of a 16-day-old chick as an enzyme and a protein from the embryonic chick aorta labeled in vivo with ^3H 6-lysine as substrate.

The tritium, set-free as ^3H_2O when enzyme and substrate were incubated in vivo, was measured. This release was inhibited by β-aminopropionitrile. In the presence of catalase, ^3H_2O is formed from the hydrogen of the α carbon atom of lysine according to the following reaction:

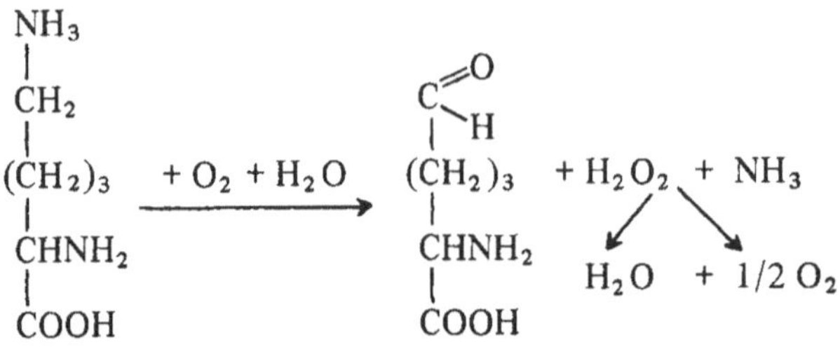

III. THE AMINE OXIDASES OF CONNECTIVE TISSUE

The discovery of lysyl oxidase prompted a great interest in this enzyme and many attempts were made to purify it. Initially, in order to show its possible lysyl oxidase activity, there was considerable interest in determining whether or not SAO was active in the production of cross-links in collagen and elastin.

Carnes et al. (see Reference 15) obtained low rates of oxygen consumption by reacting crystalline SSAO with purified proelastin. Rucker et al.[16,17] isolated an amine oxidase which was able to oxidize lysinevasopressin in chick femora and in bovine aorta. Buffoni and Della Corte[18] demonstrated by means of an immunofluorescent technique, the connective origin of SSAO.

Rucker and Goettlich-Rieman[19] found an enzyme very similar to SAO in the rabbit aorta. Nagatsu et al.[20] found an amine oxidase of the copper pyridoxal type in bovine dental pulp tissue. Buffoni et al.[21,22] isolated two nonmitochondrial oxidases in pig aorta; one similar to the plasma enzyme, the other with different pH optima and properties. Two nonmitochondrial amine oxidases were isolated from ox aorta.[23,24] One enzyme, an amine oxidase, oxidized lysinevasopressin but did not affect a number of other lysine peptides of tropoelastin, whereas the other enzyme, a true lysyl oxidase, did not act on benzylamine but oxidized tropoelastin. In agreement with the first observation made by Siegel et al.,[25] these results clearly show that LAO does not act on benzylamine and that at least one SAO is present in connective tissue, although its physiological significance is still unknown.

IV. LYSYL OXIDASE PURIFICATION

Initial attempts to purify lysyl oxidase were difficult because of the poor recovery from ion exchange or gel filtration columns.[26] The introduction of urea into the extraction procedure improved the results. In Table 1 the purifications of LAO from different sources are summarized. Enzyme activity remains stable in 6 *M* urea and can be measured upon the removal of this solvent. During purification, the enzyme was often eluted in many peaks, but some evidence suggests that both degraded and aggregated forms of the enzyme may occur, possibly accounting for some of the complicated results.

Han and Tazer[37] suggest employing nonionic detergents for the extraction of lysyl oxidase offering at least two advantages; the ability to assay the enzymic activity directly and a minimal risk of enzyme modification. Purification of lysyl oxidase is complicated by many factors; difficulty of its tissue extraction which indicates that the enzyme is not free but linked to tissue constituents, difficulty in establishing the exact degree of purification due to the presence of endogenous inhibitors in crude homogenates,[38] and difficulty in evaluating the enzymic activity as long as the endogenous substrate is present in the enzymic preparation, because of the possible competition between the different substrates when tritium release is measured.[35,38]

Results reported by Buffoni and Raimondi[35] clearly show that when lysyl oxidase is exacted with urea, substrates are also extracted, together with the enzyme. Substrates may be responsible for the tendency of LAO aggregation in the absence of urea and of the observed multiple forms of lysyl oxidase. The presence of an endogenous substrate may be easily detected by measuring auto-oxidation.[35]

Difficulties in releasing lysyl oxidase from its substrates may be indicative of the fact that a large fraction of this enzyme is not free but linked to collagen or elastin. It might be useful to treat preparations with proteases before purification. Sullivan and Kagan[39] have shown that the many forms of bovine aorta lysyl oxidase bear a remarkable resemblance to one another, both structurally and catalytically. The structural homology among the enzyme

Table 1
LAO PURIFICATION

Source	Method of extraction	Purification factors	Method of assay	Forms	Ref.
Chick embryo cartilage tibia, femur	PBS	440	P and M	1	26
Chicken aorta	4 *M* urea in PBS pH 8.4	17.4	P and M	2	27
Chick embryo cartilage	*Phosphate buffer saline pH* 7.5	1710	P and M	1	28
Bovine aorta	Phosphate buffer pH 7.4	839	P and M	1	23
Chick embryo cartilage	4 *M* urea in Tris. HCl pH 7.5	1800	P and M	1	29
Bovine lung	4 *M* urea in phosphate buffer pH 7.6	126	P and M	1	30
Chick embryo cartilage	6 *M* urea in Tris. pH 7.6	998	P and M	1	31
Bovine plasma	$CHCl_3$ treatment	—	P and M	3	32
Bovine ligamentum nuchae	4 *M* urea in phosphate buffer pH 7.7	238	P and M	4	33
Bovine aorta	4 *M* urea in phosphate buffer pH 7.7	465	P and M	3	33
Bovine aorta	4 *M* urea in PBS pH 7.7	80—146	P and M	4	34
Calf tibial cartilage	Nonidet P-40	220	P and M		
Pig aorta	H_2O	10205 B	P and M	2	35
		— A	*o*-dianisidine		
Bovine ligamentum nuchae	5 *M* guanidine hydrochloride	—	P and M	1	36

Note: P and M = Pinnell and Martin; PBS = phosphate buffer saline.

species, as revealed by peptide map analyses, suggests that the enzyme forms may be post-translational derivatives probably produced by dissociations in 4 to 6 *M* urea. However. the possibility of a separate genomic origin of each species cannot be eliminated.

Some properties of purified lysyl oxidase from different sources are reported.[23,24,26-34,36-38]

V. SPECIFIC ACTIVITY OF LYSYL OXIDASE AND PRODUCTS OF REACTION

Owing to the different substrates and assay conditions employed by investigators up to now, there is no uniformly accepted enzyme unit for LAO. Radioactivity (cpm released) is generally used as the unit. A convenient enzyme unit would be the amount of tritiated water formed which corresponds to the amount of oxidized lysine expressed in μmol (international unit).

On the other hand, it is important to point out that the amount of tritium released gives only a part of the enzymic activity (see above). Hydrogen peroxide,[35,38,40] ammonia and aldehyde[3,41,42] are the products of the reaction of LAO. The formation of hydrogen peroxide may be used for the enzymic activity determination.[35,38,40,43]

VI. LYSYL OXIDASE SUBSTRATES

Lysyl oxidase acts on collagen and elastin. To be oxidized by the enzyme, the substrates must be in their native insoluble fibrillar form.[44] Simple nonpeptidyl amines are also oxidized by LAO.[35,38,40] Trackman and Kagan[40] have shown that some diamines are oxidized by BALAO only up to 300 catalytic turnovers, before the enzyme becomes irreversibly inactivated. The presence of the insoluble aortic substrate protects the enzyme from the time-dependent development of irreversible inhibition by the nonpeptidylamines.[35,40] Lysyl oxidase can act upon lysine-containing synthetic peptides.[45]

Present evidence argues against the predominant role of the amino-acid sequence adjacent to the lysine in the LAO function; the enzyme appears to be influenced more by coacervated or fibrillar forms of polypeptide substrates.

VII. LYSYL OXIDASE INHIBITORS

The known inhibitors of LAO are reported in Table 2. They are either cupric copper chelating agents or carbonyl reagents. For other types of inhibitors such as the sulfhydryl compounds, Misiorowski and Werner[46] have shown that the requirement is a bifunctional sulfhydryl or bifunctional amine or a sulfhydryl-amine. Sodium dodecyl sulfate, fatty acid salts, and bile salts can completely inhibit the oxidation of elastin substrate.[47]

The results are consistent with a charge-based substrate-directed mechanism of modulation of lysyl oxidase activity on elastin by the elastin-bound ligands. Ligands which stimulate elastolysis inhibit LAO, whereas those inhibiting elastolysis stimulate LAO.[47]

VIII. LYSYL OXIDASE COFACTORS

Lysyl oxidase has been shown to be a cupric-copper protein.[30-38] The incorporation of copper into lysyl oxidase occurs only cotranslationally.[48]

Nutritional copper deficiency, which has been studied in several species, produces a disease characterized by multiple connective tissue manifestations.[15,49] All these changes can be attributed to an impaired cross-link of collagen and elastin.

Indirect evidence of a second cofactor in LAO comes from the observed sensitivity of the enzyme to carbonyl reagents.[27,38] It has also been demonstrated that many of these carbonyl reagents have lathyrogenic activity in vivo, resulting in the dramatically increased fragility of the mesenchymal tissues.[50]

Table 2
INHIBITORS OF LYSYL OXIDASE IN VITRO

Inhibitor	Ref.	Inhibitor	Ref.
β-Aminopropionitrile	4	1,2-Ethylenediamine	40
Semicarbazide	38	1,3-Diaminopropane	40
Phenylhydrazine	38	1,4-Diaminobutane	40
α-Aminoguanidine	38	1,5-Diaminopentane	40
Diethyldithiocarbamate	38,30	Dithiothreitol	40
α,α Dipyridyl	41	1,3-Dithio-2-propanol	46
l-Lysine	40	Dithioethane	46
l-Lysine methyl ester	40	Dithiobutane	46
N^2-Acetyl-*l*-lysine	40	β-Mercaptoethanol	46
N^2-Acetyl-*l*-ornithine	40	Penicillamine	46
N^2-Acetyl-*l*-lysine methyl amide	40	Cysteine	46
6-Aminocaproic acid	40	1,3-Diaminopropane	46
n-Butylamine	40	Putrescine	46
N^1-acetyl-*l*-lysine methyl ester	40	Cadaverine	46

Another indirect piece of evidence regarding the role of pyridoxal-phosphate (PLP) in LAO is the demonstration of a deficiency in cross-link synthesis in aortic elastin in pyridoxine-deficient chicks[51,52] and in collagen of PLP-deficient rats.[53] Murray and Levene[54] have obtained radioactivity incorporation in LAO extracted from animals treated with ^3H pyridoxine hydrochloride. This is the most direct demonstration of the presence of PLP in LAO. This has been recently confirmed.[55]

IX. LYSYL OXIDASE ACTIVITY METHODS OF ASSAY

The most significant assay method of LAO is that described by Pinnel and Martin[3] which has been recently modified by Melet et al.[56] Trackman et al.[43] have described a peroxide-coupled flourimetric method for hydrogen peroxide determination in HPLC. The formation of hydrogen peroxide may also be assayed as described by Buffoni et al.[38]

X. LYSYL OXIDASE DISTRIBUTION AND SECRETION

Fibroblasts and smooth muscle cells secrete measurable amounts of lysyl oxidase into the culture medium, a finding that is consistent with the hypothesis of an extracellular site for aldehyde formation.[57,58]

XI. LYSYL OXIDASE IN PATHOLOGY

Defects in aortic elastin are known to occur in copper deficiency and lathyrism. Miller et al.[49] observed decreased amounts of aldehyde in aorta fibers of lathyric and copper-deficient chicks. Rayton and Harris[48] found an induction of lysyl oxidase with copper which is increased by *l*-ascorbic acid,[59] thus suggesting a postabsorption role of *l*-ascorbate in copper metabolism.[59] A dietary copper deficiency decreases LAO activity in the aortic tissue of growing chickens.[60] The copper requirement for normal collagen cross-link formation in chicks is less than 2 ppm.[61] There are several hereditary disorders with distinct abnormalities in copper metabolism and connective tissue alterations: Wilson's disease, inherited copper toxicosis in Bedlington terriers and the X-linked recessive disorders such as the Menkes steely hair syndrome in man and the mottled series of mutant mice, the two forms of Ehlers-Danlos syndromes, and the curtix laxa.[62,67]

In the X-linked diseases, intestinal copper absorption is deficient, serum copper and ceruloplasmin levels are low, and cultured fibroblasts and several other cells have markedly high amounts of copper. Therefore, the deficiency in LAO activity in the Menkes syndrome, mottled mouse mutants, and Ehlers-Danlos syndrome may be secondary to functional intracellular copper deficiency.[62] Mice with "Blotchy" alleles in the X chromosomal locus have decreased levels of LAO in fibroblasts and lungs;[68,69] abnormalities in the lungs of these animals being emphysema-like changes.[68,69] Copper treatment increases the LAO level in the skin of these animals if given as a single subcutaneous injection at 7 days of age.[66] A copper-deficient, zinc-supplemented diet produces emphysema in pigs.[70]

Turkeys spontaneously develop aortic aneurysms.[71] These animals have a lower LAO activity level that that observed in the hen aorta and this may lead to the development of spontaneous aneurysms.[71] An increased formation of collagen has been observed in the blood vessels of hypertensive rats,[72] as well as an increased incorporation of lysine into the vascular protein.[73] The levels of LAO are increased in the aorta of spontaneous hypertensive rats.[74] In streptozotocin-induced diabetes the LAO activity in the rat lung is increased two- to threefold.[75] LAO and collagenase activities are increased in chronic liver damage[75,77] causing liver fibrosis.[76,77] In young granuloma tissue a high LAO activity is observed, correlated to the high number of fibroblasts.[78] LAO activity also increases the postburn scars of the human skin.[79]

XII. LYSYL OXIDASE ACTIVITY AFTER DRUG TREATMENT

LAO is a pituitary hormone-dependent enzyme.[80] In the hypophysectomized rats, aldehyde-deficient collagen does not, in fact, cross-link, indicating a lack of active LAO.[80] Rat estrogen stimulates LAO activity and accelerates the maturation of collagen and elastin.[82,86]

Cortisol increases LAO activity in human aortic smooth muscle cells in culture,[83] whereas, in the skin of newborn rats, triamcinolone produces a dose-dependent decrease in LAO activity.[84] Contrasting results have been obtained regarding the effect of cadmium. When newborn rats are given a diet containing 50 ppm cadmium, bone LAO activity is decreased,[85] whereas aerosol of cadmium in rats increases lung LAO activity.[86] The altered tissue amounts of LAO and prolyl hydroxylase after cadmium inhalation, correlates well with marked interstitial cell hyperplasia 4 to 5 days after exposure, suggesting that connective tissue protein synthesis is activated in the interstitial cell fibroblasts soon after cadmium exposure. These results may be important when we consider that industrial workers develop severe lung damage when exposed to cadmium fumes.

Bleomycin increases lung LAO and induces lung fibrosis.[87] The administration of D-penicillamine during the development of bleomycin lesions results in a normal quantity and concentration of total collagen in the lung.[88]

XIII. CONCLUSION

Lysyl oxidase is therefore an interesting enzyme with an important physiological role. It is secreted by fibroblasts and smooth muscle cells both in vivo and in vitro. It is not degraded by collagenase, but by elastase and other endopeptidases. The small subunits have a low molecular weight but the enzyme aggregates such as collagen and elastin. Many pathological conditions are characterized by a defect in the enzyme level. Some drugs may reduce this level and may therefore be useful therapeutic agents in some pathological conditions in which enzyme activity is increased.

There is still much to be learned about this enzyme from the biochemical point of view, due to the difficulty in the purification and assay of this enzyme.

Tritium assay methods may give erroneous results in crude homogenates where endogenous inhibitors and endogenous substrates are present. The best enzyme estimation comes from using different methods and dilutions to ascertain the presence of inhibitors. Some contrasting results in literature on the levels of LAO may, in fact, depend on the assay method employed, on the presence of different amounts of substrates and endogenous inhibitors, and on the varying extractability of the enzyme.

Hydralazine inhibits the copper-containing amine oxidase, LAO,[89] DAO, and SAO.[90,91] The effect of hydralazine might explain its efficacy in the treatment of essential hypertension.

Arteriosclerosis can be easily induced in rats by the administration of vitamine D_2 and a diet containing a high percentage of cholesterol, cholic acid, and methylthiouracid. In these rats, the aortic level of LAO is significantly decreased. Aortic tension and strength[92] could also be significantly reduced as well. It would be interesting to ascertain if this were also true in human vessels during the development of arterioschlerosis.

ACKNOWLEDGMENT

The original studies reported here were supported in part by the Italian National Research Council, Special Project on the "Chimica Fine", Contract No. 83.00310.95.

REFERENCES

1. **Bornstein, P., Kang, A. H., and Piez, K. A.,** The nature and location of intramolecular cross-links in collagen, *Proc. Natl. Acad. Sci. U.S.A.*, 55, 417, 1966.
2. **Bornstein, P. and Piez, K. A.,** The nature of the intramolecular cross-links in collagen. The separation and characterization of peptides from the cross-link region of rat skin collagen, *Biochemistry*, 5, 3460, 1966.
3. **Pinnell, S. R. and Martin, G. R.,** The cross-linking of collagen and elastin in enzymatic conversion of lysine in peptide linkage to α-aminoadipic-δ-semialdehyde (allysine) by an extract from bone, *Proc. Natl. Acad. Sci. U.S.A.*, 61, 708, 1968.
4. **Siegel, R. C. and Martin, G. R.,** Collagen cross-linking, enzymatic synthesis of lysine-derived aldehydes and the production of cross-linked components, *J. Biol. Chem.*, 245, 1653, 1970.
5. **Siegel, R. C.,** Lysyl oxidase, *Int. Rev. Connect. Tissue Res.*, 8, 73, 1979.
6. **Kagan, H. M. and Sullivan, K. A.,** Lysyl oxidase: preparation and role in elastin biosynthesis, in *Methods in Enzymology*, Vol. 82, Cunningham, C. W. and Frederiksen, D. W., Eds., Academic Press, New York, 1982, 637.
7. **Sandberg, L. B., Hackett, T. N., and Carnes, W. H.,** The solubilization of an elastin protein from copper-deficient porcine aorta, *Biochim. Biophys. Acta*, 181, 201, 1969.
8. **Smith, D. W., Weissman, N., and Carnes, W. H.,** Cardiovascular studies on copper deficient swine. XII. Partial purification of a soluble protein resembling elastin, *Biochim. Biophys. Res. Commun.*, 31, 309, 1968.
9. **Partridge, S. M., Elsden, D. F., and Thomas, J.,** Constitution of the cross-linkages in elastin, *Nature (London)*, 197, 1279, 1963.
10. **Partridge, S. M., Elsden, D. F., Thomas, J., Dorfman, A., Telser, A., and Ho, P. L.,** Incorporation of labeled lysine into the cross-linkages in elastin, *Nature (London)*, 209, 399, 1965.
11. **O'Dell, B. L., Elsden, D. F., Thomas, J., Partridge, S. M., Smith, R. H., and Palmer, R.,** Inhibition of the biosynthesis of the crosslinks in elastin by a lathyrogen, *Nature (London)*, 209, 401, 1966.
12. **Miller, E. J., Martin, G. R., and Piez, K. A.,** Factors influencing the biosynthesis of elastin crosslinks, *Fed. Proc.*, 24, part 1, 359, 1965.
13. **Blaschko, H., Buffoni, F., Weissman, N., Carnes, W. H., and Coulson, W. F.,** The amine oxidase of pig plasma in copper deficiency, *Biochem. J.*, 96, 4c, 1965.
14. **Martin, G. R., Pinnell, S. R., Siegel, R. C., and Goldstein, E. R.,** Lysl oxidase: The enzymatic step in collagen and elastin cross-linking, *Chemistry and Molecular Biology of the Intracellular Matrix*, Vol. 1, Bolazs, E. A., Ed., Academic Press, London and New York, 405, 1970.
15. **Carnes, W. H.,** Role of copper in connective tissue metabolism, *Fed. Proc.*, 30, 995, 1971.

16. **Rucker, R. B., Rogler, J. C., and Porker, H. E.,** The partial characterization of an amine oxidase in bone tissue, *Proc. Soc. Exp. Biol. Med.*, 130, 1150, 1969.
17. **Rucker, R. B. and O'Dell, B.,** Connective tissue amine oxidase. I. Purification of bovine aorta amine oxidase and its comparison with plasma amine oxidase, *Biochim. Biophys. Acta*, 235, 32, 1971.
18. **Buffoni, F. and Della Corte, L.,** Pig plasma benzylamine oxidase, *Adv. Biochem. Psychopharmacol.*, 5, 138, 1977.
19. **Rucker, R. B. and Goettlich-Riemann, W.,** Properties of rabbit aorta amine oxidase, *Expl. Biol. Med.*, 138, 286, 1972.
20. **Nagatsu, T., Nakano, G., Mizutani, K., and Harada, M.,** Purification and properties of amine oxidases in brain and connective tissue (dental pulp), *Adv. Biochem. Psychopharmacol.*, 5, 25, 1972.
21. **Buffoni, F., Pirisino, R., and Marino, P.,** Histamine as inhibitor of aorta amine oxidase, *Agents Actions*, 5, 429, 1975.
22. **Buffoni, F., Marino, P., and Pirisino, R.,** Partial purification of pig aorta amine oxidase, *Ital. J. Biochem.*, 25, 191, 1976.
23. **Shieh, J. J., Tamaya, R., and Yosunobu, K. T.,** A purification procedure for the isolation of homogeneous preparations of bovine aorta amine oxidase and a study of its lysyloxidase activity, *Biochim. Biophys. Acta*, 377, 229, 1975.
24. **Vidal, G. P., Shieh, J. J., and Yasunobu, K. T.,** Immunological studies of bovine aorta lysyl oxidase: evidence for two forms of the enzyme, *Biochem. Biophys. Res. Commun.*, 64, 989, 1975.
25. **Siegel, R. C., Page, R. C., and Martin, G. R.,** The relative activity of connective tissue lysyl oxidase and plasma amine oxidase on collagen and elastin substrate, *Biochim. Biophys. Acta*, 222, 552, 1970.
26. **Siegel, R. C., Pinnell, S. R., and Martin, G. R.,** Cross-linking of collagen and elastin. Properties of lysyl oxidase, *Biochemistry*, 9, 4486, 1970.
27. **Harris, E. D., Gonnerman, W. A., Sovage, J. E., and O'Dell, B. L.,** Connective tissue amine oxidase II. Purification and partial characterization of lysyl oxidase from chick aorta, *Biochim. Biophys. Acta*, 341, 332, 1974.
28. **Narayanan, A. S., Siegel, R. C., and Martin, S. R.,** Stability and purification of lysyl oxidase, *Arch. Biochem. Biophys.*, 162, 231, 1974.
29. **Stassen, F. L. H.,** Properties of highly purified lysyl oxidase from embryonic chick cartilage, *Biochim. Biophys. Acta*, 438, 49, 1976.
30. **Shieh, J. J. and Yasunobu, K. T.,** Purification and properties of lung lysyl oxidase, a copper enzyme, *Adv. Exp. Med. Biol.*, 74, 447, 1976.
31. **Siegel, R. C. and Fu, J. C. C.,** Collagen cross-linking. Purification and substrate specificity of lysyl oxidase, *J. Biol. Chem.*, 251, 5779, 1976.
32. **Shieh, J. J., Sasaki, T., Minamiura, N., and Yasunobu, K. T.,** The existence of inhibited lysyl oxidase and the presence of multiple forms, *Adv. Exp. Biol. Med.*, 79, 509, 1977.
33. **Jordan, R. E., Mitbury, P., Sullivan, K. A., Trackman, P. C., and Kagan, H. M.,** Studies on lysyl oxidase of bovine ligamentum nuchae and bovine aorta, *Adv. Exp. Biol. Med.*, 79, 531, 1977.
34. **Kagan, H. M., Sullivan, K. A., Olsson, T. A., and Cronlund, A. L.,** Purification and properties of four species of lysyl oxidase from bovine aorta, *Biochem. J.*, 177, 203, 1979.
35. **Buffoni, F. and Raimondi, L.,** A lysyl oxidase with histaminase activity in the pig aorta, *Agents Actions*, 11, 38, 1981.
36. **Serafini-Fracassini, A., Ventrella, G., Field, M. J., Hinnie, J., Onyerili, N. I., and Griffiths, R.,** Characterization of a structural glycoprotein from bovine ligamentum nuchae exhibiting dual amine oxidase activity, *Biochemistry*, 20, 5424, 1981.
37. **Han, S. and Tanzer, M. I.,** Collagen cross-linking. Purification of lysyl oxidase in solvents containing nonionic detergents, *J. Biol. Chem.*, 254, 10438, 1979.
38. **Buffoni, F., Ignesti, G., and Lodovici, M.,** Lysyl oxidase and benzylamine oxidase in pig aorta, *Ital. J. Biochem.*, 30, 179, 1981.
39. **Sullivan, K. A. and Kagan, H. M.,** Evidence for structural similarities in the multiple forms of aortic and cartilage lysyloxidase and a catalytically quiescent aortic protein, *J. Biol. Chem.*, 257, 13520, 1982.
40. **Trackman, P. C. and Kagan, H. M.,** Nonpeptidyl amine inhibitors are substrates of lysyl oxidase, *J. Biol. Chem.*, 254, 7831, 1979.
41. **Kagan, H. M., Hewitt, N. A., Salcedo, L. L., and Franzblau, C.,** Catalytic activity of aortic lysyl oxidase in an insoluble enzyme-substrate complex, *Biochim. Biophys. Acta*, 365, 223, 1974.
42. **Siegel, R. C.,** Collagen cross-linking. Synthesis of collagen cross-links in vitro with highly purified lysyl oxidase, *J. Biol. Chem.*, 251, 5786, 1976.
43. **Trackman, P. C., Zoski, C. G., and Kagan, H. M.,** Development of a peroxidase coupled fluorometric assay for lysyl oxidase, *Anal. Biochem.*, 113, 336, 1981.
44. **Siegel, R. C.,** Biosynthesis of collagen cross-links: increased activity of purified lysyl oxidase with reconstituted collagen fibrils, *J. Biol. Chem.*, 71, 4826, 1974.

45. **Kagan, H. M., Tseng, L., Trackman, P. C., Okamoto, K., Rapaka, R. S., and Urry, D. W.,** Repeat Polypeptide models of elastin as substrates for lysyl oxidase, *J. Biol. Chem.*, 255, 3656, 1980.
46. **Misiorowski, R. L. and Werner, M. J.,** Inhibition of lysyl oxidase by disulphydryl, diamines and sulphydryl-amines, *Biochem. Biophys. Res. Commun.*, 85, 809, 1978.
47. **Kagan, H. M., Tseng, L., and Simpson, D. E.,** Control of elastin metabolism by elastin ligands. Reciprocal effect of lysyl oxidase, *J. Biol. Chem.*, 256, 5417, 1981.
48. **Rayton, J. K. and Harris, E. D.,** Induction of lysyl oxidase with copper, *J. Biol. Chem.*, 254, 621, 1979.
49. **Miller, E. J. and Fullmer, H. M.,** Elastin: diminished reactivity with aldehyde reagents in copper deficiency and lathyrism, *J. Exp. Med.*, 123, 1097, 1966.
50. **Levene, C. I.,** Structural requirements for lathyrogenic agents, *J. Exp. Med.*, 114, 295, 1961.
51. **Hill, C. H. and Kim, S. C.,** The derangement of elastin synthesis in pyridoxine deficiency, *Biochem. Biophys. Res. Commun.*, 27, 94, 1967.
52. **Starcher, B. C.,** The effect of pyridoxine deficiency on aortic elastin biosynthesis, *Proc. Soc. Exp. Biol. Med.*, 132, 379, 1969.
53. **Fujii, K., Kajiwara, T., and Kurosu, H.,** Effect of vitamin B_6 deficiency on the cross-link formation of collagen, *FEBS Lett.*, 97, 193, 1979.
54. **Murray, J. C. and Levene, C. I.,** Evidence for the role of vitamin B-6 as a cofactor of lysyl oxidase, *Biochem. J.*, 167, 463, 1977.
55. **Bird, T. A. and Levene, C. J.,** Lysyl oxidase: evidence that pyridoxal phosphate is a cofactor, *Biochem. Biophys. Res. Commun.*, 108, 1172, 1982.
56. **Melet, J., Vianden, G. D. N. E., and Bachra, B. N.,** Simplified assay for lysyl oxidase activity, *Anal. Biochem.*, 77, 141, 1977.
57. **Layman, D. L., Narayanan, A. S., and Martin, G. R.,** The production of lysyl oxidase by human fibroblasts in culture, *Arch. Biochem. Biophys.*, 149, 97, 1972.
58. **Gonnerman, W. A., Ferrera, R., and Franzblau, C.,** Measurement of medium lysyl oxidase activity in aortic smooth muscle cells. Effects of multiple medium changes and inhibition of protein synthesis, *Biochemistry*, 20, 3844, 1981.
59. **Disilvestro, R. A. and Harris, E. D.,** A postabsorption effect of l-ascorbic acid on copper metabolism in chicks, *J. Nutr.*, 111, 1964, 1981.
60. **Harris, E. D. and Disilvestro, R. A.,** Correlation of lysyl oxidase activation with p-phenylenediamine oxidase activity (ceruloplasmin) in serum, *Proc. Soc. Exp. Biol. Med.*, 166, 528, 1981.
61. **Opsahl, W., Zeronian, H., Ellison, M., Lewis, D., Rucher, R. B., and Riggins, R. S.,** Role of copper in collagen cross-linking and its influence on selected mechanical properties of chick bone and tendon, *J. Nutr.*, 112, 708, 1982.
62. **Kivirikko, K. I. and Peltonen, L.,** Abnormalities in copper metabolism and disturbances in the synthesis of collagen and elastin, *Med. Biol.*, 60, 45, 1982.
63. **Kuivaniemi, H., Peltonen, L., Palotie, A., Kaitila, I., and Kivirikko, K. I.,** Abnormal copper metabolism and deficient lysyl oxidase activity in heritable connective tissue disorder, *J. Clin. Invest.*, 69, 730, 1982.
64. **Siegel, R. C., Black, C. H., and Bailey, A. J.,** Cross-linking of collagen in the X-linked Ehlers-Danlos type V, *Biochem. Biophys. Res. Commun.*, 88, 281, 1979.
65. **Byers, P. H., Siegel, R. C., Holbrook, K. A., Norayanan, A. S., Bornstein, P., and Hall, J. G.,** X-linked cutix laxa. Defective cross-link formation in collagen due to decreased lysyl oxidase activity, *N. Engl. J. Med.*, 303, 61, 1980.
66. **Royce, P. M., Camakaris, J., and Danks, D. M.,** Reduced lysyl oxidase activity in skin fibroblasts with Menkers' syndrome, *Biochem. J.*, 192, 579, 1980.
67. **Gajdos, A.,** Déficiences héréditaires en enzymes de la biosynthése du collagéne Les syndromes d'Ehlers-Danlos Nouv., *Presse Med.*, 6, 3101, 1977.
68. **Fisk, D. E. and Kuhn, C.,** Emphysema-like changes in the lungs of the Blotchy mouse, *Am. Rev. Resp. Dis.*, 113, 787, 1976.
69. **Starcher, B. C., Madaras, J. A., and Tepper, A. S.,** Lysyl oxidase deficiency in lung and fibroblasts from mice with hereditary emphysema, *Biochem. Biophys. Res. Commun.*, 78, 706, 1977.
70. **Soskel, N. T., Watanabe, S., Hammond, E., Sandberg, L. B., Benzetti, A. D., Jr., and Crapo, J. D.,** A copper-deficient, zinc-supplemented diet produces emphysema in pigs, *Am. Rev. Resp. Dis.*, 126, 316, 1982.
71. **Narayanan, A. S., Sandberg, L. B., Jones, K., Coleman, S. S., and Bagley, R. A.,** Lysyl oxidase activities of male and female turkey aorta, *Exp. Mol. Pathol.*, 36, 107, 1982.
72. **Udenfriend, S., Oshima, A., Cardinale, G., Fuller, G. C., and Spector, S.,** Increased formation of collagen in the blood vessels of hypertensive rates, *An. N.Y. Acad. Sci.*, 275, 101, 1976.
73. **Yamabe, H. and Lovenberg, W.,** Increased incorporation of ^{14}C-lysine into vascular proteins of the spontaneously hypertensive rats, *Eur. J. Pharmac.*, 29, 109, 1974.

74. **Sheridan, P. J., Kozor, L. G., and Benson, S. C.,** Increased lysyl oxidase activity in aortas of hypertensive rats and effect of β-aminopropionitrile, *Exp. Mol. Pathol.*, 30, 315, 1979.
75. **Madia, A. M., Rozovski, S. J., and Kagan, H. M.,** Changes in lung lysyl oxidase activity in streptozotocin-diabetes and in starvation, *Biochim. Biophys. Acta*, 585, 481, 1979.
76. **Carter, E. A., McCarron, J., Alpert, E., and Isselbecher, K. J.,** Lysyl oxidase and collagenase in experimental acute and chronic liver injury, *Gastroenterology*, 82, 526, 1982.
77. **Siegel, R. C., Chen, K. H., Greenspan, J. S., and Aguiar, J. M.,** Biochemical and immunochemical study of lysyl oxidase in experimental hepatic fibrosis in the rat, *Proc. Natl. Acad. Sci. U.S.A.*, 75, 2945, 1978.
78. **Chvapil, M., McCarthy, D. W., Misiorowski, R. L., Madden, J. W., and Peacock, E., Jr.,** Activity and extractability of lysyl oxidase and collagen proteins in developing granuloma tissue, *Proc. Soc. Exp. Biol. Med.*, 146, 688, 1974.
79. **Hayakawa, T., Hino, M., Fuyamada, H., Nagatsu, T., Aoyama, H., and Izawa, Y.,** Lysyl oxidase activity in human normal skin and postburns scars, *Clin. Chim. Acta*, 71, 245, 1976.
80. **Shoshan, S. and Finkelstein, S.,** Lysyl oxidase: a pituitary hormone-dependent enzyme, *Biochim. Biophys. Acta*, 439, 358, 1976.
81. **Sanada, H., Shikata, J., Hamamoto, H., Veba, Y., Yamamuro, T., and Takeda, T.,** Changes in collagen cross-linking and lysyl oxidase by estrogen, *Biochim. Biophys. Acta*, 541, 408, 1978.
82. **Ozasa, H., Tominage, T., Nishimura, T., and Takeda, T.,** Lysyl oxidase activity in the mouse uterine cervix is physiologically regulated by estrogen, *Endocrinology*, 109, 618, 1981.
83. **Järveläinen, Halme, T. and Rönnemaa, T.,** Effect of cortisol on the proliferation and protein synthesis of human aortic smooth muscle cells in culture, *Acta Med. Scand. (Suppl.)*, 660, 114, 1982.
84. **Stephen, C. B. and Luvalle, P. A.,** Inhibition of lysyl oxidase and prolylhydroxylase activity in glucocorticoids treated rats, *Biochem. Biophys. Res. Commun.*, 99, 557, 1981.
85. **Iguchi, H. and Sano, S.,** Effect of cadmium on the bone collagen metabolism of rat, *Tox. Appl. Pharmac.*, 62, 126, 1982.
86. **Chichester, C. O., Palmer, K. C., Hayes, J. A., and Kagan, H. M.,** Lung lysyl oxidase and prolyl hydroxylase: increases induced by cadmium chloride inhalation and the effect of β-aminopropionitrile in rats, *Am. Rev. Resp. Dis.*, 124, 709, 1981.
87. **Counts, D. F., Evans, J. N., Di Petrillo, T. A., Sterling, K. M., and Kelly, J.,** Collagen lysyl oxidase activity in the lung increases during bleomycin-induced lung fibrosis, *J. Pharmac. Exp. Ther.*, 219, 675, 1981.
88. **Pedullo, A. J., Karlinsky, J. B., Snider, G. L., and Goldstein, R. M.,** Lung statics and connective tissue after penicillamine in bleomycin-treated hamsters, *J. Appl. Respir. Environ. Exercise Physiol.*, 49, 1083, 1980.
89. **Numata, Y., Takei, T., and Nayakawa, T.,** Hydralazine as an inhibitor of lysyl oxidase activity, *Biochem. Pharmac.*, 30, 3125, 1981.
90. **Buffoni, F., Carpi, C., Banchelli-Soldaini, G., and Raimondi, L.,** 3-Hydrazyno-pyridazine derivatives as inhibitors of pyridoxal-phosphate dependent enzymes, *Il Farmaco, Ed. Sci.*, 35, 848, 1980.
91. **Buffoni, F., Ignesti, G., and Pirisino, R.,** 3-Hydrazion-pyridazine derivatives as inhibitors of the copper containing amine oxidases, *Il Farmaco, Ed. Sci.*, 32, 404, 1977.
92. **Hashimoto, S., Seyama, Y., Yokokura, T., and Mutai, M.,** Effects of chlorella-phospholipid on the aortic collagen and elastin metabolism and on the serum lipid content in rats with experimental arteriosclerosis, *Exp. Mol. Pathol.*, 37, 150, 1982.

Chapter 9

LOCALIZATION OF DIAMINE OXIDASE IN ANIMAL TISSUES

Maria Paola Argento-Cerù and Francesco Autuori

TABLE OF CONTENTS

I. INTRODUCTION

A. Outline of the Problem

The analysis of an enzyme follows three fundamental directions. In biochemical studies, the physical and chemical properties of the purified enzyme, as well as the stoichiometry of the enzymatic reaction, are investigated. Cytological analyses yield information about the topography of the enzyme in tissues and any interactions with other macromolecules. The function of the enzyme in living organisms is examined in physiological studies. All of these aspects, which contribute equally to the understanding of the biological role of an enzyme, have been investigated for the Cu-dependent amino oxidases; extensive information concerning these oxidases is now available.

In this chapter, we will consider only the cytological aspects of studies with AO, reviewing what is known about its localization at three subsequent levels of organization: in organs, in tissues and cells, and at the subcellular level.

B. Methods of Localization

1. In Organs

Numerous groups of workers have demonstrated the presence of AO in several organs of different animal species.[1-2] The results have allowed comparisons of the levels of enzyme activity in various organs of the same species as well as the same organ in different species. Information about DAO activity under normal circumstances has served as the basis for the examination of variations in this activity under special conditions, be they physiological (e.g., during pregnancy), experimental (e.g., during the process of regeneration), or pathological (e.g., tumors). These studies have also permitted the choice, among the organs that contain the highest levels of DAO, of those from which to purify the enzyme or to study its localization at the cellular and subcellular level.

The biochemical procedures used in these studies consist of the preparation of a crude homogenate or extract of an organ and the subsequent measurement of enzymatic activity. Since the methods that are employed are fairly simple, a direct comparison of the results obtained by different groups is often possible, at least when the experimental conditions are comparable or even identical. However, this is not always the case; in fact, several groups have used different methods for homogenization, experimental animals of various ages, sex, and nutritional state, as well as organs either freshly removed or stored frozen for various periods of time. As is discussed in Chapters 16 and 18, differences in the procedures for measuring enzyme activity also make the comparison of results obtained by various groups difficult. In fact, despite the large number of such procedures, none is without defect: the older procedures are rather insensitive and lack specificity, while the more recent ones are limited by the use of a single substrate, and are therefore difficult to compare in terms of units of enzyme activity.

2. At the Cellular Level

The localization of DAO at the cellular level can be determined either by biochemical or histochemical methods.

a. Biochemical Methods

The technical challenge lies essentially in the difficulty of obtaining an homogeneous cellular population that is large enough to allow the biochemical measurement of enzyme activity. This is actually relatively simple for certain cells, including those derived from both fluid tissues (blood and lymph) and solid tissues (liver, kidney, intestine, etc.). In fact, numerous techniques are also available for the isolation of cells that constitute parenchymal organs and for the purification of a single type of cell from a mixed population.[3-5]

b. Histochemical Methods

These methods are essentially immunohistochemical; no histochemical method for the localization of DAO has been described. This observation is surprising in that histochemical procedures for the analysis of MAO activity have been in use for some time.[6] Immunohistochemical methods (including the techniques of immunoperoxidase and immunofluorescence) have been adapted for use with some organs of the pig, human placenta, and certain types of human tumors. The relative lack of studies employing these techniques is due to the fact that only the DAOs from pig kidney and human placenta have been obtained in sufficiently pure form to allow the preparation of specific antibodies. This constitutes a serious limitation for the immunohistochemical method, since it has been shown that, despite overall similarities between the DAOs present in different tissues, often no immunological cross-reactivity exists among these enzymes.[7]

3. At the Subcellular Level

To date, the subcellular localization of DAO has been examined using only biochemical techniques. Enzyme activity has been measured in subcellular fractions obtained after differential centrifugation of crude homogenates. No cytochemical procedure for the visualization of the enzyme with the electron microscope has yet been devised, and the electron microscope has not yet been used to determine the intracellular localization of DAO with immunocytochemical techniques.

Techniques of cellular fractionation, to be acceptable, must conform to certain standards;[8-9] in the absence of these standards, the results obtained may be misleading or difficult to interpret. The critical aspects of the procedure are (1) the method used for homogenization, (2) centrifugation technique, (3) analysis of the compositions of the fractions with the electron microscope and by the measurement of enzymes used as markers, and (4) measurement of the recovery of proteins, marker enzymes, and the enzyme under study. The first two aspects must be adjusted for the various materials being studied and carefully examined in each experiment. Examinations with a phase-contrast microscope of both the crude homogenate (to check the efficiency of the homogenization) and the single fractions (to estimate the level of contamination) are essential. The third point, however, is without a doubt the most important, since it constitutes the true control of the entire procedure. Only the proper behavior of the marker enzymes can demonstrate that the preceding steps were carried out properly. The studies of cellular fractionation that fail to report the behavior of marker enzymes are all subject to criticism, and the results cannot be totally accepted. The observation of the recovery of the enzymes whose activities are followed in the fractionation steps is of considerable importance, since it constitutes the sole evidence that the enzyme under study is not subject to alterations, which could lead to its activation or inactivation. When considering DAO, it is necessary to point out that not all of the studies on its subcellular localization have been conducted with impeccable methodology, and therefore, totally reliable results in this area are relatively scarce.

II. LOCALIZATION IN ORGANS

The presence of DAO in several organs, especially in mammals, is very well documented. It is agreed that the organ containing the highest levels of DAO activity in nearly all animals is the intestine.[10-16] In the intestinal tract, the small intestine (and in particular the ileum) generally contains the highest levels of DAO activity; in fact, the activity is also high in the large intestine only in the dog.[10,17] The differences in DAO activity in the intestine of various mammals have been correlated with the diets of the animals: the activity is highest in carnivores (dog) and omnivores (man) and lowest in herbivores (guinea pig, rabbit). It has been suggested that intestinal DAO might serve to protect against biogenic amines, especially histamine, from endogenous or dietary sources.[11,18] No correlation, however, has been noted between DAO activity and intestinal flora, since germ-free rats do not have less enzyme activity than control rats[12,17] (see Chapter 15).

To complete the discussion on the gastroenteric apparatus, it is necessary to point out that the stomach contains little or no DAO in all of the species that have been examined,[10,17,19,20] with the exception of the rat, in which Beaven and de Jong[21] have demonstrated a significant DAO activity in the gastric glands. These observations seem to rule out the possibility of an important role for DAO in controlling the effect of histamine on gastric secretion. The low activity of DAO in the stomach might instead reflect the necessity for maintenance in this organ of physiological levels of histamine, which are high in all animals.

The cortical portion of the kidney contains a high level of DAO activity in certain animals, including the dog, cat,[10,22] pig,[23-25] and man.[10,26] In these mammals, the levels of renal activity are equal to or even higher than those observed in the intestine. In contrast, in other animals the renal activity levels are either low, as in the guinea pig[10,16,23] and rabbit,[1,23,27] or extremely low, as in the rat,[12,23,26,28] mouse,[10] and hen.[29] It is necessary to point out that there are major discrepancies in the reported values of renal DAO activity. In fact, DAO levels in rabbit kidney have been described by some groups as not measurable,[10,11] while other groups have noted a low but quite evident amount of this activity.[1,23,27] These discrepancies are difficult to explain, but they might be caused by certain important differences in experimental conditions, such as the various strains of animals that were used or the different methods for measurement of enzyme activity. The work of Kusche et al.[11] is relevant to this subject, since they found DAO activity only in the intestine, and in no other organ, of the rabbit. This result can be explained if one considers that this group used, as source of the enzyme from the various organs, only a supernatant after centrifugation of the crude extract at 48,000 $\times$ g for 30 min; it is now known (as will be discussed in more detail below) that in many organs DAO is a sedimentable enzyme.

In only a few animals (including the guinea pig and the hen), does the liver contain levels of DAO activity comparable to those observed in the small intestine.[10,16,29,30] In other animals, such as the rat,[10,12,23,31] the mouse,[10] the rabbit,[1,10,23,32] and the pig,[23] the hepatic activity is low, very low, or even not measurable, although there are some disagreements among various groups.

Reports in the literature on DAO levels in the lungs are also extremely conflicting. Cotzias and Dole[23] found higher levels of DAO activity in the lungs than in any other organ of the rat and the rabbit, while Beaven and de Jong[21] and Shaff and Beaven[12] reported low levels of activity in rat lung, and Waton[10] reported barely measurable levels in both animals. DAO activity has also been noted in lungs of the guinea pig and mouse,[10,23] but not in those of the hen.[29] The discrepancies in these results can be explained by considering the different sensitivities of the methods that were used for measurements of activity by the various groups or, in some cases,[23] the possibility of interference by the activity of monoamine oxidase (MAO).[22]

DAO activity has been noted in the thymus of the guinea pig[16] and the rat;[21] indeed, in the rat the level is so high that it is less only than that of the small intestine. The latter group has demonstrated, using the same animals, that DAO is also present in notable amounts in the adrenal glands and the pancreas.

DAO is present in the central nervous system of mammals, albeit not in elevated amounts and only in certain regions, such as the thalamus, the hypothalamus, and the pons.[12] This observation explains why the activity is not measurable in homogenates of the entire brain. It is worth noting that in the brain of a fish, *Cyprinus carpio,* the presence of an enzyme with the characteristics of a true DAO has been demonstrated; the activity of this enzyme is highest in the diencephalon.[1]

It is useful to specify that many organs of the rat (heart, muscle, skin, salivary glands, spleen, bone marrow, and testes) contain low or nearly no DAO activity,[12,21,33] like the skin of the guinea pig[16] and some organs (heart, skin, spleen, and ovaries) of the hen.[29]

The presence of DAO in the placenta and in the female genital apparatus of pregnant mammals deserves special mention. It has been known for a long time[34] that the maternal part of the placenta is the tissue with the highest amounts of DAO; it is the principal source of the plasma DAO (PDAO), which during pregnancy in humans and other mammals increases enormously. Furthermore, pregnancy causes significant increases in DAO activity in the uterus and, to a lesser extent, in the ovaries of the rat.[35-38] (The biological significance of these results will be considered in Chapters 14, 17, and 18).

III. CELLULAR AND SUBCELLULAR LOCALIZATION

A. Small Intestine

1. Cellular Localization

Shakir et al.[39] have shown that the DAO of the epithelium of rat small intestine is distributed according to a gradient that has its maximum in the most mature cells in the apex of the villi and its minimum in the crypt cells. The same result was obtained, using an immunohistochemical method, in the same intestinal epithelium of man.[40] These groups report that the DAO is not prevalently situated in the brush border of the intestinal epithelium, suggesting therefore that this enzyme does not play an important role in the protection of the tissue from amines originating from bacteria that are present in the intestinal lumen.

Interestingly, the distribution of DAO along the intestinal villi corresponds exactly to that observed for ODC.[41] This suggests that the role of the DAO present in intestinal cells is above all that of regulating the level of putrescine, which could, in turn, be more involved in the activities of differentiated cells than in those of proliferating cells. The mucosa of the small intestine seems to be the main source of SDAO after intravenous injection of heparin in man,[42-44] rat,[12,20,39] and in other mammals,[45] but not in the guinea pig, in which the plasma enzyme comes mainly from the liver.[46] In the rat, intravenous administration of heparin, or its perfusion in the intestinal vessels, provokes a dose-dependent release of DAO either into blood or into venous effluent, respectively, with a concomitant decrease in intestinal DAO.[12,39] A notable increase in the levels of SDAO has also been demonstrated following the injection of heparin in man; however, the relative increase is quite small in patients who show alterations in lipid metabolism[47] or suffer from ulcerative colitis.[48] Both of these pathological conditions have been associated with reduced levels of intestinal DAO.[15] The release of DAO from the intestine occurs only if the heparin is present in the intestinal vessels; in fact, this polysaccharide has no effect if it is in direct contact with the intestinal mucous membrane or with isolated epithelial cells.[39] This suggests that the DAO released into circulation does not come directly from intestinal epithelial cells, but rather from binding sites in the intestinal vasculature, which have not yet been identified.[39] In any case, the close correlation that has been observed between the levels of intestinal DAO and those of SDAO, whether during the development and post-natal maturation of the intestinal mucous membrane[14] or during the recovery process following injury to the same membrane,[49] has led to SDAO being considered as a specific indicator of the integrity of the intestinal mucous membrane (see Chapters 15 and 18).

2. Subcellular Localization

Results on the subcellular localization of intestinal DAO are difficult to interpret because of discrepancies that exist between various authors. In fact, different localizations of the enzyme have been reported not only in the various species that have been studied, but also in the same species. This is easily explainable when one considers that, as was clearly demonstrated by Hülsmann et al.,[50] results obtained with subcellular fractionation of intestinal cells are strongly influenced by the amount of goblet cells (and hence mucus) present in the starting material. It is clear, therefore, that if this factor was not taken into account during the process of fractionation, the observed differences in the intracellular localization of DAO might not be real; instead, they might simply reflect the varying influence of mucus on the fractionation procedure. The only way to evaluate the influence of this factor is that of observing both the behavior of marker enzymes in the various fractions and the morphology of the fractions under the electron microscope.

Unfortunately, none of the groups who have studied the subcellular distribution of intestinal DAO has carried out these controls. Moreover, the analysis of the results obtained by various groups is rendered much more difficult by the fact that the methods used differ in numerous and important details. The starting material is not even homogeneous, since it consists of scraped intestinal mucous membrane in some species,[13,18,22] of *in toto* fresh intestine in other species,[22,29] or even of *in toto* intestine that has been frozen and thawed.[18] The methods of homogenization also differ, be it in the type of homogenizer, the running conditions (time, rpm, number of strokes), or in the homogenization medium. In this regard, some groups use phosphate buffer, which has been known for some time to not be the most suitable medium for subcellular fractionation.[8,9] Finally, even the centrifugal fields used in fractionation vary in different studies and are often debatable.

Keeping all of these limitations in mind, we will attempt to examine the conclusions at which various groups have arrived. Backus and Kim[22] obtained results with the small intestine of the cat, dog, and rat that they themselves consider inconclusive. In the cat, 65% of the total DAO activity sediments at 600 × g in 10 min, but this does not necessarily indicate a nuclear localization for the enzyme, since the pellet, even though it contains nearly all of the nuclei, has a very heterogeneous composition regardless of tissue type. This is particularly evident in the intestine, in which mucus can cause a strong aggregation of cell organelles. In the dog and rat, DAO activity is prevalently present in a soluble form, in that 54 and 74% of the total activity, respectively, remains in the supernatant after centrifugation at 120,000 × g for 45 min. The authors, on the basis of this result, suggest that in these animals the enzyme might be easily released from particulate material during homogenization, or that it might actually be localized in the cytosol. Taylor and Lieber[13] found that, in the mucous membrane of the duodenum of the rat and macaco *(M. rhesus)*, DAO is largely soluble; this is more evident in the monkey (85.9% of the total) than in the rat (39.4%). This result, however, is somewhat unreliable in that the recoveries of both DAO activity and total proteins is low. Furthermore, in both animals (but especially in the rat), the relative specific activity of the supernatant is too low to allow a definite conclusion. Finally, Biegansky et al.[18,29] found, in the intestinal mucosa or the *in toto* intestinal wall of various animal species (dog, pig, rabbit, rat, and mouse, and man) a prevalently cytoplasmic localization for DAO (not sedimentable after 78,000 × g for 60 min). Only in the guinea pig and the chicken was the enzyme largely insoluble, in that it was sedimentable at 600 to 900 × g in 10 min. The conclusions reached by these authors are, however, unreliable since the experiments were carried out with tissues that had been frozen, thawed, and homogenized strongly in phosphate buffer; one cannot exclude the possibility that the final fraction (obtained after centrifugation at 78,000 × g for 60 min) was contaminated by fragments of organelles and especially by small vesicles derived from the endoplasmic reticulum. In addition, all of the data on the total activity of the homogenate are lacking, rendering impossible an estimation of the recovery.

A consideration of this review forces one to conclude that the problem of the subcellular localization of intestinal DAO is anything but resolved. This problem should be examined again, this time with methods that will yield indisputable results, since its resolution could be a great aid in the clarification of the relations between intestinal DAO and SDAO. A precise intracellular localization of the enzyme at the level of the intestinal epithelium could,

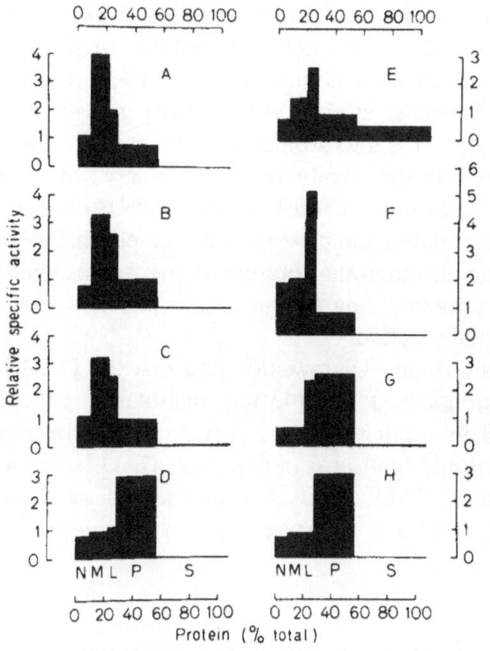

FIGURE 1. Distribution pattern of enzymes in rabbit liver. Ordinate: mean relative specific activity of fractions (% of total activity/% of total protein). Abscissa: % of total protein (related to the total homogenate). Nuclear (N), heavy mitochondrial (M), light mitochondrial (L), microsomal (P), and final supernatant (S) fractions. (A) Cytochrome oxidase; (B) monoamine oxidase; (C) diamine oxidase pH 8.8, substrate cadaverine; (D) diamine oxidase pH 7.2, substrate cadaverine; (E) acid phosphatase; (F) urate oxidase; (G) glucose-6-phosphatase; (H) diamine oxidase pH 7.6, substrate histamine. (From Argento-Cerù, M. P., Sartori, C., and Autuori, F., *Eur. J. Biochem.*, 34, 369, 1973. With permission.)

in fact, shed light on the mechanism of transfer of DAO from the cells, after its synthesis, to the intravascular binding sites from which it is released by the action of heparin.

B. Liver

1. Cellular Localization

No studies on the localization of hepatic DAO at the cellular level have been reported. The only unpublished observation deals with the liver of the rabbit, where the enzyme is present in isolated, purified hepatocytes, but is probably absent from the Kupffer cells.[89]

2. Subcellular Localization

The intracellular distribution of hepatic DAO has been studied in two mammal species, the rabbit,[23,32,51] the guinea pig,[30] and in the hen.[29] Cotzias and Dole[23] found that the DAO activity in rabbit liver is prevalently localized in the mitochondrial fraction, while Zeller et al.[51] reported, for the same organ, that the bulk of the activity is associated with the microsomal fraction. This discrepancy can be explained if one considers that the method for measurement of enzyme activity used by the first group does not allow the exclusion of an interference by MAO activity, which is known to be a marker enzyme for the external mitochondrial membrane. More recently, Argento-Cerù et al.[32] have definitively shown (Figure 1) that the DAO of rabbit liver is localized in the microsomal fraction, in which one finds about 80% of the total activity, with a relative specific activity of about three times that of the crude homogenate. The subfractionation of microsomes has shown that the enzyme is largely concentrated in the smooth endoplasmic reticulum;[52] it has the characteristics of an intrinsic microsomal membrane protein.[53] These results are in total agreement with those obtained by Schmutzler et al.[30] with guinea pig liver. In fact, in this organ too DAO is localized prevalently in the microsomal fraction (more than 80% of the total activity) and it is tightly associated with the membrane. Interestingly, endoportal injections of high doses of heparin (500 I.U./kg) cause the nearly total loss of DAO activity from all subcellular fractions; the activity is released in soluble form into the plasma, where it attains a specific activity about ten times greater than that present in the microsomal fraction. The authors interpret these results by hypothesizing that in guinea pig liver, the DAO associated with the membrane is only partially active.

In chicken liver, Biegansky and Ulatowska[29] find that the DAO is distributed between the pellet after centrifugation at 600 × g (43% of the total) and the supernatant after 9000 × g (46% of the total). This result suggests a particulate localization for the enzyme, and does not exclude the possibility that, also in this case, DAO is prevalently localized in the microsomal fraction. Finally, DAO activity has also been demonstrated in the microsomal fraction in the liver of the quail and digestive gland of the mussel.[54]

C. Kidney

1. Cellular Localization

The localization of renal DAO was determined using immunohistochemical methods, which permitted an analysis of the distribution of the enzyme in the renal cortex of the pig. Takano et al.,[55] using a direct immunofluorescence technique, observed a positive reaction in the capsular epithelium of Bowman's capsule and in the epithelium of proximal convoluted tubule. In both of these tissues the fluorescence seems diffuse in the cytoplasm but not in the nucleus, and it is particularly evident in the apical region of the tubule cells near the brush border. The authors suggest that the function of renal DAO might be that of acting on the amines that are reabsorbed by the proximal convoluted tubules, with consequent production of ammonia (which would be excreted into the lumen of the same tubules). A somewhat different localization was observed, again in the renal cortex of the pig, by Argento-Cerù et al.[56,57] In these studies, carried out both with direct and indirect immunofluorescence techniques, and with the direct immunoperoxidase method, DAO was found exclusively in the epithelium of proximal convoluted tubules; these cells showed a characteristic positive reaction with a longitudinally striated pattern under the nucleus. The slight, diffuse positive reaction observed in correspondence with the brush border can probably be explained by the fact that the preparation of renal DAO that was used to isolate the specific antibodies was actually contaminated with DOPA-decarboxylase.[7] This positivity, in fact, disappeared completely when the staining of antibody-conjugated peroxidase was not carried out by the addition of exogenous H_2O_2 to the incubation medium,[58] but by the addition of specific substrates of DAO (putrescine, cadaverine, and histamine) and consequent production of endogenous H_2O_2. On the basis of this result, one can suggest that the discrepancies between the work of Takano et al.[55] and that of Argento-Cerù et al.[56,57] can be ascribed either to the presence of contaminants in the DAO preparations that were used as antigens, or to the fact that formalin, used as a fixative by the former authors, greatly reduces the antigenicity of DAO.[59,60]

2. Subcellular Localization

The subcellular localization of DAO in the renal cortex has been examined by Backus and Kim[22] in the kidney of the cat and dog, and by Sartori et al.[27] in rabbit kidney. The former authors found that the enzyme was prevalently localized in the pellet obtained after centrifugation at 600 × g (45 and 44% of the total activity, respectively, in the dog and

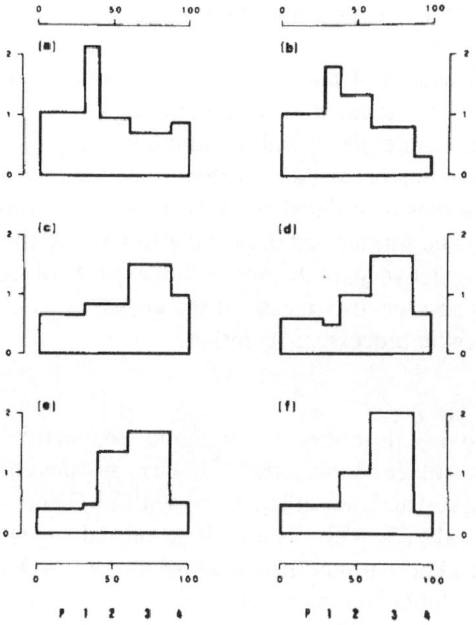

FIGURE 2. Distribution pattern of enzymes in rabbit kidney cortex microsomal subfractions. Ordinate: mean relative specific activity of subfractions (% of total activity/% of total protein). Abscissa: % of total protein (related to the total microsomal fraction). Pellet (P), microsomal subfractions (1,2,3,4). (a) (Na^+-K^+)-ATPase; (b) alkaline phosphatase; (c) NADPH-cytochrome *c* reductase; (d) NADH-cytochrome *c* reductase; (e) glucose-6-phosphatase; (f) diamine oxidase (substrate histamine). (From Sartori, C., Bargagli, A. M., and Argento-Cerù, M. P., *Biochim. Biophys. Acta*, 758, 58, 1983. With permission.)

the cat). Nevertheless, this fraction from the renal cortex, besides containing nuclei, intact cells, connective tissue, and blood cells, also contains large membranous particles that are derived from the brush border and basal infoldings. The composition of this fraction is directly dependent upon the homogenization technique and in particular upon the force of homogenization. Unfortunately, the authors did not specify the homogenization procedure; they omitted both the analysis of the fractions in the electron microscope and the measurement of marker enzyme activity, and they failed to report the specific activity of DAO in the individual fractions, so that the results obtained are not conclusive. In the renal cortex of the rabbit, Sartori et al.[27] found that more than 60% of the total DAO activity is localized in the microsomal fraction, in which the enzyme has a relative specific activity five to six times greater than that in the crude homogenate. The subsequent subfractionation of this fraction has allowed the authors to demonstrate that renal DAO is associated with the membrane of the endoplasmic reticulum (see Figure 2).

D. Placenta

As stated above, the placenta is the organ that, on the average, contains the highest levels of DAO activity among all of the normal tissues that have been studied, and it is the source of the high levels of SDAO that are observed during pregnancy.[61] The localization of this enzyme in placenta was first carried out by da Swamberg.[34] This author, by dissecting human placenta after delivery and measuring DAO activity in the single components that were mechanically isolated, demonstrated that the enzyme is most highly concentrated in the maternal part of the placenta (namely, in the *decidua basalis*). This result was later confirmed by Southren et al.[62] and by Gunther and Glick,[36] who measured enzyme activity in histological serial sections.

More recently, the purification of human placental DAO has allowed the localization of the enzyme with immunohistochemical techniques; these experiments have confirmed and broadened previous results. Lin et al.,[60] with an immunofluorescence technique, and Weisburger et al.,[63] with the immunoperoxidase method, have demonstrated that the enzyme is widespread within the cytoplasm of decidual cells, while it is completely absent from the fetal part of the placenta. The former authors have also noted a specific fluorescence, most evident in the placenta after delivery, in the intracellular spaces of the decidua. This presence could be caused either by an aspecific release of the enzyme by degenerating decidual cells or by a true and characteristic process of secretion.

E. Connective Tissue

AOs have been extensively described in numerous connective tissues such as bone,[64] derma,[65] dental pulp,[66,67] cartilage,[68] and aorta.[69-73] The properties and functions of connective tissue amine oxidases (CTAO) and their interactions with SAO and with lysyl oxidase (LAO) are not yet fully understood. We will examine here only their distributions in subcellular fractions of these tissues. The various groups agree that CTAO is present in connective tissue in two forms, one soluble and the other insoluble, which vary in relative amounts in different tissues and according to different authors. Bird et al.[69] report that in chicken aorta, homogenized in 0.15 *M* NaCl and centrifuged at 1000 × g for 15 min, less than 50% of the CTAO activity is soluble; a more complete extraction can be achieved by the addition of 1% Triton X-100 in the homogenate. Rucker et al.,[64] after fractionating an homogenate of chicken femurs in 0.1 *M* phosphate buffer (pH 7.4), found 80% of the CTAO activity in the pellet after centrifugation at 700 × g for 30 min and 20% in the supernatant. The latter percentage of activity does not sediment either at 7000 × g in 30 min (mitochondrial fraction) or at 110,000 × g in 90 min (microsomal fraction). The sedimentable CTAO activity seems to be partially latent, because it can be fully released only after sonication. Nakano et al.[66,67] found, in the dental pulp of the ox (preserved at −20°C and homogenized in 0.25 *M* sucrose) that 55% of the CTAO activity is soluble, while 25% is in the microsomal and 21% in the mitochondrial fraction. Rucker and Goettlich-Riemann[71] examined the aorta of the rabbit (frozen, crushed, and homogenized in 0.1 *M* phosphate buffer, pH 7.4) and found only 20% of the CTAO activity in the supernatant after centrifugation at 10,000 × g for 20 min; they concluded that the largest fraction of the enzyme is present in particulate form in the aorta of the rabbit. Buffoni et al.,[74] in the aorta of the pig, and Vidal et al.,[75] in the aorta of the ox, discovered two forms of CTAO: a soluble form, similar to SAO, and an insoluble form, which more closely resembles LAO. The two forms are immunologically distinct, and their relative quantities change with age.[76]

A precise evaluation of these results cannot be given without the definition of the meanings that different groups have given to the terms "soluble", "insoluble", and "particulate". The solubility of the enzyme could indicate either that the protein is loosely bound, specifically or aspecifically, to a cellular structure, or that it is actually secreted from connectival cells. Similarly, the insolubility could be caused by a tight association either with a cellular organelle or an extracellular structure that is easily sedimented, or by simple entrapment, even artifactual, in particulate structures. In any case, regarding the origin of AO in connective tissue, it seems likely that, as was demonstrated with lysyl oxidase,[77] the enzyme is synthesized in fibroblasts.[66]

A special case, in the connective tissue, is represented by the DAO of leukocytes. Zeiger et al.[78] have shown that a histaminase activity is present in the eosinophilic and neutrophilic granulocytes of human blood, while it is absent from basophilic granulocytes, lymphocytes, and monocytes. Experiments with differential centrifugation have demonstrated that DAO sediments in the fraction containing the specific granules of the granulocytes, in which the enzyme is concentrated four to five times more than in the crude homogenate.

F. Other Tissues

In rabbit lungs, Waton[10] found that the DAO activity is largely present (72.4%) in the supernatant after centrifugation at 37,000 × g for 60 min of a crude homogenate in 0.25 *M* sucrose. However, this observation does not allow a definitive conclusion concerning the localization of the enzyme in the lungs, since the centrifugal field used was not great enough to sediment all of the membranous components of the homogenate. Beaven and de Jong[21] ascertained that rat thymus DAO is present in cells other than lymphocytes, since the histaminase activity does not decrease at all when the organ is totally deprived of lymphocytes by treatment of the animal with a single dose of dexamethasone; the activity also remains high in senescent rats.

Seminal fluid contains DAO of prostatic origin,[79,80] but 5% of the total activity is associated with spermatozoa, and cannot be removed either by repeated washings with NaCl or $MgCl_2$ solutions or by treatment with 1% Triton X-100.[81] Nothing is known, however, about the localization of the enzyme in spermatozoa.

G. Tumors

The cellular distribution of DAO has been examined using immunohistochemical techniques in thyroid sections from patients with varying degrees of proliferative lesions of the parafollicular cells (C cells).[82] The enzyme is located only in some of the medullary carcinoma cells that occur in clusters and are unevenly distributed throughout the tumor mass, while it is not present in normal or hyperplastic cells. The extent of the positive reaction, however, cannot be correlated with the sizes of the tumor nodules or the morphological characteristics of the cells. The same clustered distribution of DAO was noted in metastases of medullary thyroid carcinoma, as well as in small cell carcinoma of the lungs and its metastases[82,83] and in ovarian carcinoma, in which both the distribution and the intensity of the immunohistochemical reaction cannot be correlated either with the histological pattern or the extent of tumor differentiation.[84] The study of the cellular distribution of DAO in tumors has recently been extended to many other tumor types by Nishiyama et al.[85] The enzyme, localized with the immunohistochemical method, has been noted in the cytoplasm of cells from primary tumors or metastases of the lung, thyroid, pancreas, stomach, esophagus, ovary, and pharynx, as well as in cells of carcinoids, dysgerminomas, and melanomas. In the positive cells the staining seems diffuse, scattered in the cytoplasm, with a pattern that does not correspond to any cellular organelle. Also in this work, as in the preceding studies, no correlation between the degree of the positive reaction and the histological type or extent of differentiation of the tumor was noted. The authors have not given definitive conclusions about the significance of their observations, either because of the low number of cases examined, or on account of the use of formalin as a fixative (which reduces the immunoreactivity of DAO); however, they postulate, although with caution, that the enzyme might occur more frequently in malignancies of epithelial origin (see Chapter 17).

IV. CONCLUSIONS

The results obtained in the studies on the localization of AO at various levels permit the formulation of some interesting considerations. Above all, observations that in diverse organs of the same animal the enzyme has greatly different levels of activity, and that it is particularly concentrated in certain organs (intestine, placenta, kidney) lead one to imagine that its function is in some way correlated with the specific function of the single organ. Actually, the organs richest in DAO are those with a detoxicating function in the organism, and in any case those that receive the highest concentrations of amines. For example, the intestinal mucous membrane is constantly in contact with the histamine that is produced and not metabolized by the stomach (in which the DAO activity is very low or even absent) and with the amines present in the diet or produced by the intestinal flora. Correspondingly, a substantial amount of histamine is released in the enteric vessels, together with heparin, during the process of digestion and lipid transport.[86] The kidney, on the other hand, and in particular the proximal convoluted tubule, is in contact with amines that are filtered by the renal corpuscle; these amines can even be present in considerable amounts, depending on the diet or on particular physiological (the estrous cycle, menstrual cycle, pregnancy) and pathological conditions. Finally, the placenta receives large quantities of amines produced by the developing fetus.

Conclusions of a different type can be drawn from the results obtained on the localization of AO at the cellular level. First of all, when considering the cellular types, one can say that the enzyme is not characteristic of only a single class of cells, distinctive in their embryological origin or extent of differentiation. In fact, AO is present in cells of endodermal (intestinal mucous membrane, liver), mesodermal (kidney), mesenchymal (connective tissue), and even ectodermal origin (epidermis, brain), although in the last case the levels of enzyme activity are so low as to be at the limit of detectability. On the other hand, the studies carried out with tumor tissues have not shown any correlation between the level of DAO and the degree of cellular differentiation. Instead, it should be emphasized that, at least in certain organs, the enzyme seems to have the same localization at the cellular level as ornithine decarboxylase. This parallelism has been conclusively demonstrated in the intestinal mucous membrane[41] and thymus of the rat[21] and in the kidney of the mouse;[87] it has also been well proven under certain experimental conditions in some organs that do not normally contain high levels of DAO activity. In fact, in compensatory hypertrophy of the kidney and heart of the rat[28,33] and in hepatic regeneration,[31] the remarkable increase in ornithine decarboxylase is always accompanied by an increase, although less evident, in DAO activity. Since, as is known, ornithine decarboxylase transforms ornithine into putrescine, it seems reasonable to think that DAO must play a role (the same in diverse organs) in controlling the intracellular level of this amine, which is an essential metabolite for the biosynthesis of polyamines (spermidine and spermine) (see Chapter 16).

The results on the subcellular localization of DAO lead us to the same conclusion. In fact, if one considers only those studies on cellular fractionation that were carried out with impeccable methodology, one concludes that DAO is localized in the endoplasmic reticulum in which, as has been known for some time,[88] the largest amounts of polyamines are also present. Finally, in this inquiry into the biological role of DAO at the cellular and subcellular level, it is essential that the metabolites formed as products of the oxidative deamination of substrates should not be neglected. The aldehydes and their oxidative products (e.g., γ-aminobutyric acid) or even ammonia and hydrogen peroxide could carry out important functions at the subcellular level; this subject is open to study, and might constitute an interesting topic for further research.

ACKNOWLEDGMENT

The original studies reported here were supported in part by the Italian National Research Council, Special Project on the "Chimica Fine", Contract No. 83.00310.95.

REFERENCES

1. **Zeller, E. A.,** Diamine oxidases, in *The Enzymes*, Vol. 8, Boyer, P. D., Lardy, H., and Myrbäk, K., Eds., Academic Press, New York, 1963, chap. 11.
2. **Blaschko, H.,** The natural history of amine oxidases, *Rev. Physiol. Biochem. Pharmacol.*, 70, 83, 1974.
3. **Moldéus, P., Högberg, J., and Orrenius, S.,** Isolation and use of liver cells, in *Methods in Enzymology*, Vol. 52, Part C, Fleischer, S. and Packer, L., Eds., Academic Press, New York, 1978, chap. 4.
4. **Pertoft, H., Hirtestein, M., and Kågedal, L.,** Cell separation in a new density gradient medium, Percoll, in *Cell Populations*, Vol. 8, Reid, E., Ed., Ellis Horwood, Chichester, 1979, 67.
5. **Seglen, P. O.,** Disaggregation and separation of rat liver cells, in *Cell Populations*, Vol. 8, Reid, E., Ed., Ellis Horwood, Chichester, 1979, 25.
6. **Bancroft, J. D.,** *Histochemical Techniques*, Butterworths, London, 1975, chap. 17.
7. **Oratore, A., Banchelli, G., Buffoni, F., Sabatini, S., Mondovì, B., and Finazzi-Agrò, A.,** Reaction between mammalian amine oxidases and their antibodies, *Biochem. Biophys. Res. Commun.*, 98, 1002, 1981.
8. **De Duve, C. and Berthet, J.,** The use of differential centrifugation in the study of tissue enzymes, *Int. Rev. Cytol.*, 3, 225, 1955.
9. **Allfrey, V.,** The isolation of subcellular components, in *The Cell*, Vol. 1, Brachet, J. and Mirsky, A. E., Eds., Academic Press, New York, 1959, chap. 7.
10. **Waton, N. G.,** Studies on mammalian histidine decarboxylase, *Br. J. Pharmacol.*, 11, 119, 1956.
11. **Kusche, J., Richter, H., Schmidt, J., Hesterberg, R., Friedrich, A., and Lorenz, W.,** Diamine oxidase in rabbit small intestine: separation from a soluble monoamine oxidase, properties and pathophysiological significance in intestinal ischemia, *Agents Actions*, 5, 431, 1975.
12. **Shaff, R. E. and Beaven, M. A.,** Turnover and synthesis of diamine oxidase (DAO) in rat tissues. Studies with heparin and cycloheximide, *Biochem. Pharmac.*, 25, 1057, 1976.
13. **Taylor, S. L. and Lieber, E. R.,** Subcellular distribution and properties of intestinal histamine-metabolizing enzymes from rats, guinea pigs and rhesus monkeys, *Comp. Biochem. Physiol.*, 63C, 21, 1979.
14. **Luck, G. D., Bayless, T. M., and Baylin, S. B.,** Diamine oxidase (histaminase). A circulating marker for rat intestinal mucosal maturation and integrity, *J. Clin. Invest.*, 66, 66, 1980.
15. **Hesterberg, R., Kusche, J., Stahlknecht, C. D., and Feussner, K. D.,** The start of a programme for measuring diamine oxidase activity in biopsy specimens of human rectal mucosa, *Agents Actions*, 11, 33, 1981.
16. **Fogel, W. A., Bieganski, T., and Maslinski, C.,** Gamma-amino-butyric acid (GABA) formation from putrescine in guinea pig liver during ontogenesis, *Comp. Biochem. Physiol.*, 73C, 431, 1982.
17. **Kim, K. S., Backus, B., Harris, H., and Rourke, P.,** Distribution of diamine oxidase and imidazole-*N*-methyltransferase along the gastrointestinal tract, *Comp. Biochem. Physiol.*, 31, 137, 1969.
18. **Bieganski, T., Kusche, J., Lorenz, W., Hesterberg, R., Stahlknecht, C. D., and Feussner, K. D.,** Distribution and properties of human intestinal diamine oxidase and its relevance for the histamine catabolism, *Biochim. Biophys. Acta*, 756, 196, 1983.
19. **Kobayashi, Y. and Ivy, A. C.,** Histamine metabolizing activity of dog stomach and intestine, *Am. J. Physiol.*, 195, 525, 1958.
20. **Kobayashi, Y., Kupelian, J., and Maudsley, D. Y.,** Release of diamine oxidase by heparin in the rat, *Biochem. Pharmacol.*, 18, 1585, 1969.
21. **Beaven, M. A. and de Jong, W.,** Presence of histaminase and ornithine decarboxylase activities in rat thymus, *Biochem. Pharmacol.*, 22, 257, 1973.
22. **Backus, B. and Kim, K. S.,** Subcellular distribution of diamine oxidase, *Comp. Gen. Pharmacol.*, 1, 196, 1970.
23. **Cotzias, G. C. and Dole, V. P.,** The activity of histaminase in tissues, *J. Biol. Chem.*, 196, 235, 1952.
24. **Kapeller-Adler, R. and MacFarlane, H.,** Purification and identification of hog-kidney histaminase, *Biochim. Biophys. Acta*, 67, 542, 1963.
25. **Mondovì, B., Rotilio, G., Finazzi, A., and Scioscia Santoro, A.,** Purification of pig-kidney diamine oxidase and its identity with histaminase, *Biochem. J.*, 91, 408, 1964.
26. **Kopple, J. D., Tam, C. F., Wang, M., and Swendseid, M. E.,** Diamine oxidase in renal failure, *Kidney Int.*, 13, S20, 1978.
27. **Sartori, C., Bargagli, A. M., and Argento-Cerù, M. P.,** Subcellular localization of diamine oxidase in rabbit kidney cortex, *Biochim. Biophys. Acta*, 758, 58, 1983.
28. **Desiderio, M. A., Sessa, A. and Perin, A.,** Induction of diamine oxidase activity in rat kidney during compensatory hypertrophy, *Biochim. Biophys. Acta*, 714, 243, 1982.
29. **Bieganski, T. and Ulatowska, M. A.,** Diamine oxidase in the hen, *Agents Action*, 13, 257, 1983.
30. **Schmutzler, W., Goldschmidt, O., Bethge, K. P., and Knop, J.,** The release of guinea pig liver histaminase and some of its properties, *Int. Arch. Allergy*, 36, 45, 1969.

31. **Sessa, A., Desiderio, M. A., and Perin, A.,** Diamine oxidase activity induction in regenerating rat liver, *Biochim. Biophys. Acta*, 698, 11, 1982.
32. **Argento-Cerù, M. P., Sartori, C., and Autuori, F.,** Diamine oxidase in rabbit liver. Assay conditions and subcellular localization, *Eur. J. Biochem.*, 34, 369, 1973.
33. **Perin, A., Sessa, A., and Desiderio, M. A.,** Polyamine levels and diamine oxidase activity in hypertrophic heart of spontaneously hypertensive rats and of rats treated with isoproterenol, *Biochim. Biophys. Acta*, 755, 344, 1983.
34. **Swamberg, H.,** Histaminase in pregnancy with special reference to its origin and formation, *Acta Physiol. Scand.*, 23(Suppl. 79), 1, 1950.
35. **Roberts, M. and Robson, J. M.,** The histaminase content of the rat uterus, and its relation to the decidua, *J. Physiol.*, 119, 286, 1983.
36. **Gunther, R. E. and Glick, D.,** Determination of histaminase activity in histologic samples and its quantitative distribution in intact human placenta and uterus, *J. Histochem. Cytochem.*, 15, 431, 1967.
37. **Guha, S. K. and Jänne, J.,** The synthesis and accumulation of polyamines in reproductive organs of the rat during pregnancy, *Biochim. Biophys. Acta*, 437, 244, 1976.
38. **Henningsson, A.-Ch., Henningsson, S., and Nilsson, O.,** Diamines and polyamines in mammalian reproduction, in *Advances in Polyamine Research*, Vol. 4, Bachrach, U., Kaye, A., and Chayen, R., Eds., Raven Press, New York, 1983, 193.
39. **Shakir, K. M. M., Margolis, S., and Baylin, S. B.,** Localization of histaminase (diamine oxidase) in rat small intestinal mucosa: site of release by heparin, *Biochem. Pharmacol.*, 26, 2343, 1977.
40. **Baylin, S. B. and Luk, G. D.,** Diamine oxidase (histaminase) activity in human disease, in *Polyamines in Biology and Medicine*, Morris, D. R. and Marton, L. J., Eds., Marcel Dekker Inc., New York, 1981, chap. 19.
41. **Baylin, S. B., Stevens, S. A., and Shakir, K. M. M.,** Association of diamine oxidase and ornithine decarboxylase with maturing cells in rapidly proliferating epithelium, *Biochim. Biophys. Acta*, 541, 415, 1978.
42. **Tryding, N.,** Heparin-induced diamine oxidase (DAO) activity, *Scand. J. Clin. Lab. Invest.*, 17 (Suppl. 86), 196, 1965.
43. **Hansson, R., Holmberg, C. G., Tibbling, G., Tryding, N., Westling, H., and Wetterquist, H.,** Heparin-induced diamine oxidase increase in human blood plasma, *Acta Med. Scand.*, 180, 533, 1966.
44. **Dahlbäck, O., Hansson, R., Tibbling, G., and Tryding, N.,** The effect of heparin on diamine oxidase and lipoprotein lipase in human lymph and blood plasma, *Scand. J. Clin. Lab. Invest.*, 21, 17, 1968.
45. **Hansson, R. and Thysell, H.,** Diamine oxidase in blood plasma in some vertebrates and *Anodonta cygnea* before and after injection of heparin, *Acta Physiol. Scand.*, 74, 533, 1968.
46. **Giertz, H. and Hahn, F.,** Mechanism of histamine liberation in guinea pig anaphylaxis, *Int. Arch. Allergy*, 36, 41, 1969.
47. **Baylin, S. B., Beaven, M. A., Krauss, R. M., and Keiser, H. R.,** Response of plasma histaminase to small doses of heparin in normal subjects and patients with hyperlipoproteinemia, *J. Clin. Invest.*, 52, 1985, 1973.
48. **Gäng, V., Gehring, G., and Schönborn, J.,** Combined decrease of postheparin-DAO (histaminase) and postheparin-lipoproteinlipase in inflammatory diseases, *Klin. Wschr.*, 55, 169, 1977.
49. **Luk, G. D., Vaughan, W. P., Burke, P. J., and Baylin, S. B.,** Diamine oxidase as a plasma marker of rat intestinal mucosal injury and regeneration after administration of 1-β-D-arabinofuranosylcytosine, *Cancer Res.*, 41, 2334, 1981.
50. **Hülsmann, W. C., van den Berg, J. W. O., and de Jonge, H. R.,** Isolation of intestinal mucosa cells, in *Methods in Enzymology*, Vol. 32 (Part B), Fleischer, S. and Packer, L., Eds., Academic Press, New York, 1974, chap. 67.
51. **Zeller, E. A., Barsky, J., Berman, E. R., Cherkas, M. S., and Fouts, J. R.,** Degradation of mescaline by amine oxidases, *J. Pharmacol. Exp. Ther.*, 124, 282, 1958.
52. **Argento-Cerù, M. P., Sartori, C., and Autuori, F.,** Diamine oxidase in rabbit liver. Submicrosomal distribution, *Eur. J. Biochem.*, 34, 375, 1973.
53. **Argento-Cerù, M. P., Sartori, C., and Autuori, F.,** Association of diamine oxidase with microsomal membranes in rabbit liver, *Life Sci.*, 15, 1909, 1974.
54. **Ignesti, G., Banchelli, G., Falai, C., Pirisino, R., Raimondi, L., and Buffoni, F.,** Characterization of the amine-oxidase activities of liver microsomes of different vertebrate and invertebrate species, *Biochem. Pharmacol.*, 32, 3653, 1983.
55. **Takano, K., Suzuki, T., and Yasuda, K.,** Localization of diamine oxidase and D-amino acid oxidase in kidney, demonstrated by means of immunohistochemical method, *Acta Histochem. Cytochem.*, 3, 105, 1970.
56. **Argento-Cerù, M. P., Oratore, A., Mondovì, B., and Finazzi-Agrò, A.,** Localization of diamine oxidase in pig kidney: immunofluorescence method, *Cell. Mol. Biol.*, 27, 359, 1981.

57. **Argento-Cerù, M. P., Oratore, A., Beninati, S., and Finazzi-Agrò, A.,** Localization of diamine oxidase in pig kidney: immunoperoxidase method, *Cell. Mol. Biol.*, 27, 363, 1981.
58. **Graham, R. C. and Karnovsky, M. J.,** The early stages of absorption of injected horseradish peroxidase in the proximal tubules of mouse kidney: ultrastructural cytochemistry by a new technique, *J. Histochem. Cytochem.*, 14, 291, 1966.
59. **Nishiyama, T. and Inoue, I.,** Localization of histaminase in human fetal tissues, *Acta Histochem. Cytochem.*, 12, 546, 1979.
60. **Lin, C.-W., Chapman, C. M., De Lellis, R. A., and Kirley, S.,** Immunofluorescent staining of histaminase (diamine oxidase) in human placenta, *J. Histochem. Cytochem.*, 26, 1021, 1978.
61. **Kapeller-Adler, R.,** Histamine catabolism in vitro and in vivo, *Fed. Proc.*, 24, 757, 1965.
62. **Southren, A. L., Kobayashi, Y., Brenner, P., and Weingold, A. B.,** Diamine oxidase activity in human maternal and fetal plasma and tissues at parturition, *J. Appl. Physiol.*, 20, 1048, 1965.
63. **Weisburger, W. R., Mendelsohn, G., Eggleston, J., and Baylin, S. B.,** Immunohistochemical localization of histaminase (diamine oxidase) in decidual cells of human placenta, *Lab. Invest.*, 38, 703, 1978.
64. **Rucker, R. B., Rogler, J. C., and Parker, H. E.,** The partial characterization of an amine oxidase in bone tissue, *Proc. Soc. Exp. Biol. Med.*, 130, 1150, 1969.
65. **Lovemberg, W., Dixon, E., Keiser, H. R., and Sjoerdsma, A.,** Enzymic activities during the transformation of crypt to columnar intestinal cells. *Biochem. Pharmacol.*, 13, 1117, 1968.
66. **Nakano, G. and Nagatsu, T.,** Purification and properties of an amine oxidase in bovine dental pulp, *Experientia*, 27, 1399, 1971.
67. **Nakano, G., Harada, M., and Nagatsu, T.,** Purification and properties of an amine oxidase in bovine dental pulp and its comparison with serum amine oxidase, *Biochim. Biophys. Acta*, 341, 366, 1974.
68. **Pinnel, S. R. and Martin, G. R.,** The cross-linking of collagen and elastin: enzymatic conversion of lysine in peptide linkage to α-aminoadipic-δ-semialdehyde (allysine) by an extract from bone, *Proc. Natl. Acad. Sci. U.S.A.*, 61, 708, 1968.
69. **Bird, D. W., Savage, J. E., and O'Dell, B. L.,** Effect of copper deficiency and inhibitors on the amine oxidase activity of chick tissues, *Proc. Soc. Exp. Biol. Med.*, 123, 250, 1966.
70. **Rucker, R. B. and O'Dell, B. L.,** Connective tissue amine oxidase. I. Purification of bovine aorta amine oxidase and its comparison with plasma amine oxidase, *Biochim. Biophys. Acta*, 235, 32, 1971.
71. **Rucker, R. B. and Goettlich-Riemann, W.,** Properties of rabbit aorta amine oxidase, *Proc. Soc. Exp. Biol. Med.*, 139, 286, 1972.
72. **Harris, E. D., Gonnerman, W. A., Savage, J. E., and O'Dell, B. L.,** Connective tissue amine oxidase. II. Purification and partial characterization of lysyl oxidase from chick aorta, *Biochim. Biophys. Acta*, 341, 332, 1974.
73. **Buffoni, F. and Della Corte, L.,** Immunofluorescence histochemistry of porcine tissues using antibodies to pig plasma amine oxidase, *Proc. R. Soc. London B*, 195, 417, 1977.
74. **Buffoni, F., Pirisino, R., and Marino, P.,** Histamine as inhibitor of pig aorta amine oxidase, *Agents Actions*, 5, 429, 1975.
75. **Vidal, G. P., Shieh, J. J., and Yasunobu, K. T.,** Immunological studies of bovine aorta, lysyl oxidase: evidence for two forms of the enzyme, *Biochem. Biophys. Res. Commun.*, 64, 989, 1975.
76. **Misiorowsky, R. L., McCarthy, D. W., and Ulreich, J. B.,** Some physical properties of lysyl oxidase and its substrate, *Fed. Proc.*, 33, 618, 1974.
77. **Layman, D. L., Narayanan, A. S., and Martin, G. R.,** The production of lysyl oxidase by human fibroblast in culture, *Arch. Biochem. Biophys.*, 149, 97, 1972.
78. **Zeiger, R. S., Yurdin, D. L., and Colten, H. R.,** Histamine metabolism. II. Cellular and subcellular localization of the catabolic enzymes, histaminase and histamine methyl transferase, in human leukocytes, *J. Allergy Clin. Immunol.*, 58, 172, 1976.
79. **Mann, T.,** *The Biochemistry of Semen and of the Male Reproductive Tract*, Methuen, London, 1964, 1.
80. **Jänne, J., Hölttä, E., Haaranen, P., and Elfving, K.,** Polyamines and polyamine-metabolizing enzymes activities in human semen, *Clin. Chim. Acta*, 48, 393, 1973.
81. **Pulkkinen, P., Kanerva, S., Elfving, K., and Jänne, J.,** Association of spermine and diamine oxidase activity with human spermatozoa, *J. Reprod. Fertil.*, 43, 49, 1975.
82. **Mendelsohn, G., Eggleston, J. C., Weisburger, W. R., Gann, D. S., and Baylin, S. B.,** Calcitonin and histaminase in C-cell hyperplasia and medullary thyroid carcinoma, *Am. J. Pathol.*, 92, 35, 1978.
83. **Baylin, S. B., Weisburger, W. R., Eggleston, J. C., Mendelsohn, G., Beaven, M. A., Abeloff, M. D., and Ettinger, B. S.,** Variable content of histaminase, L-dopa decarboxylase and calcitonin in small cell carcinoma of the lung: biological and clinical implications, *N. Engl. J. Med.*, 299, 105, 1978.
84. **Ettinger, D. S., Rosenshein, N. B., Parmley, T. H., Mendelsohn, G., and Baylin, S. B.,** Tumor cell origin of histaminase activity in ascites fluid from patients with ovarian carcinoma, *Cancer*, 45, 2568, 1980.

85. **Nishiyama, T., Hayashi, I., Inoue, I., Lin, C.-W., Akagi, G., and Sano, T.,** Distribution of histaminase in human tumor tissues. An immunohistochemical study, *Oncodev. Biol. Med.*, 4, 197, 1983.
86. **Mahadoo, J., Wright, C. J., and Jaques, L. B.,** Effect of intrapulmonary heparin on plasma diamine oxidase (histaminase) activity in mice, *Agents Actions*, 11, 335, 1981.
87. **Persson, L., Rosengren, E., Sundler, F., and Uddman, R.,** Immunocytochemical demonstration of ornithine decarboxylase, in *Advances in Polyamine Research*, Vol. 4, Bachrach, U., Kaye, A., and Chayen, R., Eds., Raven Press, New York, 1983, 571.
88. **Raina, A. and Telaranta, T.,** Association of polyamines and RNA in isolated subcellular particles from rat liver. *Biochim. Biophys. Acta*, 138, 200, 1967.
89. **Autuori, F.,** unpublished observations.

Chapter 10

PLASMA AMINE OXIDASE

G. Pettersson

TABLE OF CONTENTS

I. INTRODUCTION

Copper-containing amine oxidases have been found in a wide range of eukaryotic cells of fungal, plant, and mammalian origin.[1-3] The Cu-AOs isolated from different tissues or phylae have many properties in common, but are distinctly different enzymes with respect to molecular characteristics and substrate specificity. This chapter will examine what is known about the Cu-AO which occurs in the plasma of mammals. Emphasis is placed on the mechanism of action of the enzyme.

II. PHYSICOCHEMICAL PROPERTIES

A. Catalytic Activity

Plasma (serum) amine oxidase (SAO) catalyzes the oxidative deamination of primary aliphatic or aromatic amines according to the equation:

$$RCH_2NH_2 + O_2 + H_2O \rightarrow RCHO + H_2O_2 + NH_3$$

Secondary and tertiary amines, and amines with a substituent at the α-carbon, are not catalytically affected.[3,4]

Substrates such as benzylamine and heptylamine are readily oxidized by SAO from all sources examined. The enzymes occurring in nonruminants (e.g., horse,[5] pig,[6] rabbit,[7] and human[8]) are often referred to as benzylamine oxidase. They usually act on biogenic amines such as histamine, tyramine, tryptamine, dopamine, and serotonin, but show negligibly low activity towards diamines. The converse specificity pattern applies for SAO from ruminants (e.g., sheep[9] and beef[10]). The latter type of enzyme has been described as spermine oxidase because of its high activity towards spermine and spermidine. Detailed data on substrate and inhibitor specificities of SAO from different sources can be found in the review of Buffoni[4] (see Chapter 13).

B. Isolation and Protein Structure

Highly purified SAO has been isolated from rabbit,[7] human,[8] pig,[5] and bovine[10] plasma. The pig (SSAO) and bovine (BSAO) enzymes were obtained in the crystalline state, and their physicochemical and catalytic properties have been extensively studied. Modified methods are now available for the purification of BPAO[11-13] and PPAO[14-17] to homogeneity.

Reported molecular weights are 186,000 to 196,000 for SSAO[6,17,18] and 170,000 to 190,000 for BSAO.[19,20] Both enzymes are dimers consisting of two apparently identical subunits;[18,19] BSAO shows a tendency to undergo association into higher polymers.[19] Amino acid compositions of BSAO[21] and SSAO[17,22] are similar, and both enzymes contain substantial amounts (7 to 10%) of carbohydrate.[17,22,23] The chromatographically distinct fractions of SSAO separated during enzyme preparation[6,14,16] appear to differ mainly with regard to the carbohydrate content,[17] and such is likely to be the case also for the observed multiple forms of BSAO.[24]

SSAO (benzylamine oxidase) and BSAO (spermine oxidase) are certainly not identical, but seem to be sufficiently closely related to be considered as source variants of one and the same enzyme.

C. Metal Content

SAO contains two copper atoms per enzyme molecule.[20,24-26] No other metals have been detected in significant amounts. The copper is all EPR detectable (i.e., cupric)[18,27] and remains cupric during the interaction of the enzyme with substrates or substrate analogs such as phenylhydrazine and amphetamine.[28-32] Enzyme-bound copper can be reduced to the cuprous state with dithionite under anaerobic conditions.[29,30] Inhibitory concentrations of ammonia,[33] cuprizone,[34] cyanide,[18] and azide[18,35,36] give rise to EPR spectral changes indicative of direct ligation of the reagents to copper. Inhibition through ligation to copper appears to be found also with metal chelators such as 1,10-phenanthroline, 2,2-dipyridyl, and diethyldithiocarbamate.[4,28]

Yamada et al.[28] reported that copper can be removed from BSAO by dialysis of the enzyme against diethyldithiocarbamate (DDC). The inactive enzyme derivative thus prepared showed an intense 380-nm absorption band and could be 65 to 70% reactivated by dialysis against cupric copper. A similar claim for reversible copper removal by DDC treatment has been made in the case of SSAO,[4] but could neither be confirmed by Knowles and co-workers,[2,37] nor by our group. We reisolated SSAO after DDC treatment under a variety of experimental conditions, including those of Yamada et al.,[28] and the reisolated enzyme species were invariably found by atomic absorption spectrometry to contain two equivalents of copper. BSAO, treated with DDC as described by Yamada et al.,[28] has been similarly shown by atomic absorption spectrometry to retain about 80% of its original copper content,[38] indicating that the copper-reversible inactivation of SAO by DDC involves formation of a tight enzyme-DDC complex rather than copper extraction. Spectral data reported by Suzuki et al.[38] provide clear evidence that the inactivated enzyme species described by Yamada et al.[28] represents a Cu-containing enzyme-DDC complex, the 380-nm absorption band being attributable to Cu-S charge transfer between DDC and enzyme-bound copper. Since chelation of enzyme-bound copper by DDC is likely to have interfered with the chemical methods for copper analysis used by Yamada et al.,[28] evidence presented for the preparation of reversibly Cu-depleted SAO by DDC treatment[4,28] cannot be accepted as convincing. A recent preliminary report[39] on the preparation of BSAO half depleted of Cu by DDC treatment would seem to require confirmation.

Copper-free BSAO has been prepared by Suzuki et al.[40,41] by dialysis of the dithionite-reduced enzyme against cyanide under anaerobic conditions. The apoenzyme (which lacks a 380-nm absorption band) is catalytically inactive. Copper can be dialytically reincorporated to be about 90%, and the reconstituted enzyme is more than 80% active and spectrally indistinguishable from native enzyme.[27] Zinc, nickel(II), and cobalt(II) substituted derivatives of the apoenzyme have been prepared and described.[40,41]

EPR Spectral data (35 GHz) presented by Knowles and co-workers provide evidence that the two copper sites in SSAO are nonequivalent and differently affected by substrate during catalysis.[18,32] Azide binding abolishes the heterogeneity of the 35-GHz EPR spectrum[18] and gives 9-GHz spectra with at least 9 Cu-N superhyperfine lines in the g region.[35] Nitrogen superhyperfine splitting can be detected also in the absence of azide,[35] indicating that nitrogenous protein ligands contribute to copper binding. EPR spectra (9 GHz) of BSAO show a similar nitrogen superhyperfine structure when recorded in the presence of high concentrations of ammonium sulfate.[40] The number of metal-coordinated nitrogen atoms cannot be estimated reliably from the observed superhyperfine patterns (the EPR spectrum derives from ^{63}Cu and ^{65}Cu in two different chemical environments, i.e., from four different signals), but is likely to exceed one for both of the copper sites.

Further information on the nature of the copper ligands in SAO comes from NMR measurements. SSAO drastically enhances the rate of water proton spin relaxation, which demonstrates that at least one of the two copper sites is accessible to solvent.[42] Barker et al.[18] found that the relaxation rate enhancement decreases by a factor of 0.5 with saturating concentrations of azide and cyanide, indicating that these inhibitory anions displace 50% (more cautiously interpreted: at least one but not all) of the copper-bound water molecules. Barker et al.[18] proposed that each copper ion coordinates two water molecules (one equatorial and one axial), the equatorial ones being displaced by the inhibitors. Olsson et al.[36] proposed from the observed pH dependencies of azide and cyanide binding that the inhibitors are bound with displacement of a protein ligand showing a pK_a close to 9 in the displaced state. Both proposals were based partly on the assumption of equivalent copper sites, and the experimental evidence does not seem to exclude the possibility that azide and cyanide may displace an equatorial water from one of the copper ions and a protein ligand from the corresponding position in the coordination sphere of the second copper ion. Such an interpretation of the experimental data would make it easier to understand why the copper sites are basically nonequivalent, and why they become equivalent following inhibitor binding (see Chapter 12).

D. Carbonyl Cofactor

SAO contains a catalytically essential prosthetic group which interacts strongly with a variety of carbonyl reagents.[3,8,22,43] The reaction of this group with hydrazine derivatives is virtually irreversible and leads to formation of intensely colored inactive enzymic complexes. Titration of enzyme activity[15,44] or absorbance[15,26,45] against increasing concentrations of hydrazine derivatives has shown that 1 mol of inhibitor is bound per mole of enzyme, indicating the presence of a single reactive prosthetic group per enzyme dimer. This prosthetic group will be referred to as the carbonyl cofactor.

Strong evidence has accumulated which indicates that the carbonyl cofactor may be a pyridoxal derivative. Pronase digestion and acid hydrolysis of SSAO,[22] or freezing-thawing in the presence of 4 *M* urea,[46] has been reported to release a substance which reactivates pyridoxal phosphate-depleted tyrosine decarboxylase. Blaschko and Buffoni found that this substance shows the optical absorption and fluorescence properties of pyridoxal, and that the enzyme and its hydrolysate restores growth to rats put on a vitamin B_6-deficient diet.[22] Yamada and Yasunobu hydrolyzed BSAO with pronase and isolated a chromophoric peptide fragment with absorbance and fluorescence properties closely similar to pyridoxal phosphate.[43] In particular, absorbance spectra of the peptide fragment and authentic pyridoxal phosphate were found to exhibit the same pronounced and characteristic dependence on pH. Additional spectral features of SAO, consistent with the presence of a 3-hydroxypyridine chromophore, will be considered in the following section.

Borohydride reduction experiments indicate that the carbonyl cofactor is reducible,[43,47] but have failed to provide convincing evidence for a reductive trapping of substrate-pyridoxal intermediates.[47,48] Watanabe et al.[49] isolated a chromophoric peptide fragment from hydrolysates of phenylhydrazine-treated BSAO. Since this fragment neither contained phosphate, nor reacted in chromatographic spray tests for pyridoxal, the authors concluded that another cofactor may be present. Unsuccessful attempts to detect pyridoxal in hydrolysates of BSAO by a microbiological assay were reported by Abeles and co-workers, who strongly questioned the presence of a pyridoxal cofactor also on mechanistic grounds.[50-52] Alternative structures for the carbonyl cofactor have been proposed.[51-54]

Evidence identifying the carbonyl cofactor as a pyridoxal derivative is not conclusive, but appears considerably stronger than the evidence so far reported in support of alternative proposals (see Chapter 11).

E. Spectral Properties

Optical absorption spectra of SAO show a broad maximum at about 470 nm (Figure 1), which renders the enzyme pink in color. The color disappears on reduction of the enzyme with dithionite or amine substrates under anaerobic conditions, and reappears in the presence of oxygen.[6,7,10,31] The reducible absorption band is centered at a slightly higher wavelength than the absorption maximum (cf. Table 2) and will be referred to as the 480-nm band. Copper-depleted and nickel- or cobalt-substituted preparations of BSAO show a reducible absorption maximum closely similar to that of native enzyme,[20] indicating that the 480-nm

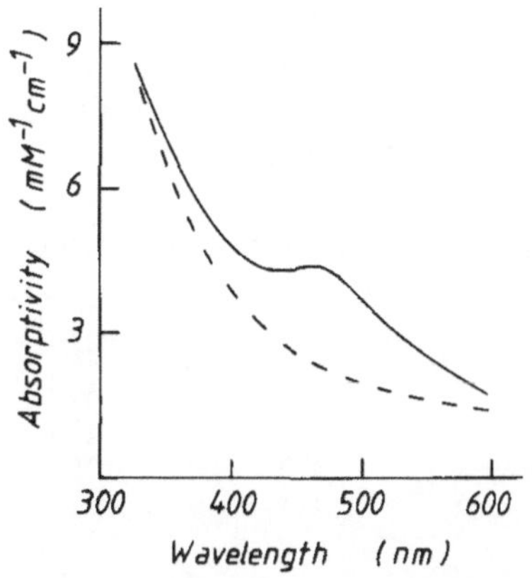

FIGURE 1. Absorption spectrum of native (—) and reduced (----) pig SAO.

absorption band is associated primarily with the chromophoric carbonyl cofactor. Confirmatively, full reduction of the band is achieved with 1 mol of substrate per mole of enzyme under anaerobic conditions.[26]

Circular dichroism measurements have provided evidence that the carbonyl cofactor is associated also with an optically active electronic transition at 360 to 370 nm.[20,24,55] The corresponding absorption band is likely to contribute significantly to the optical spectra and could account for a major part of the absorption of reduced enzyme in the region above 350 nm. The intensity of the 360 to 370 nm absorption band appears to be drastically enhanced in one of the enzyme-substrate complexes formed during catalysis.[55]

Electronic transitions associated with the metal in native and cobalt- or nickel-substituted BSAO have been discussed by Suzuki et al.[20] Their spectral data establish that the main chromophoric properties of SAO are attributable to the carbonyl cofactor rather than to copper. The spectral contributions provided by the metal are more difficult to evaluate, considering the weakness of the corresponding absorption bands and the evidence indicating that the two copper sites are nonequivalent.

Absorption bands attributable to copper-ligand charge transfer appear following azide[35] and diethyldithiocarbamate[38] binding to SAO. Binding of the copper-chelating reagent cuprizone leads to reaction with the carbonyl cofactor and shifts the 480-nm absorption band to 350 nm, agreeing with the absorption maximum for the product formed btween cuprizone and pyridoxal phosphate.[34] Absorption spectra for enzymic derivatives obtained with more typical carbonyl reagents are similar to, but not identical with, those of the corresponding pyridoxal phosphate derivatives.[22,43] Thus, the phenylhydrazine derivatives of SSAO and BSAO show characteristic hydrazone absorption bands which are more intense and red-shifted by 20 to 40 nm compared to that of the pyridoxal phosphate phenylhydrazone in aqueous solution at low pH (Table 1). These spectral differences must be considered as minor in view of the environmental (hydrophobicity and enzyme-bound copper) effects on the chromophores documented by data in Table 1.

The latter data, as well as the spectral properties of native enzyme and of the chromophoric peptide fragment isolated by Yamada et al.[43] provide clear evidence that the carbonyl cofactor contains a free or derivativized carbonyl group conjugated with a structure allowing for extensive π-electron delocalization (e.g., an aromatic ring structure). This would seem to exclude that the carbonyl cofactor is a sulfenic acid residue,[53] and –S–S– group,[51] an aliphatic carbonyl compound such as pyruvate, oxobutyrate, or dehydroalanine,[24] or a carbonyl group provided by one of the protein carbohydrate residues. The 360- to 370-nm and the 480-nm

Table 1
SPECTRAL DATA FOR PHENYLHYDRAZONE DERIVATIVES OF SAO AND PYRIDOXAL PHOSPHATE

Carbonyl group donor	λ_{max} (nm)	ϵ(mM^{-1}cm^{-1})	Ref.
Bovine SAO	448	38	20
Copper depleted	432	28	27
Pig SAO	430	38	15
In 4 *M* urea	430	23	15
Pyridoxal phosphate	410	25	56
In 50% isopropanol	414	34	57

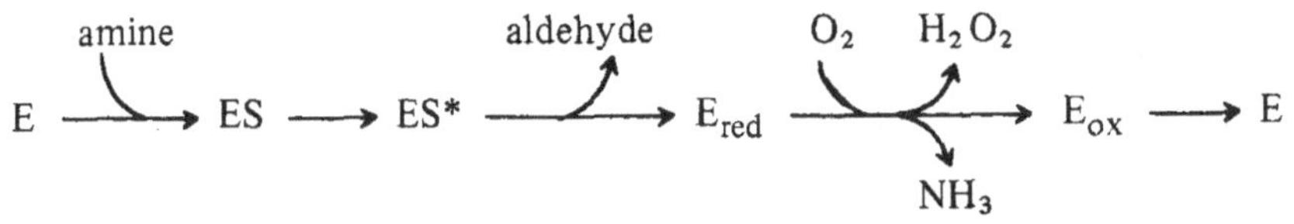

SCHEME 1

absorption bands of the carbonyl cofactor are within the characteristic regions of maximum absorption of nonprotonated (355 to 375 nm) and carbanionic semiquinonoid (480 to 500 nm) pyridoxal Schiff bases.[58,59] The presence of a positive Cotton effect deriving from the cofactor is considered as a fundamental characteristic of pyridoxal enzymes, and several pyridoxal-dependent transaminases and decarboxylases show this effect (i.e., a positive CD band) in the 360- to 370-nm region.[59] Early proposals that SAO contains a pyridoxal cofactor[22,43] are obviously consistent with, and in several respects strongly supported by, the available spectral evidence. The spectral properties of PAO might be consistent also with the pyrimidine type of cofactor discussed by Hamilton,[54] but this remains to be shown (see Chapter 11).

III. MECHANISTIC PROPERTIES

A. Kinetic Behavior

Steady-state kinetic studies indicate that SAO operates by an enzyme-substitution (ping-pong) mechanism,[14,31,60] but the order of substrate binding and product release cannot be reliably deduced from reported product inhibition data. The catalytic reaction is virtually irreversible, and product inhibitions observed may (in the case of ammonia[33] are known to) reflect effects other than kinetic reversal of the catalytic steps of product dissociation. The addition of benzylamine to BSAO under anaerobic conditions, however, has been shown to result in formation of 1[60,61] or 2[24,50] mol of benzaldehyde per mole of enzyme. The corresponding reaction of SSAO yields 1 mol of aldehyde per mole of enzyme,[26,62,63] while no hydrogen peroxide is formed.[64] These observations provide direct evidence that SAO operates by a substitution mechanism in which the interaction of enzyme with amine substrate leads to formation of a reduced enzyme intermediate which is subsequently reoxidized by molecular oxygen.

Stopped-flow studies of SSAO have led to the detection and characterization of additional enzymic reaction intermediates (Scheme 1). Formation of the reduced enzyme species (E_{red}) proceeds minimally via two enzyme-substrate complexes, ES and ES*, the former being spectrally similar to free enzyme (E) and the latter showing an absorption difference band at about 360 nm.[55,65] This difference band disappears on conversion of ES* into E_{red}, and

Table 2
SPECTRAL PROPERTIES OF ENZYME SPECIES IN SCHEME 1

Species	λ_{max} (nm)	ϵ (M^{-1} cm^{-1})	Ref.
E (or ES)	470	4,400	35
ES* - E	360	3,000	55
E_{ox} - E	290	11,000	66
E - E_{red}	480	2,000	55

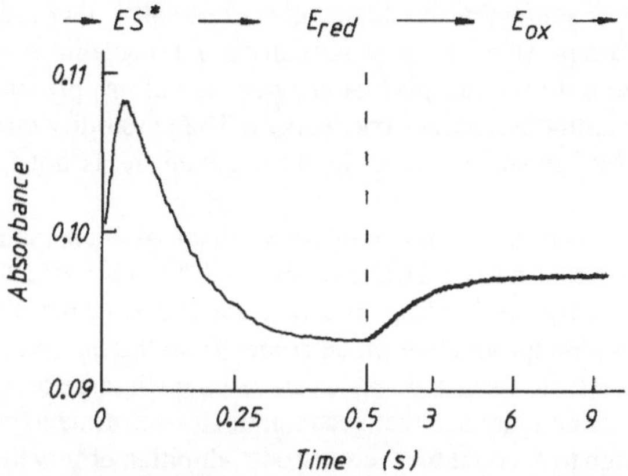

FIGURE 2. Pre-steady-state absorbance changes recorded at 380 nm during the reaction of 20 μM pig PAO with 5 mM benzylamine in 0.1 M phosphate buffer solution equilibrated with air (0.25 mM oxygen).

the reduced enzyme species accumulating during catalysis is indistinguishable from that formed under anaerobic conditions.[31,55] Reduction of the 480-nm absorption band may be associated with the ES* → E_{red} step,[55,56] but is more likely to occur already in the ES → ES* step and to be masked by the appearance of the broad and comparatively intense 360-nm band. The 480-nm band reappears in an oxygen-dependent step leading to formation of an intermediate (E_{ox}) showing drastically enhanced absorption in the 280- to 320-nm region.[66] Table 2 summarizes reported spectral data for species in Scheme 1. The spectrally modified intermediates are formed at readily distinguishable rates (Figure 2), and the effect of pH and substrate structure on these rates has been examined.[55,65,66]

Transient-state kinetic methods have been applied also to identify the catalytic steps of product dissociation in the SSAO reaction. Aldehyde release, as monitored by stopped-flow techniques, occurs at a rate indistinguishable from that of conversion of ES* into E_{red};[62,63] the characterization of species ES and ES* as enzyme-substrate complexes is based on that observation. Hydrogen peroxide has been similarly shown to dissociate in the oxygen-dependent reaction step associated with reappearance of the 480-nm absorption band, i.e., in the E_{red} → E_{ox} step.[64] Quenched-flow data recently presented by Knowles and co-workers[16,67] establish that the third product, ammonia, is released at an oxygen-dependent rate agreeing with that of hydrogen peroxide release and reappearance of the 480 nm band (cf. Scheme 1). The amount of transiently formed product was in all three cases found to correspond to the presence of a single active site per enzyme dimer.[16,62-64]

The product release pattern thus determined for SSAO cannot be readily reconciled with the mechanistic proposals of Benitez et al.[53] and Abeles and co-workers,[50-52] according to which the oxidative deamination of amine substrate leads to primary formation of an imine product; this product could not be experimentally detected,[52] and was therefore assumed to be rapidly hydrolyzed by the enzyme to form aldehyde and ammonia during the anaerobic step of enzyme reduction. The anaerobic reaction of benzylamine with SSAO,[14,64] and BSAO,[50] when carried out in the presence of a NADH/oxoglutarate/glutamate dehydrogenase assay system for ammonia detection, has been observed to give rise to NADH oxidation, possibly attributable to glutamate formation from ammonia. Lindström et al. found,[62] however, that such apparent ammonia release neither requires the presence of catalytically active SAO, nor leads to any consumption of the amine substrate. Catalytic ammonia production could be detected only when oxygen was admitted to the anaerobic reaction solution after 40-min reaction time. The latter observations, which are consistent with the transient-state ammonia release data of Knowles and co-workers,[16] establish that reduced enzyme retains the amino portion of amine substrate in a virtually nondissociable (i.e., covalently bound) form. Results obtained with the pig plasma enzyme, therefore, provide clear evidence that SAO is reduced by an aminotransferase mechanism. The possibility that SAO from different sources may operate by fundamentally different mechanisms is not favoured by presently available evidence.

Scheme 1 does not account for the inhibitory effect of high concentrations of amine substrate on the steady-state rate of SAO catalysis.[7,60,68-70] This effect has been interpreted in terms of inhibitory substrate binding to a noncatalytic enzymic site.[68-70] Evidence that substrates may bind in a nonproductive mode comes from the observation that benzylamine affects low-temperature EPR spectra of the inactive phenylhydrazine derivative of BSAO[45] and SSAO,[57] but there is no evidence that such ligation of substrate to copper affects enzyme activity or is tight enough to account for the substrate inhibition observed at room temperature. Transient-state kinetic measurements have established the absence of inhibitory effects of amine substrate on the process of enzyme reduction[31,62] and on the oxygen-dependent step of the reoxidation process.[66] This argues strongly against the presence of inhibitory noncatalytic sites for amine binding and indicates that substrate inhibition is due to formation of an abortive or weakly productive complex between amine substrate and the reoxidized species E_{ox} in Scheme 1. Formation of an inhibitory E_{ox}-substrate complex is consistent also with the observation that substrate inhibition becomes more pronounced at alkaline pH[69] where accumulation of E_{ox} is kinetically favored.[66]

B. Mechanism of Enzyme Reduction

Differences in substrate specificity between SAO from different sources may be physiologically important,[3] but probably reflect only minor structural modifications of the amine binding site. Observed correlations between K_m values and chain length of aliphatic amines indicate that substrate binding is stabilized by hydrophobic interactions.[7,68] Reported fluorescence data for kynuramine binding provide evidence in the same direction and confirm that a single amine binding site is available.[71] The activity of SSAO towards benzylamine is not detectably affected by high concentrations of benzylalcohol or *N*-methylbenzylamine,[65] while α-substituted primary amines such as amphetamine act as comparatively tightly bound competitive inhibitors.[37] Hydrophobic interactions, therefore, appear to be of secondary importance compared to the binding contribution provided by the unsubstituted amino group of the substrate.

If the substrate amino group participates strongly in binding, then complex formation between enzyme and amines should be strongly dependent on the protonation state of the amino group. Examination of the effect of pH on K_m for benzylamine oxidation by SAO from different sources has confirmed that such is the case, but has led to conflicting conclusions as to what protonation state is the binding one.[7,24,68,72] This is not surprising; pH dependence data for a single ionizing ligand can only be tentatively interpreted in the latter

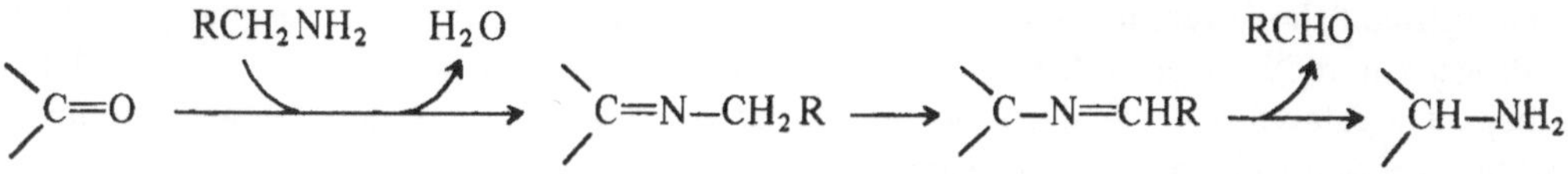

SCHEME 2

respect. Steady-state,[63] and transient-state[65] kinetic data obtained with substrates of different acidity provide unambiguous evidence that amine binding to SSAO requires the unprotonated form of the substrate and the protonated form of an enzymic group with an estimated pK_a of 8.8. Hence, it may be concluded that the substrate amino group contributes to binding through its nucleophilic properties, i.e., through interaction with an electrophilic group at the amine binding site.

Suva and Abeles[52] proposed that amines are bound as sulfenamides to an enzymic –S–X– group with one-electron reduction of a chromophore X, subsequent oxidation of the substrate being associated with reduction of the sulfur to a thiol. This proposal was supported by evidence showing that substrate-reduced BSAO incorporates 1.3 to 1.9 equivalents of *N*-ethylmaleimide more than free enzyme, due possibly to formation or unmasking of sulfhydryl groups. Using spin-labeled maleimide derivatives, Yasunobu and co-workers[73] found that the thiol reactivity of reduced enzyme is induced also by phenylhydrazine binding and thus cannot be related to catalytic reduction of an –S–X– group. Spin-labeling of the groups unmasked by enzyme reduction had no effect on the binding of phenylhydrazine, and phenylhydrazine binding did not affect EPR spectra of the spin-labeled groups. The increased reactivity of reduced or phenylhydrazine-treated enzyme towards maleimide derivatives, therefore, appears to involve enzymic groups which are not at the amine binding site and which may not even be thiols. Native BSAO contains no groups showing reactivity towards the thiol reagent *p*-chloromercuribenzoate,[70] but incorporated about one[52] or four[73] equivalents of the different maleimide derivatives tested.

There is no doubt, however, that SAO does contain a strongly electrophilic prosthetic group (the carbonyl cofactor) which interacts with and is reduced by amine substrates. Early proposals[74,75] that amines are directly bound to the carbonyl cofactor were based on the observed reactivity of the enzyme towards inhibitory substrate analogs such as hydrazine derivatives and other nucleophilic reagents containing a primary amino group.[3,22,43] Later studies have confirmed the mechanistic analogy between catalytic substrate binding and the binding of hydrazine derivatives to the carbonyl cofactor. The latter process is competitively affected by substrate binding,[44,45,76] involves the unprotonated form of the inhibitor,[76] and requires protonation of the pK_a-8.8 group, which controls substrate binding to SSAO.[76] Chromophoric properties of the enzyme-inhibitor complexes formed with hydrazine derivatives are characteristic of hydrazone formation, indicating that substrate binding leads to Schiff base formation. This is consistent also with the absence of strong competitive effects of alcohol analogs on the binding of amine substrates[65] and offers a reasonable explanation for the absolute specificity of SAO for primary amines.

The evidence indicating that SAO contains pyridoxal led Blaschko[75] to propose that the enzyme is reduced by a transamination between amine substrate and enzyme-bound pyridoxal. A transamination mechanism for enzyme reduction (Scheme 2) would seem plausible irrespective of the identity of the cofactor which carries the carbonyl reactive group, and was proposed also for the pyrimidine cofactor discussed by Hamilton.[54] Suva and Abeles[52] assumed that direct carbonyl group reduction, as in Scheme 2, must lead to incorporation of nonexchangeable hydrogen from substrate or water into the reduced enzyme; their observation that no nonexchangeable hydrogen is incorporated on reduction of BSAO, therefore, was taken as evidence against a transamination mechanism. Since neither the structure, nor the mode of binding of the carbonyl cofactor has been definitely established, it would seem difficult, however, to make any reliable predictions as to the actual exchange rate of hydrogens introduced on reduction of the cofactor. Hydrogens on the reactive methylene group of a catalytically involved enzyme-bound pyridoxamine derivative would certainly not be expected to be nonexchangeable.[58,77]

The conversion of amine substrate into aldehyde product requires breaking of a C–H bond at the α-carbon of the substrate. Obvious predictions of Scheme 2 are that such bond breaking should occur in the primary Schiff base, limit the rate of Schiff base tautomerization, and proceed by a mechanism of proton dissociation rather than hydrogen radical or hydride ion dissociation. Direct evidence for a proton dissociation mechanism has been presented by Abeles and co-workers,[51,78] who demonstrated that BSAO catalyzes deprotonation of the α-carbon of primary amines with great efficiency. Data reported for the enzymic nonoxidative conversion of β-Cl-phenylethylamine into phenylacetaldehyde through HCl elimination[51] are particularly convincing in that respect and establish that nonoxidative amine binding is sufficient to trigger proton abstraction. Summers et al. found that the oxidative deamination of selectively tritiated dopamine leads to nonstereospecific loss of tritium from the β- as well as the α-carbon,[79] which suggests that proton abstraction is facilitated primarily through substrate methylene group hyperconjugation with the π-electron system of the carbonyl cofactor.[77] The oxidation of benzylamine, however, leads to stereospecific loss of the pro-*S*-hydrogen at the α-carbon,[80] indicating that specific basic groups at the active site may participate in the proton abstraction process. Olsson et al. noted a primary isotope effect of about 3 on the K_m value for α,α-bisdeuterobenzylamine oxidation by SSAO[81] and showed by transient-state kinetic measurements that this isotope effect derives from a decreased rate of formation of species ES* in Scheme 1.[55] Hence, it may be concluded that C-H bond breaking at the amine α-carbon limits the rate of isomerization of an early enzyme-substrate complex (ES) to give a spectrally modified enzyme-substrate complex (ES*) which is capable of decomposing into reduced enzyme and the aldehyde product. Considering the latter results and the evidence indicating that the C-H bond is broken through catalytic proton abstraction, that substrates are bound as Schiff bases to a carbonyl cofactor, and that amine oxidation proceeds by an aminotransferase mechanism, it would seem sufficiently well established that SAO is reduced by the reaction in Scheme 2 or a closely related transamination process.

C. Mechanism of Enzyme Reoxidation

Reduced SAO (E_{red}) is reoxidized in two kinetically distinguishable reaction steps (Scheme 1).[14,60,66] The interaction of E_{red} with oxygen in the first step is rate limiting for the release of ammonia and hydrogen peroxide,[16,64] as well as for reappearance of the 480-nm absorption band.[66] This characterizes the first step as a carbonyl cofactor reoxidation process, and the oxidized enzyme species (E_{ox}) formed in this process might well show carbonyl group reactivity (e.g., interact inhibitorily with amine substrate). The subsequent reconversion of E_{ox} into the resting form of the enzyme (E) is kinetically first-order or pseudo first-order, i.e., may reflect an enzyme isomerization step or a reaction involving water consumption or production. The activating effect of ammonia and imidazole on the latter reaction step[33] could be attributable to relief of substrate inhibition.

The fact that SAO contains copper suggests that the metal may have a catalytic function in the oxygen-dependent carbonyl cofactor reoxidation step. Experimental evidence for such a functional role of enzyme-bound copper came from inhibition studies showing that ligation of azide to copper prevents SSAO reoxidation without affecting the binding of substrate or phenylhydrazine to the enzyme.[35] Subsequent kinetic measurements located the effect of azide to the carbonyl cofactor reoxidation step, and this step was shown to be strictly inhibited also by cyanide binding to copper.[18,36] Suzuki et al.[20] reported that copper-depleted BSAO can be reduced by amine substrate, but not reoxidized by oxygen. Color was oxidatively restored to the copper, nickel, and cobalt reconstituted enzyme derivatives, and spectral studies of the latter derivatives, native enzyme, and apoenzyme led to the conclusion that the electronic state of the carbonyl cofactor is greatly influenced by the enzyme-bound metal ions.[20] The influence of the carbonyl cofactor on enzyme-bound copper is demonstrated by the EPR spectral changes occurring in the reaction of SAO with amine substrates.[28-32] Grant et al.[32] found that these spectral changes are most pronounced in the reduced enzyme species, and that they involve only one of the two copper sites. Since SAO contains a single carbonyl cofactor per enzyme dimer, the latter observations provide confirmatory evidence for a catalytic and electronic interdependence of the carbonyl cofactor and the copper ion at the active-site subunit.

The above observations are obviously consistent with the idea that the carbonyl cofactor provides at least one ligand atom for the active-site copper ion.[82,83] This would also explain the nonequivalence of the two copper sites, water being a plausible ligand substituting for the carbonyl cofactor at the "noncatalytic" copper ion. Displacement of such water and the cofactor ligand atom(s) by azide or cyanide could account for the EPR spectral effects of these inhibitors and would then relate their inhibitory effect to Cu-cofactor bond breaking. Knowles and co-workers[18,33] proposed that azide and cyanide inhibition is due to displacement of equatorially metal-bound water, which ionizes with a pK_a of 8 to 9, and which acts in the ionized state as a nucleophile facilitating hydride transfer from the reduced carbonyl cofactor to oxygen. The NMR spectral evidence[18] indicating ionization of all copper-bound water molecules with a pK_a value close to 8 might require reinterpretation, however, since such ionization cannot be readily reconciled with the observed pH independence of cyanide binding between pH 7 and 9.[18,36] EPR spectra of SSAO are not significantly affected by pH over the pH range 6 to 9, but change drastically in a fully reversible transition with a pK_a value (about 10.5[57]) which could be more typical for a copper-bound water molecule.[33]

Although attempts to demonstrate oxygen binding to copper have been unsuccessful,[35] it seems reasonable to conclude from the available evidence for copper coordination of the carbonyl cofactor (Ccf) that the metal is likely to function as a mediator of electron transfer from the reduced cofactor ($CcfH_2$) to oxygen. Since copper is not detectably reduced during catalysis,[32] the actual electron transfer step should conform formally (proton transfer steps are omitted) to the equation:

$$\mathrm{Ccf}^{2-} \cdot \mathrm{Cu}^{2+} \cdot \mathrm{O}_2 \rightarrow \mathrm{Ccf} \cdot \mathrm{Cu}^{2+} \cdot \mathrm{O}_2^{2-}$$

The catalytic principles for such a metal-linked reoxidation process have been discussed by Hamilton,[54,83-85] who pointed out that ligation of oxygen to copper eliminates the spin problem associated with reduction of triplet oxygen, thus increasing the chemical reactivity of the oxygen molecule. Electron transfer between complexed cofactor and oxygen can be anticipated to proceed rapidly via the metal-ligand orbitals, and quantum mechanistic studies would seem to be required to decide whether or not the catalytic reaction may involve reduced (e.g., $\mathrm{Ccf}^{-} \cdot \mathrm{Cu}^{+} \cdot \mathrm{O}_2$) or oxidized (e.g., $\mathrm{Ccf}^{2-} \cdot \mathrm{Cu}^{3+} \cdot \mathrm{O}_2^{-}$) transitory states of the resting cupric form of the metal. The electron transfer step must be associated with proton transfer (via solution or system of hydrogen bonds at the active site) from the reduced cofactor to oxygen, and such proton transfer would be expected to trigger C–N bond breaking, resulting in ammonia release from the cofactor. Detailed mechanisms have been proposed for copper-linked reoxidation of a pyridoxal cofactor,[83,86] but these proposals require modification in view of present knowledge about the actual order of substrate binding and product release. If reduction of the carbonyl cofactor is depicted as in Scheme 2, Scheme 3 could be adequate for simplified description of the cofactor reoxidation step.

IV. CONCLUDING REMARKS

Even though the main mechanistic features of SAO catalysis might be well understood

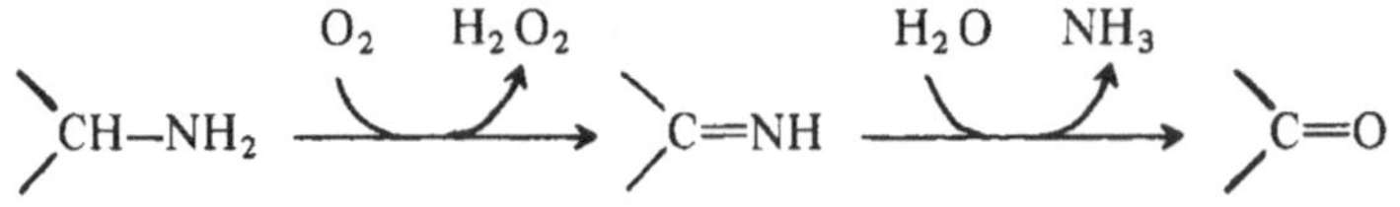

SCHEME 3

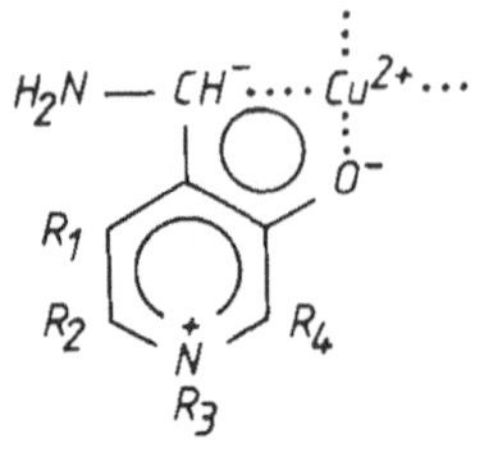

FIGURE 3. Possible structure of the reduced carbonyl cofactor.

(Schemes 1 to 3), it seems appropriate to conclude that the detailed mechanism for carbonyl cofactor reoxidation is unknown and that most details of the mechanism of action of the enzyme will remain obscure until the structure and mode of binding of the cofactor has been more firmly established. It seems likely that the carbonyl cofactor is bound to copper, but we do not know how many or what ligand atoms the cofactor provides, nor do we know if there is any ligand exchange at copper during catalysis. The cofactor may be bound as an internal Schiff base,[55] which makes it uncertain whether ammonia is released by an hydrolytic or an aminolytic mechanism. Catalytic proton transfer steps are probably facilitated by ionizing enzymic groups, but essentially no information is available as to the identity of such groups or about the actual protonation state of the carbonyl cofactor in different enzyme species. The "deprotonated" pyridoxamine structure in Figure 3 is presented to illustrate this and to emphasize the fact that enzymic carbonyl group reduction need not necessarily lead to introduction of an hydrogen atom at the reduced carbonyl carbon;[52] carbanionic reaction intermediates could be stabilized by enzyme-bound copper or through electronic conjugation[77] with the carbonyl cofactor. The presence or catalytic formation of carbanionic cofactor derivatives could explain many of the unexpected observations made by Suva and Abeles,[52] including the substrate-induced appearance of a group showing high reactivity towards the thiol reagent *N*-ethylmaleimide.[58]

The steady-state turnover of SSAO is known to be rate limited by the process of enzyme reoxidation, and attention should be drawn to the pronounced difference in catalytic efficiency of the enzyme with respect to this process and the transamination process by which the enzyme is reduced (Figure 2). Under physiological conditions, the two reaction steps of the aerobic reoxidation process show first-order rate constants of about 0.5 sec^{-1},[66] while the two first-order steps of the anaerobic transamination process appear to be about 100-fold more rapid irrespective of the detailed structure of the amine substrate.[55,63,65] This is not what would be expected from an evolutionary point of view; evolution should act on the slowest catalytic step and so tend to eliminate rate differences between all first-order steps of the catalytic reaction. The striking kinetic disparity between the reductive and oxidative parts of the SSAO reaction could be taken to indicate that the enzyme originates from an ancestor capable of carrying out transamination reactions with considerable efficiency. The selection of pyridoxal phosphate as a cofactor for aminotransferases and related enzymes in the amino acid metabolism was certainly made at an early evolutionary stage, and highly effective pyridoxal enzymes are likely to have been operative long before the appearance of atmospheric oxygen. There is strong reason to believe, therefore, that SAO may originate from a pyridoxal-containing ancestor with transaminase or decarboxylase activity; the latter alternative seems particularly attractive since presently available pyridoxal-dependent decarboxylases form amines as enzyme-bound products and have been found in several instances to contain a catalytically essential binding site for divalent metal ions.[58]

The chemical reactivity, mechanistic behavior, and spectral properties of SAO are likewise indicative of the presence of an enzyme-bound pyridoxal cofactor. Nevertheless, further attempts should be made to isolate the carbonyl cofactor and to establish its identity by unambiguous analytical methods. Such studies should provide valuable inferences also as to the mode of binding of the cofactor, information which is of crucial importance for our understanding of the spectral and catalytic properties of the enzyme.

The presence of a single active site in SSAO has been firmly established, and recent investigations indicate that there is a similar 2:1 correspondence between copper and carbonyl cofactor content in BSAO. The nonequivalence of the two copper sites could present problems in spectral as well as mechanistic studies of SAO, and attention has been drawn to the possibility that both copper sites might be catalytically involved.[2] However, new approaches to the study of properties and functional roles of the copper sites seem possible now that methods are available for the preparation of copper-depleted and metal-substituted enzyme, and future investigations will undoubtedly lead to a better understanding of the structure and function of the copper-containing amine oxidase from mammalian blood plasma.

ACKNOWLEDGMENT

I wish to thank P. F. Knowles, B. Mondovì, T. Okuyama, and S. Suzuki for helpful discussions.

REFERENCES

1. **Malmström, B. G., Andréasson, L.-E., and Reinhammar, B.,** Copper containing oxidases and superoxide dismutase, in *The Enzymes*, Vol. 12B, 3rd ed., Boyer, P. D., Ed., Academic Press, London, 1975, 507.
2. **Knowles, P. F. and Yadav, K. D. S.,** Amine oxidases, in *Copper Proteins and Copper Enzymes*, Vol. 2, Lontie, R., Ed., CRC Press, Boca Raton, 1984, 103.
3. **Kapeller-Adler, R.,** *Amine Oxidases and Methods for their Study*, Wiley-Interscience, New York, 1970, 124.
4. **Buffoni, F.,** Histaminase and related amine oxidases, *Pharmacol. Rev.*, 18, 1163, 1966.
5. **Bergeret, B., Blaschko, H., and Hawes, R.,** Occurrence of an amine oxidase in horse serum, *Nature (London)*, 180, 1127, 1957.
6. **Buffoni, F. and Blaschko, H.,** Benzylamine oxidase and histaminase: purification and crystallization of an enzyme from pig plasma, *Proc. R. Soc. (London) B*, 161, 153, 1964.
7. **McEwen, C. M., Cullen, K. T., and Sober, A. J.,** Rabbit serum monoamine oxidase: purification and characterization, *J. Biol. Chem.*, 241, 4544, 1966.
8. **McEwen, C. M.,** Human plasma monoamine oxidase. I. Purification and identification, *J. Biol. Chem.*, 240, 2003, 1965.
9. **Hirsch, J. G.,** Spermine oxidase: amine oxidase with specificity for spermine and spermidine, *J. Exp. Med.*, 97, 345, 1953.
10. **Yamada, H. and Yasunobu, K. T.,** Monoamine oxidase. I. Purification, crystallization, and properties of plasma monoamine oxidase, *J. Biol. Chem.*, 237, 1511, 1962.
11. **Ishizaki, H. and Yasunobu, K. T.,** Bovine plasma amine oxidase interactions with concanavalin A in solution and with concanavalin A sepharose, *Biochim. Biophys. Acta*, 611, 27, 1980.
12. **Oda, O., Manabe, T., and Okuyama, T.,** Oxidation of ε-amino group of lysyl peptides by bovine serum amine oxidase, *J. Biochem. (Tokyo)*, 89, 1317, 1981.
13. **Turini, P., Sabatini, S., Befani, O., Chimenti, F., Casanova, C., Riccio, P. L., and Mondovì, B.,** Purification of bovine plasma amine oxidase, *Anal. Biochem.*, 125, 294, 1982.

14. **Taylor, C. E., Taylor, R. S., Rasmussen, C., and Knowles, P. F.**, A catalytic mechanism for the enzyme benzylamine oxidase from pig plasma, *Biochem. J.*, 130, 713, 1972.
15. **Lindström, A. and Pettersson, G.**, Active-site titration of pig plasma benzylamine oxidase with hydrazine derivatives, *Eur. J. Biochem.*, 34, 564, 1973.
16. **Rius, F. X., Knowles, P. F., and Pettersson, G.**, The kinetics of ammonia release during the catalytic cycle of pig plasma amine oxidase, *Biochem. J.*, 220, 767, 1984.
17. **Falk, M. C., Staton, A. J., and Williams, T. J.**, Heterogeneity of pig plasma amine oxidase: molecular and catalytic properties of chromatographically isolated forms, *Biochemistry*, 22, 3746, 1983.
18. **Barker, R., Boden, N., Cayley, G., Charlton, S. C., Henson, R., Holmes, M. C., Kelly, I. D., and Knowles, P. F.**, Properties of cupric ions in benzylamine oxidase from pig plasma as studied by magnetic-resonance and kinetic methods, *Biochem. J.*, 177, 289, 1979.
19. **Achee, F. M., Chervenka, C. H., Smith, R. A., and Yasunobu, K. T.**, Amine oxidase XII: the association and dissociation, and number of subunits of beef plasma amine oxidase, *Biochemistry*, 7, 4329, 1968.
20. **Suzuki, S., Sakurai, T., Nakahara, A., Manabe, T., and Okuyama, T.**, Effect of metal substitution on the chromophore of bovine serum amine oxidase, *Biochemistry*, 22, 163, 1983.
21. **Yamada, H., Gee, P., Ebata, M., and Yasunobu, K. T.**, Monoamine oxidase. VI. Physicochemical properties of plasma monoamine oxidase, *Biochim. Biophys. Acta*, 81, 165, 1964.
22. **Blaschko, H. and Buffoni, F.**, Pyridoxal phosphate as a constituent of the histaminase (benzylamine oxidase) of pig plasma, *Proc. R. Soc. London B*, 163, 45, 1965.
23. **Watanabe, K. and Yasunobu, K. T.**, Carbohydrate content of bovine plasma amine oxidase and isolation of a carbohydrate-containing fragment attached to asparagine, *J. Biol. Chem.*, 245, 4612, 1972.
24. **Yasunobu, K. T., Ishizaki, H., and Minamiura, N.**, The molecular, mechanistic, and immunological properties of amine oxidases, *Mol. Cell. Biochem.*, 13, 3, 1976.
25. **Boden, N., Charlton, S. C., Holmes, M. C., and Knowles, P. F.**, Structural studies on the copper sites in benzylamine oxidase from pig plasma, *Biochem. Soc. Trans.*, 1, 1008, 1973.
26. **Lindström, A. and Pettersson, G.**, Active-site titration of pig plasma benzylamine oxidase, *Eur. J. Biochem.*, 83, 131, 1978.
27. **Suzuki, S.**, personal communication, 1983.
28. **Yamada, H. and Yasunobu, K. T.**, Monoamine oxidase. II. Copper, one of the prosthetic groups of monoamine oxidase, *J. Biol. Chem.*, 237, 3077, 1962.
29. **Yamada, H., Yasunobu, K. T., Yamano, T., and Mason, H. S.**, Copper in plasma amine oxidase, *Nature (London)*, 198, 1092, 1963.
30. **Buffoni, F., DellaCorte, L., and Knowles, P. F.**, The nature of copper in pig plasma benzylamine oxidase, *Biochem. J.*, 106, 575, 1968.
31. **Lindström, A., Olsson, B., and Pettersson, G.**, Kinetics of the interaction between pig plasma benzylamine oxidase and substrate, *Eur. J. Biochem.*, 35, 70, 1973.
32. **Grant, J., Kelly, I., Knowles, P. F., Olsson, J., and Pettersson, G.**, Changes in the copper centres of benzylamine oxidase from pig plasma during the catalytic cycle, *Biochem. Biophys. Res. Commun.*, 83, 1216, 1978.
33. **Yadav, K. D. S. and Knowles, P. F.**, A catalytic mechanism for benzylamine oxidase from pig plasma: stopped-flow kinetic studies, *Eur. J. Biochem.*, 114, 139, 1981.
34. **Lindström, A. and Pettersson, G.**, The mechanism of inhibition of pig plasma amine oxidase by the copper-chelating reagent cuprizone, *Eur. J. Biochem.*, 48, 229, 1974.
35. **Lindström, A., Olsson, B., and Pettersson, G.**, Effect of azide on some spectral and kinetic properties of pig plasma benzylamine oxidase, *Eur. J. Biochem.*, 48, 237, 1974.
36. **Olsson, B., Olsson, J., and Pettersson, G.**, Effects on enzyme activity of ligand binding to copper in pig plasma benzylamine oxidase, *Eur. J. Biochem.*, 87, 1, 1978.
37. **Rasmussen, C.**, Inhibition Studies and the Catalytic Mechanism of Pig Plasma Benzylamine Oxidase, *M. Phil. thesis*, University of Leeds, 1975.
38. **Suzuki, S., Sakurai, T., Nakahara, A., Oda, O., Manabe, T., and Okuyama, T.**, Copper binding site in serum amine oxidase treated with sodium diethyldithiocarbamate, *Chem. Lett. (Tokyo)*, 487, 1982.
39. **Mondovi, B., Sabatini, S., and Befani, O.**, The active site of copper amine oxidases, *J. Mol. Catalysis*, 23, 325, 1984.
40. **Suzuki, S., Sakurai, T., Nakahara, A., Oda, O., Manabe, T., and Okuyama, T.**, Spectroscopic aspects of copper binding site in bovine serum amine oxidase, *FEBS Lett.*, 116, 17, 1980.
41. **Suzuki, S., Sakurai, T., Nakahara, A., Oda, O., Manabe, T., and Okuyama, T.**, Preparation and characterization of cobalt(II)-substituted bovine serum amine oxidase, *J. Biochem.*, 90, 905, 1981.
42. **Boden, N., Holmes, M. C., and Knowles, P. F.**, Binding of water to types I and II Cu^{2+} in proteins, *Biochem. Biophys. Res. Commun.*, 57, 845, 1974.
43. **Yamada, H. and Yasunobu, K. T.**, Monoamine oxidase. IV. Nature of the second prosthetic group of plasma monoamine oxidase, *J. Biol. Chem.*, 238, 2669, 1963.

44. **Hucko-Haas, J. E. and Reed, D. J.,** Hydrazines as substrates for bovine plasma amine oxidase, *Biochem. Biophys. Res. Commun.*, 39, 396, 1970.
45. **Rinaldi, A., Floris, G., Sabatini, S., Finazzi-Agrò, A., Giartosio, A., Rotilio, G., and Mondovì, B.,** Reaction of beef plasma and lentil seedlings Cu-amine oxidases with phenylhydrazine, *Biochem. Biophys. Res. Commun.*, 115, 841, 1983.
46. **Buffoni, F. and DellaCorte, L.,** Pig plasma benzylamine oxidase, *Adv. Biochem. Psychopharmacol.*, 5, 133, 1972.
47. **Buffoni, F.,** Pyridoxal catalysis in pig plasma benzylamine oxidase (histaminase), in *Pyridoxal Catalysis: Enzymes and Model Systems*, Snell, E. G., Braunstein, A. C., Severin, E. S., and Torchinsky, Y. M., Eds., Wiley-Interscience, London, 1968, 363.
48. **Inamasu, M., Yasunobu, K. T., and Konig, W. A.,** Cofactor investigation of bovine plasma amine oxidase, *J. Biol. Chem.*, 249, 5265, 1974.
49. **Watanabe, K., Smith, R. A., Inamasu, M., and Yasunobu, K. T.,** Prosthetic group of beef plasma amine oxidase, *Adv. Biochem. Psychopharmacol.*, 5, 107, 1972.
50. **Berg, K. A. and Abeles, R. H.,** Mechanism of action of plasma amine oxidase: products released under anaerobic conditions, *Biochemistry*, 19, 3186, 1980.
51. **Neumann, R., Hevey, R., and Abeles, R. H.,** The action of plasma amine oxidase on β-haloamines, *J. Biol. Chem.*, 250, 6362, 1975.
52. **Suva, R. H. and Abeles, R. H.,** Studies on the mechanism of action of plasma amine oxidase, *Biochemistry*, 17, 3538, 1978.
53. **Benitez, L. V. and Allison, W. S.,** The inactivation of acylphosphatase activity catalyzed by the sulphenic acid form of glyceraldehyde-3-phosphate dehydrogenase by dimedone and olefins, *J. Biol. Chem.*, 249, 6234, 1974.
54. **Hamilton, G. A.,** Copper containing amine oxidases, in *Copper Proteins*, Spiro, T. G., Ed., Wiley, New York, 1981, 205.
55. **Olsson, B., Olsson, J., and Pettersson, G.,** Stopped-flow spectrophotometric characterization of enzymic reaction intermediates in the anaerobic reduction of pig plasma benzylamine oxidase by amine substrates, *Eur. J. Biochem.*, 71, 375, 1976.
56. **Wada, H. and Snell, E. S.,** The enzymic oxidation of pyridoxine and pyridoxamine phosphates, *J. Biol. Chem.*, 236, 2089, 1961.
57. **Lindström, A. and Pettersson, G.,** unpublished results.
58. **Fasella, P.,** Pyridoxal phosphate, *Annual Rev. Biochem.*, 36, 185, 1967.
59. **Dunathan, H. C.,** Stereochemical aspects of pyridoxal phosphate catalysis, *Adv. Enzymol.*, 35, 79, 1971.
60. **Oi, S., Inamasu, M., and Yasunobu, K. T.,** Mechanistic studies of beef plasma amine oxidase, *Biochemistry*, 9, 3378, 1970.
61. **Reed, D. J. and Swindell, R.,** Molecular oxygen and the reaction between soluble amine oxidases and their amine substrates, *Fed. Proc. Fed. Am. Soc. Exp. Biol.*, 28, 891, 1969.
62. **Lindström, A., Olsson, B., and Pettersson, G.,** Transient kinetics of benzaldehyde formation during the catalytic action of pig plasma benzylamine oxidase, *Eur. J. Biochem.*, 42, 377, 1974.
63. **Lindström, A., Olsson, B., Pettersson, G., and Szymanska, J.,** Kinetics of the interaction between pig plasma benzylamine oxidase and various monoamines, *Eur. J. Biochem.*, 47, 99, 1974.
64. **Lindström, A. and Pettersson, G.,** The order of substrate addition and product release during the catalytic action of pig plasma benzylamine oxidase, *Eur. J. Biochem.*, 84, 479, 1978.
65. **Lindström, A., Olsson, B., Olsson, J., and Pettersson, G.,** Effect of pH on the transient reduction of pig plasma benzylamine oxidase by benzylamine derivatives, *Eur. J. Biochem.*, 64, 321, 1976.
66. **Olsson, B., Olsson, J., and Pettersson, G.,** The kinetics of reoxidation of reduced benzylamine oxidase, *Eur. J. Biochem.*, 74, 329, 1977.
67. **Knowles, P. F., Lowe, D. J., Peters, J., Thorneley, R. N. F., and Yadav, K. D. S.,** Kinetic and magnetic resonance studies on amine oxidases, in *The Coordination Chemistry of Metalloenzymes*, Bertini, I., Drago, R. S., and Luchinat, C., Eds., Reidel, Dordrecht, 1983, 159.
68. **McEwen, C. M.,** Human plasma monoamine oxidase. II. Kinetic studies, *J. Biol. Chem.*, 240, 2011, 1965.
69. **Kelly, D., Knowles, P. F., Yadav, K. D. S., Barsley, W. G., Leff, P., and Waight, R. D.,** Steady-state kinetic studies on benzylamine oxidase from pig plasma, *Eur. J. Biochem.*, 114, 133, 1981.
70. **Wang, T. M., Achee, F. M., and Yasunobu, K. T.,** Amine oxidase. XI. Beef plasma amine oxidase, a sulfhydryl protein, *Arch. Biochem. Biophys.*, 128, 106, 1968.
71. **Massey, J. B. and Churchich, J. E.,** Kynuramine, a fluorescent substrate and probe of plasma amine oxidase, *J. Biol. Chem.*, 252, 8081, 1977.
72. **Buffoni, F., DellaCorte, L., and Ignesti, C.,** Pig plasma benzylamine oxidase (histaminase): kinetic studies, *Pharmacol. Res. Commun.*, 4, 99, 1972.

73. **Zeidan, H., Watanabe, K., Piette, L. H., and Yasunobu, K. T.,** Electron spin resonance studies of bovine plasma amine oxidase, *J. Biol. Chem.*, 255, 7621, 1980.
74. **Davison, A. N.,** The mechanism of the irreversible inhibition of rat-liver monoamine oxidase by iproniazid, *Biochem. J.*, 67, 316, 1957.
75. **Blaschko, H.,** The amine oxidases of mammalian blood plasma, *Adv. Comp. Physiol. Biochem.*, 1, 67, 1962.
76. **Lindström, A., Olsson, B., and Pettersson, G.,** Kinetics of the interaction between pig plasma benzylamine oxidase and hydrazine derivatives, *Eur. J. Biochem.*, 42, 177, 1974.
77. **Pullman, B.,** Quantum mechanistical investigation of pyridoxal-dependent reactions, in *Chemical and Biological Aspects of Pyridoxal Catalysis*, Snell, E. E., Fasella, P. M., Braunstein, A., and Rossi Fanelli, A., Eds., Pergamon, Oxford, 1963, 103.
78. **Hevey, R. C., Babson, J., Maycock, A. L., and Abeles, R. H.,** Highly specific enzyme inhibitors: inhibition of plasma amine oxidase, *J. Am. Chem. Soc.*, 95, 6125, 1973.
79. **Summers, M. C., Markovic, R., and Klinman, J. P.,** Stereochemistry and kinetic isotope effects in the bovine plasma amine oxidase catalyzed oxidation of dopamine, *Biochemistry*, 18, 1969, 1979.
80. **Battersby, A. R., Staunton, J., Klinman, J., and Summers, M. C.,** Stereochemistry of oxidation of benzylamine by the amine oxidase from beef plasma, *FEBS Lett.*, 99, 297, 1979.
81. **Olsson, B., Olsson, J., and Pettersson, G.,** Kinetic isotope effects on the catalytic activity of pig plasma benzylamine oxidase, *Eur. J. Biochem.*, 64, 327, 1976.
82. **Finazzi Agrò, A., Guerrieri, P., Costa, M. T., and Mondovì, B.,** The nature of chromophore in pig kidney diamine oxidase, *Eur. J. Biochem.*, 74, 435, 1977.
83. **Hamilton, G. A.,** A possible model reaction for pyridoxal phosphate dependent amine oxidases, in *Pyridoxal Catalysis: Enzymes and Model Systems*, Snell, E. E., Braunstein, A. E., Severin, E. S., and Torchinsky, Y. M., Eds., Interscience, New York, 1968, 375.
84. **Hamilton, G. A.,** Mechanisms of two- and four-electron oxidations catalyzed by some metalloenzymes, *Adv. Enzymol.*, 32, 55, 1969.
85. **Hamilton, G. A.,** Proton in biological redox reactions, *Prog. Bioorg. Chem.*, 1, 83, 1971.
86. **Buffoni, F. and Della Corte, L.,** Mechanism of the enzymic reaction of benzylamine oxidase (histaminase) in pig plasma, *Boll. Soc. Ital. Biol. Sper.*, 43, 1395, 1967.

Chapter 11

OPTICAL SPECTROSCOPIC PROPERTIES OF COPPER-CONTAINING AMINE OXIDASES

A. Finazzi-Agrò

All but one (1) copper-containing amine oxidase (Cu-AO) have now been isolated as chromoproteins. Their color has been described as pink-yellow to frankly pink and is related to a broad absorption (width at half-height ≈ 200 nm), peaking between 460 and 510 nm, with an extinction coefficient ranging between 2000 and 6500 M^{-1} cm^{-1} (see Table 1). The highest value has been reported for the *Aspergillus niger* AO, which at variance with other AOs shows two carbonyl-like reacting groups per mole.[15] This broad band is the envelope resulting from several electronic transitions as demonstrated by CD spectra. Beef serum AO shows at least five CD bands in the visible at 350(+), 450(−), 540(−), 660(+), and 810(−) nm.[7] Lentil seedling AOs show two major CD bands at 360(−) and 520(+) nm with two shoulders at about 430 and 580 nm (see Chapter 6, Figure 1). Pig plasma AO has positive bands at 372 and above 600 nm and a broad negative band at 435 nm with shoulders at 470 and 500 nm.[17] The visible absorption is not related to copper, though the CD bands at the longer wavelengths might be attributed to the d-d transitions of Cu^{2+}.

From this point of view, AOs might recall other "colorless" copper-containing natural or artificial copper proteins like galactose oxidase, dopamine β-hydroxylase, carbonic anhydrase (Cu-substituted), etc.[18] The removal of the metal under nonreducing conditions, i.e., by treatment with diethyldithiocarbamate, does not significantly affect the position or the intensity of the absorption band (Table 2). CD experiments, however, show that the position, sign, and intensity of the component peaks are significantly modified in the copper-free AOs. The removal of copper by the dithionite-cyanide method yields instead an almost colorless enzyme which regains its color upon readdition of copper in the presence of O_2.

Nickel and cobalt can substitute for Cu in the reconstitution of the enzyme, giving essentially the same spectral features.[7] On the other hand, Cu-free LSAO is irreversibly bleached by substrate or by dithionite. Some highly purified animal AOs show an absorption peak at 400 to 410 nm which seems attributable to a minor contamination with a heme-containing protein. In the presence of Cu ligands like N_3^- and SCN^-, native AO forms absorption bands typical of Cu adducts without affecting the 480-nm absorption.[3,5,20] This absorption is instead affected by several "carbonyl reagents" like phenylhydrazine, semicarbazide, isoniazide, etc. The reaction of AOs with phenylhydrazine has been studied in detail since this compound behaves as an irreversible inhibitor. This reaction can be followed spectrophotometrically since the adduct is deeply yellow ($\epsilon_{430\text{-}445}$ = 30 to 40 mM^{-1} cm^{-1}). This adduct is formed at the expense of the enzyme chromophore as demonstrated by the presence of an isosbestic point in the titration with phenylhydrazine. The binding of phenylhydrazine to AOs is irreversible, allowing the titration of the active sites. Generally, AOs show one reacting site per molecule, i.e., one every two Cu atoms, the only exception being the enzyme from *Aspergillus niger*. Contradictory results showing more than one site per molecule in some AOs seem to be due to experimental drawbacks. However, recent evidence might indicate the presence of two active sites per enzyme molecule. This finding can be reconciled with the phenylhydrazine titration experiments by assuming either a half-of-the sites reactivity or the binding of phenylhydrazine to a site distinct from the active site. Similarly, semicarbazide and isoniazide give rise to characteristic absorption spectra with

Table 1
VISIBLE ABSORPTION SPECTRA OF Cu-AOs

Source	λ_{max}(nm)	ϵ(mM^{-1}cm^{-1})	Ref.
Swine kidney	500	3.5—4	2
	470—480	2	3
Swine serum	470	3.4	4
	470	4.3	5
Bovine serum	460	3.3	6
	476	3.8	7
Bovine aorta	465—470	—	8
Pea seedlings	500	2.7	9
	500	2.7	10
	495	3.14	11
	525	1.2	12
Lentil seedlings	495	2.6	13
E. characias latex	490	2	14
Aspergillus niger	490	6.5	15
Micrococcus rubens putrescine oxidase (FAD-containing)	460, 370	10	16

Table 2
SPECTROSCOPIC PROPERTIES OF Cu-FREE AOs

Enzyme source	Chelating agent	λ_{max} (nm)	Ref.
Pea seedlings	DDC[a]	480	9
Pea seedlings	DDC	395, 410	12
Lentil seedlings	DDC	480	13
Bovine serum	DDC	380	19
Bovine serum	KCN	330, 370, 415, 550 (CD)	7

[a] DDC = Diethyldithiocarbamate.

various AOs. The spectra obtained are similar to, but not identical with, those of adducts with authentic pyridoxal.[20]

Other chemical compounds can form irreversible adducts with AOs.

For instance, Br-ethylamine is able to inactivate several amine oxidases. ANAO forms a covalent adduct characterized by an absorption at 320 nm (ϵ_{mM} = 3.2), catalytically inactive but still able to react with phenylhydrazine. However, if the adduct is reduced with $NaBH_4$ its absorption shifts to 380 nm and it becomes unreactive to phenylhydrazine.[21]

Cycloserine, a known inhibitor of pyridoxal phosphate enzymes,[22] also strongly inhibits AOs and gives spectrophotometrically observable adducts. After addition of cycloserine, BSAO and PKDAO form species with absorption at 370 nm very much like those observed with authentic pyridoxal enzymes.[20,23] These adducts could be at least partially reversed by addition of substrates. Upon the removal of Cu, cycloserine-treated PKDAO showed a fluorescence similar to that observed with the cycloserine adduct of aspartate aminotransferase, a typical pyridoxal-containing enzyme.[20]

Azide, cyanide, thiocyanate, and sulfide give rise to new absorption bands when added to AOs[3,24] that can be attributed to their reaction with copper. However, an additional reaction with the organic cofactor cannot be excluded at least for some of them.

BSAO forms an adduct absorbing at 415 nm with glycine which reversibly inhibits the enzyme.[23]

The visible absorption of AOs is particularly useful in order to study the kinetic reaction of these enzymes with their substrates. In fact, they are reversibly bleached by substrate both in the presence and absence of air. The bleaching pattern in the presence of air depends on the concentration of substrate added. If enough substrate is added to aerobic, nonstirred solutions of AO, the system rapidly goes to a quasi-anaerobic situation where the enzyme is practically colorless. Instead LSAO, PSAO, and ANAO, when reacted with their substrates in the absence of air, yield a yellow intermediate characterized by two sharp absorption bands at 470 to 480 and 440 to 430 nm, and by a broader band at 370 nm. This absorption spectrum is quite similar in the three enzymes and independent of the substrate nature.[13] Dithionite is able to bleach both the "resting" and the "yellow intermediate" AO, indicating that the latter is not a truly "reduced" form of the enzyme. Interestingly, other AOs may generate the same yellow intermediate in the presence of substrate, but only when the catalytic activity is slowed by subzero temperatures[12] or in the presence of CN^-.[20] This finding seems to indicate that the yellow intermediate can be formed within every AO, but its steady-state concentration is different in each enzyme. The yellow intermediate is bleached by the addition of Cu chelating agents[9] and is not observed with Cu-free AO.[13] Copper seems, therefore, to be needed for the generation of this intermediate that is formed after the release of aldehyde. Copper-free AO is still able to release aldehyde but is irreversibly bleached by substrate addition.[13]

Despite several efforts, the nature of the chromophore responsible for the visible absorption in AOs is still uncertain. The oldest and never fully disproved hypothesis suggests that a special form of pyridoxal phosphate is the organic cofactor present in amine oxidases. This hypothesis rests on the following data. First, a pyridoxal-based mechanism, proposed by Braunstein,[25] fits most of the reported mechanistic data. Second, a number of spectroscopic features could be reconciled with the formation of pyridoxal or pyridoxal derivatives with hydrazines, semicarbazide, cycloserine, etc. However, the spectra of the resting and of the substrate-treated AOs are different from those of typical pyridoxal enzymes. Therefore, it appears that if pyridoxal phosphate is the coenzyme of AOs, it should neither be present in the usual aldamine nor in the aldimine structure.[20] The optical spectroscopic features shown by AOs during their catalytic cycle seem rather similar to intermediates observed with pyridoxal model systems.[26,27] Third, various biological and chemical evidences[18,28,29] have indicated that pyridoxal derivatives were in fact present in various purified AOs, though the relative stoichiometry has not been reported.

On the other hand, several reports disclaimed some of the above-mentioned results or proposed mechanisms not requiring pyridoxal as the second prosthetic group.[30-32]

Alternatively, FAD or other flavin derivatives[33] have been postulated in the active site of AOs by analogy with the related enzymes, MAO and PAO. Just recently, Ameyama et al.[34] were able to isolate from the hydrolysate of three different AOs, namely pig kidney, beef plasma, and *Aspergillus niger*, two yellow compounds, one of which gave fluorescence spectra very similar to the coenzyme present in methylamine dehydrogenase from *Pseudomonas*. This coenzyme has been identified as pyrroloquinoline quinone (methoxatin).[35]

Whatever the organic moiety involved, it confers on AOs their spectroscopic and catalytic properties which make them a new distinct class of enzymes.

ACKNOWLEDGMENT

The original studies reported here were supported in part by the Italian National Research Council, Special Project on the "Chimica Fine", Contract No. 83.00310.95.

REFERENCES

1. **Mehrabian, Z. B. and Nalbandyan, R. M.,** Benzylamine oxidase from brain microvessels, *FEBS Lett.*, 164, 89, 1983.
2. **Rinaldi, A., Vecchini, P., and Floris, G.,** Diamine oxidase from pig kidney: new purification method and amino acid composition, *Prep. Biochem.*, 12, 11, 1982.
3. **Dooley, D. M. and Golnik, K. C.,** Spectroscopic and kinetic studies of the inhibition of pig kidney diamine oxidase by anions, *J. Biol. Chem.*, 258, 4245, 1983.
4. **Buffoni, F. and Blaschko, H.,** Benzylamine oxidase and histaminase: purification and crystallization of an enzyme from pig plasma, *Proc. R. Soc.*, 161, 153, 1964.
5. **Lindström, A., Olsson, B., and Pettersson, G.,** Effect of azide on some spectral and kinetic properties of pig plasma benzylamine oxidase, *Eur. J. Biochem.*, 48, 237, 1974.
6. **Suzuki, S., Sakurai, T., Nakahara, A., Oda, O., Manabe, T., and Okuyama, T.,** Spectroscopic aspects of copper binding site in bovine serum amine oxidase, *FEBS Lett.*, 116, 17, 1980.
7. **Suzuki, S., Sakurai, T., Nakahara, A., Manabe, T., and Okuyama, T.,** Effect of metal substitution on the chromophore of bovine serum amine oxidase, *Biochemistry*, 22, 1630, 1983.
8. **Bird, T. A. and Levene, C. I.,** Lysyl oxidase: evidence that pyridoxal phosphate is a cofactor, *Biochem. Biophys. Res. Commun.*, 108, 1172, 1982.
9. **Hill, J. M. and Mann, P. J. G.,** Further properties of the diamine oxidase of pea seedlings, *Biochem. J.*, 91, 171, 1966.
10. **McGowan, R. E. and Muir, R. M.,** Purification and properties of amine oxidase from epicotyls of *Pisum sativum, Plant Physiol.*, 47, 644, 1971.
11. **Macholàn, L. and Haubrova, J.,** Isolation and some characteristics of diamine oxidase from etiolated pea seedlings, *Coll. Czechoslov. Chem. Commun.*, 41, 2987, 1976.
12. **Kluetz, M. D., Adamsons, K., and Flynn, J. R., Jr.,** Cryoenzymology and spectrophotometry of pea seedling diamine oxidase, *Biochemistry*, 19, 1617, 1980.
13. **Rinaldi, A., Giartosio, A., Floris, G., Medda, R., and Finazzi-Agrò, A.,** Lentil seedlings amine oxidase: preparation and properties of the copper-free enzyme, *Biochem. Biophys. Res. Commun.*, 120, 242, 1984.
14. **Rinaldi, A., Floris, G., and Finazzi-Agrò, A.,** Purification and properties of diamine oxidase from *Euphorbia* latex, *Eur. J. Biochem.*, 127, 417, 1982.
15. **Suzuki, H., Ogura, Y., and Yamada, H.,** Stoichiometry of the reaction by amine oxidase from *Apergillus niger, J. Biochem. (Tokyo)*, 69, 1065, 1971.
16. **Swain, W. F. and Desa, R. J.,** Mechanism of action of putrescine oxidase. Binding characteristics of the active site of putrescine oxidase from Micrococcus rubens, *Biochim. Biophys. Acta*, 429, 331, 1976.
17. **Falk, M. C.,** Stoichiometry of phenylhydrazine inactivation of pig plasma amine oxidase. *Biochemistry*, 22, 3740, 1983.
18. **Lontie, R., Ed.,** *Copper Proteins and Copper Enzymes*, Vol. 2, CRC Press, Boca Raton, 1984.
19. **Yamada, H. and Yasunobu, K. T.,** Monoamine oxidase. IV. Nature of the second prosthetic group of plasma monoamine oxidase, *J. Biol. Chem.*, 238, 2669, 1983.
20. **Finazzi-Agrò, A., Guerrieri, P., Costa, M. T., and Mondovì, B.,** On the nature of chromophore in pig kidney diamine oxidase, *Eur. J. Biochem.*, 74, 435, 1977.
21. **Kumagai, H., Uchida, H., and Yamada, H.,** Reaction of fungal amine oxidase with -bromoethylamine, *J. Biol. Chem.*, 254, 10913, 1979.
22. **Karpeisky, M. Y., Khomutov, R. M., Severin, E. S., and Brensov, Y. N.,** The investigation on the interaction of cycloserine and related compounds with aspartate-glutamate transaminase, in *Chemical and Biological Aspects of Pyridoxal Catalysis*, Snell, E. E., Fasella, P. M., Braunstein, A. E., and Rossi-Fanelli, A., Eds., Pergamon, Oxford, 1963, 323.
23. **Costa, M. T., Sabatini, S., Turini, P., and Mondovì, B.,** Inhibition of beef plasma amine oxidase activity by glycine. *FEBS Lett.*, 94, 112, 1978.
24. **Dooley, D. M. and Coté, C. E.,** Inactivation of beef plasma amine oxidase by sulfide, *J. Biol. Chem.*, 259, 2923, 1984.
25. **Braunstein, A. E.,** Pyridoxal phosphate, in *The Enzymes*, Boyer, P. D., Lardy, H., and Myrbäck, K., Eds., Academic Press, New York, 1960, chap. 6.
26. **Schirch, L.-V. and Slotter, R. A.,** Spectral properties of Schiff-bases of amino acid esters with pyridoxal and pyridoxal N-methochloride in ethanol, *Biochemistry*, 5, 3175, 1966.
27. **Walters, L. S., Cornish, T. J., Askins, H. W., and Ledbetter, J. W.,** Observation of nanosecond transient states in laser-induced reactions of pyridoxal-5-phosphate, *Anal. Biochem.*, 127, 361, 1982.
28. **Goryachenkova, E. V.,** Nature of the active groups in diamine oxidase (histaminase), *Biochemistry (USSR)*, 21, 249, 1956.
29. **Mondovì, B., Costa, M. T., Finazzi-Agrò, A., and Rotilio, G.,** Pyridoxal phosphate as a prosthetic group of pig kidney diamine oxidase, *Arch. Biochem. Biophys.*, 119, 373, 1967.

30. **Inamasu, M., Yasunobu, K. T., and Konig, W. A.,** Cofactor investigation of bovine plasma amine oxidase. $NaBH_4$ reduction of enzyme-substrate mixture, *J. Biol. Chem.*, 249, 5265, 1974.
31. **Benitez, L. V. and Allison, W. S.,** The inactivation of the acyl phosphatase activity catalyzed by the sulfenic acid form of glyceraldehyde 3-phosphate dehydrogenase by dimedone and olefins, *J. Biol. Chem.*, 249, 6234, 1974.
32. **Neumann, R., Jevey, R., and Abeles, R. H.,** The action of plasma amine oxidase on β-haloamines. Evidence for proton abstraction in the oxidative reaction, *J. Biol. Chem.*, 250, 6262, 1975.
33. **Hamilton, G. A.,** Oxidases with monocopper reactive sites. Galactose oxidase and copper-containing amine oxidases, in *Copper Proteins*, Spiro, T. G., Ed., John Wiley & Sons, New York, 1981, chap. 5.
34. **Ameyama, M., Hayashi, M., Matsushita, K., Shinagawa, E., and Adachi, O.,** Microbial production of pyrroloquinoline quinone, *Agric. Biol. Chem.*, 48, 561, 1984.
35. **Salisbury, S. A., Forrest, H. S., Cruse, W. B. T., and Kennard, O.,** A novel coenzyme from bacterial primary alcohol dehydrogenase, *Nature (London)*, 280, 843, 1979.

Chapter 12

SPECTROSCOPIC AND CHEMICAL PROPERTIES OF THE AMINE OXIDASE COPPER

G. Rotilio

TABLE OF CONTENTS

I. INTRODUCTION

On the basis of the evidence so far available for AOs from various sources, copper is the only inorganic cofactor known for these enzymes and is reproducibly present in all enzyme preparations as 2 g-atoms per mole of protein, likely to be one copper for each subunit of enzyme. Reports on manganese being an additional essential metal[1-3] are not sufficiently supported by the degree of purity of the preparations used or by adequate control of contamination by adventitious metal. In spite of its essential nature, no clear mechanistic role has been proposed for the copper center of AOs. The purpose of this chapter is to summarize the chemical and physicochemical properties of the AO copper, as compared to the copper of other copper proteins. The matter dealt with includes spectroscopy of amine oxidases (such as EPR spectra, NMR studies, optical absorption, and CD studies) and chemical reactions such as redox titrations, ligand binding, and metal removal and substitution.

II. SPECTROSCOPY

A. EPR Spectra: Parameters, Intensity and Line Shape (Figure 1)

The EPR parameters of all Cu-containing AOs fall in the category of the so-called Type 2 Cu-EPR spectra.[4] This means that they are similar to those of small copper complexes. In fact, all the spectra reported are in the limits of axial symmetry, with $g_{\|}$ values between 2.29 and 2.35, $g_{\perp}$ between 2.06 and 2.07, and $A_{\|}$ in the 15 to 16 mT range. Practically all the copper is accounted for by the integrated intensity of the EPR signal, indicating that all the metal is in the Cu(II) state with no spin or dipolar interaction between the two centers ($\geq$ 14 Å apart).

A single report[5] on the *Aspergillus niger* enzyme indicated that only two of the three coppers assayed by chemical analysis per enzyme molecule are EPR detectable. However, the unusual copper-to-protein ratio suggests that the EPR-silent copper may be of adventitious nature. In fact, a similar report was published in an early EPR study of the beef plasma enzyme,[6] while later reports established the usual 1:1 stoichiometry. Significant changes of the peak-to-peak amplitude of the derivative absorption in the $g_{\perp}$ region are detectable upon addition of substrate.[7] However, the integrated intensity of the signal seems not to vary significantly under these conditions. Conformational changes that affect the line shape of the copper spectrum are likely to be responsible for such changes. This view is supported by other modifications of the EPR spectrum in the presence of substrate. Pig kidney DAO shows different nitrogen hyperfine patterns with different substrates, which are unequivocally due to protein nitrogens, as demonstrated by isotopic substitution.[8] Beef plasma and lentil seedling AOs show a larger copper $A_{\|}$ after anaerobic addition of substrates,[9] while a narrower coupling has been observed under similar conditions in the case of the pea seedling enzyme.[2] Depending on the different enzyme sources and experimental conditions, the EPR spectrum of AOs is not homogeneous. In part, heterogeneity is of trivial origin as it can be minimized by removal of adventitious copper with extensive dialysis and/or chelating agents.[10a] However, two species of copper were detected by EPR in clean preparations of pig kidney DAO,[11] and they responded as independent entities to various conditions, such as addition of substrate and thiol reagents. Considerable heterogeneity can be detected by high-frequency EPR in the beef plasma AO,[12] and it has been interpreted in terms of structural diversity of the copper site on each enzyme subunit. In particular, it has been proposed that one copper should have more rhombic distortion than the other one. Addition of anionic ligands like N_3^- and CN^- makes the two sites identically axial. This, and other evidence obtained with substrate,[13] suggests that the nonidentity of the two copper sites in the native enzyme may have functional significance.

Nitrogen hyperfine lines are observed in the $g_{\perp}$ region of the EPR spectra of AOs either

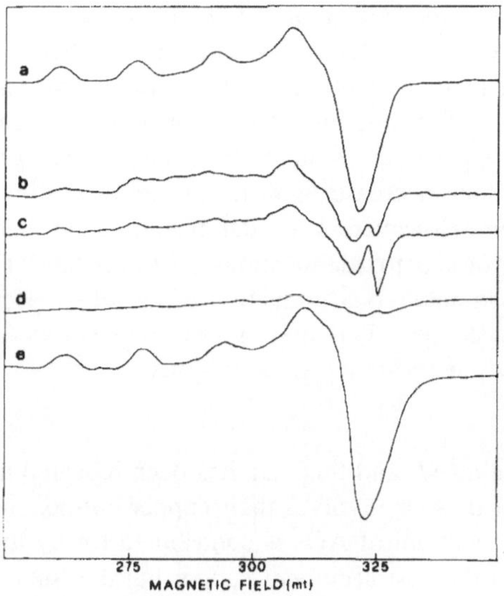

FIGURE 1. EPR Spectra of beef plasma AO (1.2×10^{-4} *M* copper) in 0.1 *M* phosphate buffer, pH 7.4. (a) Enzyme as such; anaerobic addition of large excesses of either ferrocyanide or ascorbate or NADH had no effect on the spectrum. (b) To the enzyme solution (as in a) 5×10^{-4} *M* NADH was added anaerobically in the presence of 50 μM phenazine methosulfate; the solution was frozen for spectrum after 5 min incubation at 25°C; after 24 hr no EPR signal was detectable. Similar results were obtained with ascorbate. (c) The enzyme solution, completely reduced after 24 hr incubation under the conditions described in b, was open to air and frozen after 5 min standing at 25°C. In both b and c the radical of phenazine methosulfate is apparent. (d) The enzyme solution (as in a) from 1 min after anaerobic addition of stoichiometric dithionite. (e) The solution as in (d) was opened to air and frozen for spectra after 2 min at 25°C. EPR conditions: microwave frequency, 9.14 GHz; microwave power, 15 mW field modulation amplitude, 10 gauss; temperature, − 160°C.

in native enzyme preparations[8] or under certain experimental conditions, including alkaline pH under nondenaturating conditions,[14] high ionic strength,[15] and substrate[8] or inhibitor[16] addition. No doubt they reflect nitrogen coordination from protein side chains, probably imidazoles.

B. NMR Relaxation

Paramagnetic copper centers of proteins may have large enhancing effects on the nuclear relaxation rates of water protons when at least one water molecule rapidly exchanges between the metal first coordination sphere and the bulk solvent. This is the case of the AOs so far investigated by this technique. The pig kidney and the beef plasma enzymes have been extensively studied in this respect. In the pig kidney enzyme,[10a] a substrate analog, 3-methyl-4-(β-dimethylammonium)ethyl imidazole chloride, reduces the water relaxation rates at very high concentrations, but it is not able to abolish water exchange completely. Since the limiting value is not one half of the original value, the result can be interpreted in terms of more than one water bound to the copper in a nonequivalent fashion rather than in terms of two nonidentical copper sites. Moreover, the high concentrations of substrate analog needed for the effect to be shown confirm that the effect is an indirect one, as previously demonstrated by EPR data with isotopically substituted diamines.[8] Later, ^1H NMR relaxation dispersion measurements showed that the two Cu(II) ions are also intrinsically different,[10b] because both the T_1 and T_2 frequency dependencies show diphasic behaviors.

From the values of the proton relaxation rate of beef plasma AO,[12] it was inferred that two water molecules were coordinated to the copper, one being equatorial and the other axial. Measurements were made with N_3^- and CN^-, which indicated displacement by anions of the equatorial water only. The functional significance of these data will be discussed later on. Deprotonation of the copper-bound water with an approximate pK = 8 has been suggested to be the cause of the sudden decrease of water proton relaxation rate of the beef plasma enzyme in this pH region. However, a similar phenomenon leads to an increase of the relaxation rate in the case of superoxide dismutase.[17] In the latter situation it was proposed, on the basis of a parallel rhombic-axial transition of the EPR spectrum line shape, that the OH^- became close to the copper. This may not be the case of AOs, whose EPR spectra show almost tetragonal symmetry under all conditions.

C. Optical Spectra

It has generally been believed, and this idea has been repeated in most recent reviews on AOs,[18] that the pink color of AOs involves their copper cromophore. However, there is no evidence that the visible spectrum of AOs is contributed for by the inorganic cofactor (see also Chapter 11). Perhaps the best argument supporting this statement is that bleaching of the 500-nm absorption by substrate is associated with minimal changes of the copper EPR spectrum.[8] On the other hand, drastic alterations of the copper EPR spectrum by N_3^- or CN^- leave the 500-nm band practically unmodified.[16] Recently, copper removal by apparently nonreducing procedures have given rise to copper-free enzyme with substantially identical visible absorption spectra to those of the native copper-containing enzyme[19] (see Section III.C). If these results can be confirmed, the AO copper should be considered optically silent and not even indirectly associated with the organic cofactor as far as the origin of the pink color is concerned. Therefore, also the attribution[15] of the CD bands at 350, 450, and 540 nm of BSAO to charge transfer bands from the ligands (hypothetically sulfur side chains) to the copper has to be discussed in the light of the previous considerations. The assignment of the CD peaks at 650 and 800 nm of the same enzyme to Cu(II) d-d transitions appears, on the other hand, to be less questionable.

III. REACTIONS WITH LIGANDS

A. Redox Reagents

The AO copper seems to have a rather low redox potential and is rather inert kinetically in redox reactions. Ferrocyanide is not able to reduce the copper of BSAO (Figure 1), contrary to the case of another type "Type 2-copper" enzyme, superoxide dismutase.[20] The enzyme copper is reduced by stoichiometric NADH and excess ascorbate only in the presence of suitable mediators, and then very sluggishly (Figure 1). Thus the E_0' of this copper site should be considerably negative.

Also, reoxidation by air occurs rapidly only in the presence of mediators. However, the reoxidation rate is dependent on the nature of the reductant. The copper reduced by stoichiometric dithionite is reoxidized by oxygen very rapidly (Figure 1), supporting a possible role of the copper in the oxidase activity of the enzyme (see below). Reoxidation by O_2 of the dithionite-reduced enzyme does not inactivate the enzyme or alter the EPR spectrum as for instance in the case of superoxide dismutase.[21] This is probably due to the lack of sensitivity of resting AOs to H_2O_2,[22] which is a product of the reoxidation process of Cu(I) enzymes by O_2.[21] On the other hand, the sulfide-reduced BSAO is inactivated upon reoxidation.[23] The reason for this discrepancy is not clear.

B. Substrate

It has already been mentioned that substrates decolorize the visible spectrum of AOs, but perturb their EPR spectra only in a fashion suggesting conformational changes affecting the copper geometry. By an isotopic substitution approach[8] it was shown that the substrate amino groups to be oxidized do not bind to the copper in the pig kidney enzyme and this has been considered to be generally true for all AOs. In fact the result[24] that high concentrations of ammonium ion change the EPR spectrum of the cupric centers has to be considered as just another indication of conformational changes of the enzyme and not an evidence for direct amino group binding to the copper, as suggested by the authors. Positively charged molecules usually do not coordinate directly to protein-bound copper. Also, in the case of superoxide dismutase, drastic changes of the copper EPR spectra were seen in the presence of Ca(II), but they were shown to be due to conformational effects brought about by electrostatic interactions with negatively charged amino acid side chains.[25] In fact, NMR relaxation data with ^{15}N-labeled molecules clearly show that neither NH_3 nor the amino end of substrates displace the copper-coordinated water,[10a,b] implying a more indirect role, if any, for water coordination to the copper in enzyme catalysis (see below). As already mentioned, high concentrations of a substrate analog actually displace coordinated water, but water substitution occurs via coordination of the opposite substrate end.[10a,b] This would suggest that the copper-bound equatorial waters is the site of the inhibition by excess substrate, which is observed in most AOs,[26] rather than that of binding of the substrate amine group to be oxidized. In fact, the optical data also agree in indicating that the organic chromophore is the site of binding by substrate and substrate analogs through their reactive end (see Chapters 6, 7, 10, 11, and 13).

An interesting feature of substrate-perturbed EPR spectra of AOs is that in the beef plasma enzyme a reaction of only half the copper sites has been demonstrated by 35-GHz EPR analysis.[13] Also, in the case of the substrate analog phenylhydrazine, only half of plasma AO copper seems to be involved.[9] This may be related to the presence of only one substrate-reacting site in AOs[27] (see Chapters 10 and 11).

C. Inhibitors and Chelating Agents

Beside competitive inhibitors and carbonyl-reagent-like inhibitors, it is assumed that most other AO inhibitors and inactivators bind to copper. Azide and cyanide[12,16] strongly perturb the EPR spectrum of the copper, with no significant effect on the visible absorption spectrum of the enzyme. Interestingly, the enzyme is still active at inhibitor concentrations which completely change the EPR spectrum of the resting enzyme.[16] Thus, either copper binding to inhibitors does not interfere with the enzyme activity or, more likely, as already observed with other copper oxidases,[28] azide and cyanide bind with different affinity to the working enzyme. The already-mentioned conformational changes produced by substrate binding are in line with this hypothesis.

An additional fact complicates the picture even more. At higher concentrations than those affecting the low temperature EPR spectrum, azide binding leads to the formation of a 400-nm charge transfer band in parallel with progressive inhibition of enzyme activity[16] and of proton relaxation rates of the equatorial water.[12] Although comparison of data obtained from frozen solution EPR with room temperature data should always be considered with some caution,[29] this would again suggest that only one of the two copper sites is functionally important and that replacement of its equatorially bound water molecule impairs the involvement of this copper in the catalytic act. In fact, 35-GHz EPR spectra in the presence of azide and cyanide show no longer copper heterogeneity.[12]

The changes of the copper EPR spectrum which are observed in the presence of other inhibitors, such as cuprizone[30] and phenylhydrazine,[9] are rather similar to those induced by substrate and this perhaps indicates indirect conformational effects through binding to other sites, probably the substrate-binding site. The EPR spectra in the presence of N_3^-,[16] CN^-,[12] and S^-[23] are instead indicative of stronger ligand fields around the copper and are in line with inner sphere coordination of the copper, probably through binding to the equatorial water-binding site. Thus, a substrate-binding site and a water-binding site seem to be the regions where different inhibitors approach the enzyme. In both cases the copper environment is perturbed, but only in the case of direct copper binding at the water-binding site is its coordination geometry extensively altered.

IV. METAL REMOVAL AND SUBSTITUTION

CN^- is able to reversibly remove the copper from the enzyme only in the case of prior copper reduction, e.g., by dithionite.[31] The apoenzyme lacks the 500-nm band, which is restored by copper addition. This evidence may suggest involvement of copper in the visible absorption of the enzyme; however, it is now clear (see below) that this is due to lack of reoxidation of the reduced organic cofactor.

The most complicated case of ligand interaction with AOs is that of the well known metal chelating agent, diethyldithiocarbamate (DDC). Early reports[32,33] generally agreed that DDC was able to reversibly remove the copper from these enzymes, thus explaining the loss of enzyme activity in the presence of DDC. More recent data[34] indicate that a ternary DDC-Cu-protein complex is formed, with optical and EPR spectra similar to those described in the case of Cu-carbonic anhydrase.[35] No clear evidence is available for actual separation of a Cu-DDC complex from the enzyme, as in the case of superoxide dismutase,[36] at least for the well characterized BSAO. Data with plant AOs seem to indicate that DDC is actually able to remove the copper with no disappearance of the visible absorption,[19,37] thus confirming that copper is not involved in the visible spectrum of these enzymes.

The copper-free enzyme, as obtained by the dithionite-CN^- treatment, can accept either Co(II) or Ni(II) at a site very likely to be the original Cu(II) site.[31] These two derivatives are interesting insofar that they show that metal binding, even though the metal ion is not the native one, is able to reoxidize the dithionite-reduced visible chromophore, giving back the original optical spectrum of the copper enzyme. Moreover, when the Ni-enzyme was depleted of nickel under nonreducing conditions by treatment with dimethylglyoxine and CN^-, the 500-nm band stayed unaltered. This result again, and conclusively, confirms that copper per se is not necessary to create the visible spectrum of the enzyme, but it is required for the reoxidation of the reduced organic chromophore. It is important to recall that only copper was able to reoxidize the substrate-bleached optical absorption.[31] A sound interpretation of these results can be that metal binding results in a conformational change of the enzyme molecule which facilitates access of O_2 to the reduced chromophore. However, when substrate is bound only copper is effective in catalyzing electron transfer to O_2, perhaps according to the recently suggested mechanism,[24] in which a hydroxyl coordinated to copper acts as a nucleophile to facilitate hydride ion transfer to oxygen during the reoxidation process.

V. FINAL REMARKS

There are two major questions concerning the AO copper, which have repeatedly been addressed throughout the book and which I would like to reconsider in the light of the present chapter. One question is the possible vicinity of the two types of cofactors, and the other is the functional role of copper in these enzymes. As far as the first question is concerned I should say that there is no conclusive evidence for the two cofactors being close to each other in AOs. Copper appears to be not involved in the optical spectrum of the organic chromophore. All the cross-effects seen by different researchers between the substrate binding site and the copper may well be due to conformational changes mediated by even long-range protein rearrangement. This is the case of the EPR changes in the presence of substrate,[8] phenylhydrazine,[9] and cuprizone.[30] When copper is unequivocally bound to ligands (such as N_3^- and CN^-), no effects on the visible spectrum are observed.[12,16] The only suitable approach I see in this context is to spin label the substrate or a substrate analog and see the effects on its EPR spectrum by varying the valence state of the copper.

On the other hand, there is considerable evidence that copper plays a specific role in the reoxidation of the substrate-reduced enzyme, though no valence change can be seen by traditional EPR experiments, even with rapid freeze techniques.[13] This is not, however, an obstacle to accepting the idea that copper is required in the reoxidation step of the enzyme catalysis. The proposal[24] that copper may catalyze oxygen reduction by facilitating hydride transfer from substrate via a nucleophilic mechanism shows that other explanations can be suggested as alternative to simplistic redox shuttles. Recent work from our laboratory[38] has shown that when copper is bound by CN^- a free radical intermediate can be detected by EPR in plant AO even in the presence of oxygen. A similar case can be the free radical reported[31] to appear when the reduced copper-depleted BSAO is reoxidized by $Fe\ (CN)_6{}^{3-}$. In both cases the organic cofactor might have been trapped in a transient state which is rapidly turned over by oxygen only in the presence of copper.

ACKNOWLEDGMENT

The original studies reported here were supported in part by the Italian National Research Council, Special Project on the "Chimica Fine", Contract No. 83.00310.95.

REFERENCES

1. **Mann, P. J. G.,** Purification and properties of the amine oxidase of pea seedlings, *Biochem. J.*, 79, 623, 1961.
2. **Nylen, U. and Syzbek, P.,** Kinetics and other characteristics of diamine oxidase of pea seedlings, *Acta Chem. Scand.*, B28, 1153, 1974.
3. **Crabe, M. J. C., Waight, R. D., Bardsley, W. G., Barker, R. W., Kelly, I. D., and Knowles, P. F.,** Human placental diamine oxidase: improved purification and characterization of a copper and manganese containing amine oxidase with novel substrate specificity, *Biochem. J.*, 155, 679, 1976.
4. **Malmström, B. G., Andreasson, L-E., and Reinhammar, B.,** Copper containing oxidases and superoxide dismutase, in *The Enzymes*, Vol. 12, 3rd ed., Boyer, P. D., Ed., Academic Press, New York, 1975, 507.
5. **Yamada, H., Adachi, O., and Yamano, T.,** Electron paramagnetic resonance spectra of amine oxidase from *Aspergillus niger*, *Biochem. Biophys. Acta*, 191, 751, 1969.
6. **Yamada, H., Yasunobu, K. T., Yamano, T., and Mason, H. S.,** Cu in plasma amine oxidase, *Nature (London)*, 198, 4885, 1963.
7. **Mondovì, B., Rotilio, G., Finazzi Agrò, A., Vallogini, M. P., Malström, B. G., and Antonini, E.,** Copper reduction by substrate in diamine oxidase, *FEBS Lett.*, 2, 182, 1969.
8. **Mondovì, B., Rotilio, G., Costa, M. T., Finazzi-Agrò, A., Chiancone, E., Hansen, R. E., and Beinert, H.,** Diamine oxidase from pig kidney; improved purification and properties, *J. Biol. Chem.*, 242, 1160, 1967.
9. **Rinaldi, A., Floris, G., Sabatini, S., Finazzi-Agrò, A., Giartosio, A., Rotilio, G., and Mondovì, B.,** Reaction of beef plasma and lentil seedlings Cu-amine oxidases with phenylhydrazine, *Biochim. Biophys. Res. Commun.*, 115, 841, 1983.

10a. **Kleutz, M. D. and Schmidt, P. G.,** Proton relaxation study of the hog kidney diamine oxidase active centre, *Biochemistry*, 16, 5191, 1977.

10b. **Kluetz, M. D. and Schmidt, P. G.,** Proton magnetic relaxation dispersion in solutions of the cuproprotein diamine oxidase, *Biophys. J.*, 29, 283, 1980.

11. **Mondovì, B. and Calabrese, L.,** Copper heterogeneity in pig kidney histaminase. An EPR study on the effect of substrate, guanidine and mersalyl, *Agents Action*, 3, 187, 1973.

12. **Barker, R., Boden, N., Cayley, G., Charlton, S. C., Henson, R., Holmes, M. C., Kelly, I. D., and Knowles, P. F.**, Properties of cupric ions in benzylamine oxidase from pig plasma as studied by magnetic resonance and kinetic methods, *Biochem. J.*, 177, 289, 1979.
13. **Grant, J., Kelly, I., Knowles, P., Olsson, J., and Pettersson, C.**, Changes in the copper centres of benzylamine oxidase from pig plasma during the catalytic cycle, *Biochem. Biophys. Res. Commun.*, 83, 1216, 1978.
14. **Olsson, B. and Pettersson, G.**, personal communication.
15. **Suzuki, S., Sakurai, T., Nakahara, A., Oda, O., Manabe, T., and Okuyana, T.**, Spectroscopic aspects of copper binding site in bovine serum amine oxidase, *FEBS Lett.*, 116, 17, 1980.
16. **Lindstrom, A., Olsson, B., and Pettersson, G.**, Effect of azide on some spectral and kinetic properties of pig plasma benzylamine oxidase, *Eur. J. Biochem.*, 48, 237, 1974.
17. **Terenzi, M., Rigo, A., Franconi, C., Mondovì, B., Calabrese, L., and Rotilio, G.**, pH dependence of the nuclear magnetic relaxation rate of solvent water protons in solutions of bovine superoxide dismutase, *Biochim. Biophys. Acta*, 351, 230, 1974.
18. **Knowles, P. P. and Yadov, K. D. S.**, Amine oxidase, in *Copper Proteins and Copper Enzymes*, Vol. II, Lontie, R., Ed., CRC Press, Boca Raton, 1984, 103.
19. **Rinaldi, A., Giartosio, A., Floris, G., Medda, R., and Finazzi-Agrò, A.**, Lentil seedlings amine oxidase: preparation and properties of the copper-free enzyme, *Biochem. Biophys. Res. Commun.*, 120, 242, 1984.
20. **Rotilio, G., Morpurgo, L., Calabrese, L., and Mondovì, B.**, On the mechanism of superoxide dismutase. Reaction of the bovine enzyme with H_2O_2 and ferrocyanide, *Biochim. Biophys. Acta*, 302, 229, 1973.
21. **Viglino, P., Scarpa, M., Coin, F., Rotilio, G., and Rigo, A.**, submitted.
22. **Mondovì, B., Rotilio, G., Finazzi-Agrò, A., and Costa, M. T.**, Diamine oxidase inactivation by hydrogen peroxide, *Biochim. Biophys. Acta*, 132, 521, 1967.
23. **Dooley, D. M. and Coté, C. E.**, Inactivation of beef plasma amine oxidase by sulfide, *J. Biol. Chem.*, 259, 2923, 1984.
24. **Yadav, K. D. S. and Knowles, P. F.**, A catalytic mechanism for benzylamine oxidase from pig plasma — stopped flow kinetic studies, *Eur. J. Biochem.*, 114, 139, 1981.
25. **Cocco, D., Calabrese, L., Finazzi-Agrò, A., and Rotilio, G.**, Electrostatic interactions in Cu,Zn superoxide dismutase, *Biochim. Biophys. Acta*, 746, 71, 1983.
26. **Zeller, E. A.**, Diamine oxidases, in *The Enzymes*, Vol. 8, 2nd ed., Boyer, P. D., Ed., Academic Press, New York, 1963, 313.
27. **Mondovì, B., Sabatini, S., and Befani, O.**, The active site of copper amine oxidases, *J. Mol. Cat.*, 23, 325, 1984.
28. **Morpurgo, L., Rotilio, G., Finazzi-Agrò, A., and Mondovì, B.**, Anion binding to Rhus vernicifera laccase, *Biochim. Biophys. Acta*, 336, 324, 1974.
29. **Morpurgo, L., Calabrese, L., Desideri, A., and Rotilio, G.**, Dependence on freezing of the geometry and redox potential of Type 1 and Type 2 copper sites of Japanese lacquer free laccase, *Biochem. J.*, 193, 639, 1981.
30. **Lindstrom, A. and Pettersson, C.**, The mechanism of inhibition of pig-plasma benzylamine oxidase by the copper-chelating reagent cuprizone, *Eur. J. Biochem.*, 48, 229, 1974.
31. **Suzuki, S., Sakurai, T., Nakahara, A., Manabe, T., and Okuyama, T.**, Effect of metal substitution on the chromophore of bovine serum amine oxidase, *Biochemistry*, 22, 1630, 1983.
32. **Mann, P. J. C.**, Purification and properties of the amine oxidase of pea seedlings, *Biochem. J.*, 79, 623, 1961.
33. **Yamada, H. and Yasunobu, K. T.**, Mono amine oxidase. II. Copper, one of the prosthetic groups of plasma monoamine oxidase, *J. Biol. Chem.*, 237, 3077, 1962.
34. **Suzuki, S., Sakurai, T., Nakahara, A., Oda, O., Manabe, T., and Okuyama, T.**, Copper binding site in serum amine oxidase treated with sodium diethyldithiocarbamate, *Chem. Lett.*, 487, 1982.
35. **Morpurgo, L., Desideri, A., Rigo, A., Viglino, P., and Rotilio, G.**, Reaction of diethyldithiocarbamate and other bidentate ligands with Zn, Co and Cu carbonic anhydrase, *Biochim. Biophys. Acta*, 746, 183, 1983.
36. **Cocco, D., Calabrese, L., Rigo, A., Argese, E., and Rotilio, G.**, Reexamination of the reaction of diethyldithiocarbamate with the copper of superoxide dismutase, *J. Biol. Chem.*, 256, 8983, 1981.
37. **Kluetz, M. D., Adamsons, K., and Flynn, J. E.**, Cryoenzymology and spectrophotometry of pea seedlings diamine oxidase, *Biochemistry*, 19, 1617, 1980.
38. **Finazzi-Agrò, A., Rinaldi, A., Floris, G., and Rotilio, G.**, A free radical intermediate in the reaction of plant Cu-amine oxidases, *FEBS Lett.*, 176, 378, 1984.

Chapter 13

INHIBITORS OF COPPER AMINE OXIDASES

W. G. Bardsley

TABLE OF CONTENTS

I. INTRODUCTION AND DEFINITION OF TERMS TO BE USED

The enzymes to be discussed in this chapter are those from such sources as kidney, placenta, pregnancy plasma, serum, semen, aorta, intestine, pea seedling, or *Aspergillus niger*.

These copper-containing AOs (subsequently referred to as copper AO) catalyze the reaction:

$$RCH_2NH_2 + O_2 + H_2O \rightleftharpoons RCHO + H_2O_2 + NH_3 \qquad (1)$$

where RCH_2NH_2 is a mono- or diamine and RCHO is an aldehyde product. In homogeneous, well-stirred aqueous solutions at constant pressure and temperature, the rate or velocity of the enzyme-catalyzed reaction, v, can be defined as:

$$v = \frac{d\,[RCHO]}{dt} = \frac{d\,[H_2O_2]}{dt} = \frac{d\,[NH_3]}{dt}$$

$$= -\frac{d\,[RCH_2NH_2]}{dt} = -\frac{d\,[O_2]}{dt} \qquad (2)$$

where square brackets indicate concentrations. Similar expressions can also be written down for the additional reactions catalyzed by polyamine oxidases (PAOs) and in all cases the rate of enzyme-catalyzed reaction depends upon time, t, enzyme, E, substrate, and product concentrations. Also, the rate can be altered by the presence of modifier substances, $M_1, M_2, \ldots, M_n$. These modifiers could be protons, buffers, inorganic salts, nonspecific organic compounds, substrate analogs, specific reactive agents, i.e., any substances that can alter the rate of reaction. So we can write a general expression for v in the form:

$$v = f\,(t, [E], [RCH_2NH_2], [O_2], [RCHO], [NH_3],$$

$$[H_2O_2], [M_1], [M_2], \ldots [M_n]) \qquad (3)$$

where f represents a functional relationship connecting v and the independent variables. It is useful to have appropriate formulae for the function, f, when this is possible in order that changes in v can be expressed quantitatively. This approach makes definitions more unambiguous and interpretation in molecular terms more certain as statistical treatment of experimental data and discrimination between alternative models becomes possible.

Classic reviews of the literature concerning the effects of inhibitors on AO are to be found in books[1,2] and in the definitive article by Yasunobu et al.[3] These studies established that substrate analogs can compete reversibly with amine substrates for binding sites on the enzymes and the study of this phenomenon has illuminated the nature of the substrate binding sites. Slow inhibition by carbonyl-group reagents has suggested the presence of a carbonyl group at the active site; chelating agents have confirmed the importance of copper and substrate inhibition has also been reported. However, in recent years these well-known results have been subjected to some criticism and also to more quantitative investigations. Product inhibition has been exploited to identify molecular mechanisms; new phenomena such as suicide inhibition have been discovered, and extensive in vivo studies have been performed. So in this chapter I shall concentrate on reviewing the quantitative studies carried out since 1970 where inhibition studies have been used to understand the chemistry and in vivo function of copper AO.

Compounds that inhibit AO can often be classified into alternative types such as substrate analogs, time-dependent enzyme inactivators, copper-chelating reagents, substrate inhibitors, product inhibitors, or suicide substrates. We shall see that in some cases the inhibition falls clearly into one of the above classes but we shall also find that some compounds have more than one type of effect. For each class of inhibitor I shall give a brief discussion of the general principles involved and draw attention to key references and results that have been secured.

II. RAPID INHIBITION BY SUBSTRATE ANALOGS

It often happens that a modifier substance is not an obligatory reactant or cofactor, but is an effector that can bind rapidly to an enzyme thereby altering its kinetic characteristics. Enzyme-modifier complexes of this type are presumed to result from hydrogen bonds, Van der Waal's forces, or electrostatic attraction, but not from covalent bond formation. Even simple mechanisms for modifiers of enzyme action can lead to inhibition, activation, or other complex results, and a theoretical study of some model kinetic schemes (together with rate equations to account for these effects) has been published.[4] An appropriate formulation of Equation 3 is obtained by keeping all reactants constant and measuring the steady-state rate at different concentrations of modifier. Then the v[M] profile takes the form:

$$\frac{v}{E_o} = \frac{a_o + a_1[M] + a_2[M]^2 + \ldots + a_n[M]^n}{b_o + b_1[M] + b_2[M]^2 + \ldots + b_n[M]^n}; \quad a_i, b_j \geqslant 0 \tag{4}$$

accounting for the nonobligatory effect of M, i.e., $a_o \neq 0$. Equation 4, even with $n = 2$, can give activation, inhibition, activation followed by inhibition, inhibition followed by activation, and other more complex patterns are possible when $n > 2$, all resulting from binding of M to a single binding site which changes the kinetic behavior of the enzyme when occupied.[4]

Such effects are not simply of academic interest but have been observed with copper AO. For instance, substances such as chlorpromazine with PSAO,[5] alcohols with rabbit serum AO,[6] 2-oxosuccinamic, or malonamic acids with PKDAO,[7] can lead to activation and simple compounds such as NH_3 or imidazole, cause activation followed by inhibition with PSAO.[8] Indeed, it was this last result which stimulated the theoretical studies referred to previously.[4] However, the presence of artifacts should always be borne in mind, especially when high concentrations of modifiers are employed. For instance, the supposed activation of placental AO by Mn^{2+} and hemoglobin has been shown to be an artifact[9] in a paper where the numerous problems associated with the use of coupled peroxidase assays for H_2O_2 to measure AO kinetics are also emphasized.

Usually, however, interest has centered on compounds that are nitrogenous bases having structural features like the amine substrates. The aim has been to study the topology of the active site by measuring the extent to which these substrate analogs can displace amine substrates from the substrate binding sites. A characteristic feature of this type of effect is that it should be rapidly and quantitatively reversed by removal of the modifier, but this has not always been confirmed. However, Macholan's group has studied the inhibition of the pea seedling and hog kidney enzymes by a series of diamino ketones.[10] These keto analogs of diamines are potent inhibitors, but it was shown that extensive reactivation occurs after passing the enzyme fully inhibited by 2×10^{-3} *M* 1,4-diamino-2-butanone through a Sephadex® G-25 column.

If it can be established that a substance produces an immediate effect which can be fully reversed by dialysis, chromatography or some other way of removing the ligand then, providing the enzyme obeys Michaelis-Menten kinetics,[11] the appropriate form for Equation 3 is

$$\frac{1}{v} = \frac{k}{V}\left(1 + \frac{[I]}{K_{i\ slope}}\right)\frac{1}{[RCH_2\ NH_2]} + \frac{1}{V}\left(q + \frac{[I]}{K_{i\ intercept}}\right) \qquad (5)$$

Here the approach of Cleland is being used[12] where v is a steady-state rate. Almost invariably the amine has been the varied substrate at fixed oxygen tension in these types of studies due to the difficulties involved in varying oxygen concentration with amine fixed. The slopes and intercepts of linear double reciprocal plots should be replotted in order to determine the K_i values and also to test for curvature which would indicate higher-order stoichiometry. According to Cleland's nomenclature, slope effects (i.e., $K_{i\ slope}$ measurable) are referred to as competitive inhibition, intercept effects (i.e., $K_{i\ intercept}$ measurable) as uncompetitive inhibition and simultaneous occurrence of both effects (i.e., $K_{i\ slope}$ and $K_{i\ intercept}$ both measurable but not necessarily equal) is called noncompetitive inhibition.

We have used this technique with the enzyme from hog kidney,[13] human placenta,[14] and human kidney[15] and a homologous series of compounds of the form:

$$X - (CH_2)_n - Y$$

as substrates or inhibitors where n was in the range 2 to 12 and X and Y were such groups as $-CH_3$, $-NH_2$, $-N(CH_3)_2$, $N^+(CH_3)_3$, $S^+(CH_3)_2$, $S^+{=}C(NH_2)_2$, or $N^+H{=}C(NH_2)_2$. In addition, a number of miscellaneous compounds were studied, including some with benzene rings or imidazoles replacing the methylene chains. From these studies it was concluded that the three enzymes were very similar; slope effects were interpreted as resulting from the competition between amine substrate and inhibitor for an oxidized form of the enzyme, while intercept effects were proposed to be due to binding to a reduced form of the enzyme. A number of general conclusions were reached as now listed.

1. Sulfonium and trimethylammonium compounds tended to give competitive inhibition.
2. Other *bis*-onium and *p*-xylylene-compounds gave noncompetitive inhibition.
3. The inhibitory potency in any series was approximately isothiuronium = guanidinium >> dimethylsulfonium > trimethylammonium.
4. Usually a chain length of 5 to 8 methylene units or a benzene ring gave optimum separation in *bis*-onium compounds leading to maximum inhibition.
5. In any series, *bis*-onium compounds were not generally more inhibitory than *mono*-onium compounds.
6. Replots of slopes and intercepts were usually linear although there were a few exceptions.

From the K_i values measured it was possible to give estimates of the standard free energies of interaction between the various amine analogs and the oxidized and reduced forms of the enzymes and it was concluded that *bis*-onium compounds became bound end-on to the enzyme at a negatively charged substrate binding site some 0.6 to 0.9 nm from the uncharged oxidizing site. This confirmed earlier views resulting from substrate specificity studies[16,17] and did not support a mechanism requiring the inhibitor to bind in an extended manner to two negatively charged sites.

The picture of the substrate binding region in AO that emerges from these and related studies suggests that almost any water-soluble nitrogenous base will be able to act as a competitive inhibitor. The presence of sp^2 hybridized moieties as in guanidines, imidazoles, isothiuronium compounds, etc. will usually result in stronger inhibition, often of a noncompetitive type.

III. SLOW INHIBITION BY ENZYME INACTIVATORS

Some important modifiers of enzyme action are reagents that react chemically with the enzyme leading to partial or total loss of activity as a result of covalent bond formation. Recovery of enzyme activity on dialysis is usually slow, sometimes too slow to measure depending upon the reagent used. However, experiments with these reagents can give strong indications concerning the nature of groups involved in the catalysis. In the case of AO the carbonyl group reagents such as phenylhydrazine, hydroxylamine, semicarbazide, or aminoguanidine fall into this category, being reagents that lead to loss of activity due to a scheme such as:

$$\text{E–CHO} + H_2NHNR \underset{k_{-1}}{\overset{k_1}{\rightleftharpoons}} X \underset{k_{-2}}{\overset{k_2}{\rightleftharpoons}} \text{E–CH=N–HN–R} + H_2O \tag{6}$$

Here X is an enzyme-inhibitor complex or possibly a carbinolamine which is rapidly formed within seconds and then, with a slower time scale of minutes, the Schiff base is produced. However, a very slow solvolysis step (i.e., $1 >> k_{-2} > 0$ in Equation 6) may lead to recovery of enzyme activity. Although this is the classical explanation for inhibition by carbonyl group reagents, we shall see shortly that this interpretation has been both challenged[18,19] and defended recently[20,21] and must therefore be regarded as controversial at present.

Phenylhydrazine gives pronounced spectral changes during the reaction with AO while reagents such as semicarbazide, hydroxylamine, and aminoguanidine, which universally inhibit the copper AO, do not produce such strongly characteristic chromophores, but have been used merely to reinforce the belief that the copper AOs have active-site carbonyl groups. For instance, titration studies with phenylhydrazine have been reported for the AO from kidney,[20] plasma,[22-24] pea seedling,[25] and *Aspergillus niger*.[26] For the interpretation of results obtained with phenylhydrazine or cycloserine,[20,27] there has been a tendency to identify the chromophore as resulting from the presence of a cofactor such as pyridoxal phosphate at the active site. However, despite determined attempts, pyridoxal phosphate has not yet been unambiguously identified by chemical or microbiological techniques.

One way to formulate Equation 3 for the inhibition of AO by carbonyl group reagents is the method of time-dependent inhibition or affinity labeling where enzyme is incubated with inhibitor; the reaction is then quenched by the addition of substrate and activity is measured as a function of time. Over short periods of time the inactivation can sometimes be approximately modeled as a first-order process but AOs reach equilibrium positions due to the reverse solvolysis reaction and this has stimulated a theoretical treatment of the origins of curvature and occurrence of horizontal asymptotes in the semilogarithmic plots.[27] According to this treatment, appropriate expressions for the affinity labeling of AO where $[I] = [RNHNH_2]$ are of the form:

$$\frac{v}{v_o} = \frac{k_{-2}[H_2O](k_1[I] + k_{-1})}{\lambda_1 \lambda_2} + \frac{k_1[I]k_2}{(\lambda_2 - \lambda_1)} \left\{ \frac{1}{\lambda_1} e^{\lambda_1 t} - \frac{1}{\lambda_2} e^{\lambda_2 t} \right\} \tag{7}$$

where λ_1 and λ_2 satisfy the characteristic equation:

$$\lambda^2 + (k_1[I] + k_{-1} + k_2 + k_{-2}[H_2O])\lambda + k_1[I](k_2 + k_{-2}[H_2O]) + k_{-1}k_{-2}[H_2O] \quad (8)$$

When k_1 and k_{-1} are large compared to k_2 and k_{-2}, then species X is in rapid equilibrium with ECHO and $RNHNH_2$. The fast relaxing exponential term in Equation 7 decays very rapidly and over the measurable time course the approach to equilibrium depends upon the single exponential: $\exp-\{k_2 + k_{-2}(1 + k_{-1}/k_1[I])\}$.

We used this approach for aminoguanidine and some other carbonyl group reagents with amine oxidases from hog kidney, human placenta, and human kidney,[15,29] and in all instances the semilogarithmic plots were nonlinear, a fact confirmed recently by others[30] and hence justifying a double exponential equation such as Equation 7 rather than a first order model. Interpretation of the nonlinearity in terms of a reversible reaction was confirmed by addition of pyruvate and other carbonyl compounds which promoted recovery of enzyme activity. It was observed that the inhibition was complicated by the fact that most of the carbonyl group reagents had an instantaneous as well as a time-dependent effect. For instance, at least two molecules of aminoguanidine combined with the AO, giving curvature (i.e., at least quadratic terms) in the replots of double reciprocal plots, and this instantaneous reversible inhibition was taking place at the same time as the time-dependent approach to a final equilibrium position.

A number of possible artifacts should be borne in mind when studying time-dependent inhibition by active site reagents. For instance, the velocity is not strictly proportional to enzyme concentration with AO. These enzymes tend to aggregate[31,32] in strong solutions giving concave down $v([E_0])$ curves, and this could complicate the interpretation of v/v_0 data. Results with hydrazines have to be interpreted with caution since phyenylhydrazine is unstable,[33] and some hydrazines have been shown to act as substrates and are oxidized by AO.[34] Also, there have been other controversial results as in the case of SSAO where titration of one active site per mole of enzyme with phenylhydrazine seems to be established now,[22,33] although three had been claimed previously.[35]

Perhaps the best-known traditional examples of affinity labelings with other enzymes are for those with active site imidazoles, serines, and cysteines. Classically, the copper AOs were not reported to be sensitive to reagents that covalently modified $-SH$, $-OH$, or $-NH_2$ groups. However, exceptions are known, such as the *Aspergillus niger* AO.[36] Recently it has become clear that the BSAO, though insensitive to iodoacetate and iodoacetamide, does have active site SH groups that are exposed after reacting with substrate and may be involved in the reaction mechanism.[37] The fact that two essential $-SH$ groups in AO from *Euphorbia characias* latex are unreactive in the native enzyme but can be readily titrated by 4,4'-dithiodipyridine after denaturation or anaerobically in the presence of amine substrate[38] is another example. Similarly, active site histidines in bovine plasma, *Aspergillus niger*, and pea seedling AO were suggested by reaction with ethoxyformic acid anhydride and photooxidation of the plasma AO in the presence of Rose Bengal.[39] So it seems now that the classical results with such reagents as iodoacetate, iodoacetamide, NEM, pCMB, etc. remain true, but recent advances confirm that active site groups with copper AO may be rather more difficult to label than with some other enzymes. Clearly a searching reexamination of this area is required.

IV. THE EFFECT OF CHELATING AGENTS

The presence of copper in AO has been established by both chemical and spectroscopic analysis and strong confirmation of the necessity of copper for enzyme activity came from the fact that incubation with DDC gave precipitation of the copper chelate complex and loss of enzyme activity.[40] In many instances partial or complete recovery of enzyme activity was demonstrated after replacing the copper ions.

A theoretical treatment of a possible form for Equation 3 for use in these circumstances was developed, taking into account ligand substitution, chelate complex desorption, and possible dissociation of the enzyme into subunits.[41] When this was applied to the hog kidney[42] and human placental AOs,[43] it was found that the inhibitory effect of 1,10-phenanthroline, DDC, 8-hydroxyquinoline, 2,2′-bipridyl, and dithio-oxamide was very rapid in onset. The kinetic behavior was rather complex but was claimed to be consistent with the formation of chelate complexes in solution rather than simple ligand substitution. Stability constants were then estimated but one important line of argument used was based upon the nonlinearity of replots of slopes and intercepts in the presence of the chelating agents. Since this was also subsequently found with other nonchelating agents and since more detailed investigations have shown that the double reciprocal plots are not really linear, it now seems that a rapid ligand substitution reaction giving instantaneous inhibition followed by slow dissociation of ligand-copper complex, particularly at high concentration of chelating agent cannot be ruled out.

Ligands such as azide or cyanide seem to undergo ligand substitution reactions. They can be regarded as chelating agents for copper at the active site in that they seem to displace water from coordination sites with copper, and this has been extensively studied with the pig plasma enzyme.[44,45] A surprising but interesting result obtained by Pettersson[46] was that cuprizone, a reagent previously thought of as a straightforward chelating agent, was actually inhibiting SSAO by acting as a carbonyl group reagent without extracting any copper. In the light of this experiment and the fact that many chelating agents are chemically reactive substances or can act as substrate analogs, it is clear that precise interpretation of binding of kinetic data with chelating agents presents special problems.

V. INHIBITION BY EXCESS OF SUBSTRATE

It has been suggested that many enzymes would show deviation from Michaelis-Menton kinetics if v(S) data were obtained over a wide range of substrate concentration[11] and analyzed by computer fitting[47] and statistical tests.[48] AOs are no exception to this generalization; indeed, study of these enzymes was partly responsible for the formulation of this point of view. AOs from hog kidney, human placenta, human kidney,[49] and pig plasma[8,47] have been shown to give nonlinear double reciprocal plots if extended ranges of substrate concentration are used. The deviations could be due to the presence of two cooperatively linked active sites per enzyme molecule since most AOs have more than one subunit. Such deviations are unlikely to be due entirely to mixtures of alternative enzyme forms or enzyme polymerization but such effects could well contribute to the kinetic complexity of AO. However, all AOs do show one marked deviation from Michaelis-Menton kinetics: substrate inhibition, i.e., inhibition by excess substrate, which is more pronounced with some substrates than others.[6,17,50-52]

The form of Equation 3 necessary to study this phenomenon is a rational function[53] such as:

$$\frac{v}{E_o} = \frac{a_1[S] + a_2[S]^2 + \ldots + a_n[S]^n}{b_o + b_1[S] + b_2[S]^2 + \ldots + b_n[S]^n}; \quad a_i, b_j \geq 0 \tag{9}$$

where the a_i and b_j are calculated by well known rules, S is the amine (other effectors being fixed), n is the stoichiometry of the reaction involving the substrate varied, and $a_n b_{n-1} < a_{n-1} b_n$. A review of substrate inhibition and a new theoretical treatment have recently appeared[54] and attempts have been made to interpret the phenomenon with SKDAO.[17,50,51] With this enzyme, as with most copper AOs, substrate inhibition is more pronounced for substrates having delocalized π - electron systems such as isothiuronium, guanidinium, or imidazole groups. It tends to be much weaker with monoamines. Using Equation 9 with $n = 2$ and $a_2 = 0$, inhibition binding constants were estimated and the dead-end inhibition was interpreted in terms of a specific inhibitory binding site which could be copper[55] or a site near to copper.[56]

Substrate inhibition has usually been recorded with amine as the varied substrate, but it also occurs with oxygen varied as, for instance, with HKDAO.[57] Some caution is needed if the rate is measured spectrophotometrically using benzylamine derivatives since concentrations of amine over 5 m*M* can cause spectral changes with aldehyde products due to Schiff base formation.[8] This artifact can be minimized by following the reaction at 240 rather than 250 nm, but it is better to confirm the presence of substrate inhibition independently by oxygen uptake.

The presence of substrate inhibition in a kinetic curve is often regarded as a nuisance and is simply dismissed as being an artifact occurring outside the range of physiological concentration. This attitude cannot be accepted uncritically with AO. It is possible that local concentrations of substrates could become sufficiently high to make the phenomenon of regulatory significance. So eventually, some satisfying explanation must be advanced to explain why the phenomenon is much more pronounced with compounds such as histamine than with others (such as monamines) and experiments should be performed to see if the phenomenon has any biological significance.

VI. INHIBITION BY PRODUCTS

In any enzyme-catalyzed reaction, it is arbitrary which reactants are called substrates and which are called products since reactions are facilitated in both directions. Accordingly, addition of either H_2O_2, RCHO, or NH_3 will slow the reaction down at fixed concentrations of RCH_2NH_2 and O_2 by backing up the reaction sequence. The possible effects of products on the double reciprocal plots given by substrates can then be used to decide between possible molecular mechanisms.[12] The products are used at several different fixed concentrations and at each concentration the slope and intercept of a double reciprocal plot with amine or O_2 varied are measured. Replots of these slopes and intercepts as functions of the product concentrations then confirm the type of mechanism as being, say, ordered or ping pong and suggest the order of addition of substrates and release of products.

This technique has been used with considerable success to establish ping pong mechanisms for the AOs from bovine plasma,[58] pig plasma,[59] pea seedling,[27] hog kidney,[60,61] and human placenta.[43] The results obtained are probably good first approximations to reality but it must be borne in mind that AOs do not actually obey Michaelis-Menten kinetics and furthermore there are considerable experimental difficulties involved in performing product inhibition studies with AOs. It is usually necessary to employ a direct, continuous spectrophotometric method with benzylamine or *p*-dimethylaminomethylbenzylamine[62,63] as substrates in order to obtain a sufficient quantity of data and this then limits the range of aldehyde product that can be used due to the high background absorbance. Also, it is difficult to vary the oxygen concentration accurately over a sufficiently wide range for model discrimination, and artifacts can occur such as the substrate inhibition referred to previously, the time-dependent destruction of enzyme by H_2O_2[64] or complex inhibition patterns with NH_3,[8,59] which has to be used in high concentration since it is a rather weak inhibitor.

VII. SUICIDE INHIBITION

There is one characteristic difference between inhibitors in this category and the affinity labeling compounds discussed earlier: suicide inhibitors have to be altered by enzyme catalysis before they can react covalently with the enzyme leading to inactivation. Some of these reagents are converted into products that are first released from the enzyme but then reassociate, perhaps at some site other than the active site. However, many products can react with the enzyme directly they are formed and often there is a partition between the two possible outcomes: dissociation of product or enzyme inactivation. In all instances the inhibition is time-dependent, substrate protection can often be demonstrated by nonsuicide substrates, and the inactive enzyme-product complex involves covalent bond formation. Mathematical expressions for Equation 3 can be obtained as infinite series but, as these are of little value experimentally, simplified treatments involving steady state or first order kinetic assumptions are employed.

Rando and De Mairena[65] studied the inhibition of beef plasma and hog kidney AO by propargylamine (1-amino-2-propyne). They demonstrated an inactivation with a time scale of minutes which was diminished by substrate protection with benzylamine and abolished by prior inactivation of the enzyme following copper removal. Interpretation involved imine formation, isomerization of the acetylene to a reactive allene, and finally irreversible inactivation by nucleophilic attack on the allene involving unknown groups on the enzyme. Abeles and co-workers then reported a number of detailed studies concerning the use of suicide inhibitors to probe the active site chemistry with BSAO.

They demonstrated[66] that 2-chloroallylamine, 2-butynylamine, and 2-propynylamine (propargylamine) were oxidized with concomitant inactivation of the enzyme, but allylamine just behaved as a substrate. Protection from inactivation was found using benzylamine and covalent modification of the enzyme was established using tritiated 2-propynylamine. The results were again interpreted as being due to allene formation followed by addition of nucleophile from the enzyme to form a covalent adduct. However, the findings with 2-chloroallylamine were explained by base-catalyzed elimination of an ∝-methylene proton as follows:

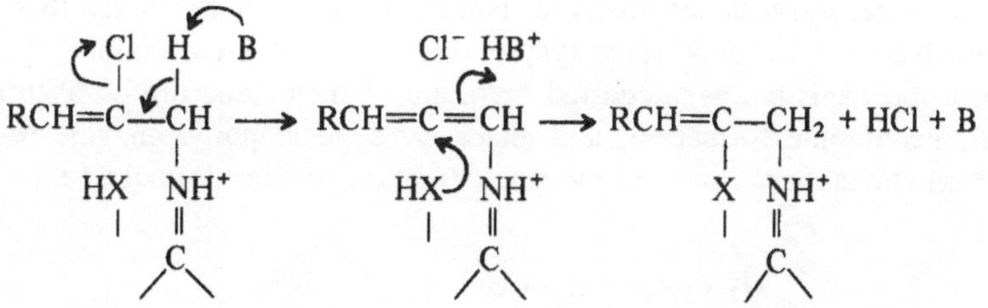

Subsequently,[67] it was shown that glycine esters with good leaving groups could act as suicide substrates but other glycine esters were just substrates. Proton abstraction from the ∝-methylene followed by elimination from the carbanion to give ketenes was invoked as in the sequence:

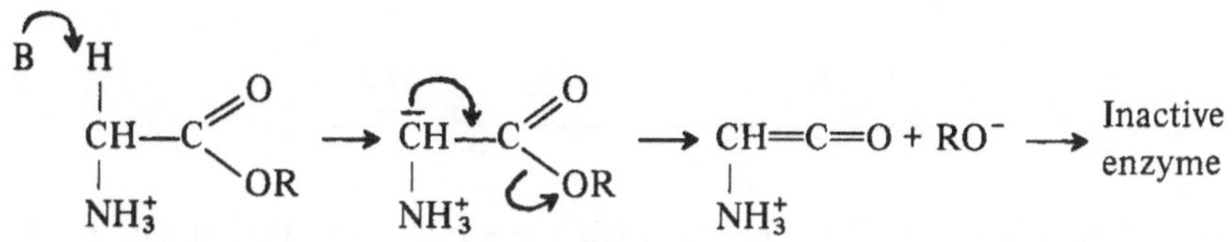

Aminoacetonitrile, a well-known lathyrogen, was not a substrate but it caused covalent inactivation as shown by carbon 14 labeling and confirmed by substrate protection. Proton abstraction followed by ketenimine formation was considered to be a likely explanation. The elimination of HCl from β-chlorophenylethylamine to produce phenylacetaldehyde was at a rate comparable to normal oxidation,[18] again supporting a proton abstraction mechanism. In addition, it was proved that in approximately one out of every nine turnovers the β-bromoacetaldehyde resulting from β-bromoethylamine led to enzyme inactivation by alkylation. Serious doubts were expressed about the carbonyl nature of the active site and the –S–S– system was suggested as an electron acceptor. This theme was further developed[19] in a paper containing evidence that imine formation occurs first but that only the aldehydes are released from the enzyme. It was shown that reduction of the enzyme makes two –SH groups available to NEM, a sulfenic acid mechanism proposed earlier for the active site was criticized and it was demonstrated that inactivation by β-bromoethylamine led to nonspecific labeling of several groups, including some cysteine residues.

The suicide inhibitors just discussed are largely synthetic compounds that have been specially selected to exploit elimination reactions that can occur with good leaving groups following proton abstraction. It is therefore surprising to note that suicide inhibition has been reported by Kagan et al.[68] using lysyl oxidase and simple mono- and diamine substrates. Product inhibition by ammonia, aldehyde, or H_2O_2 was eliminated as a possible explanation and [1,5-^{14}C]-diaminopentane covalently labeled the enzyme. The kinetics were rather complicated and were interpreted as discounting a simple partitioning effect between productive catalysis and a single inactivation mechanism. It was estimated that something like 100 catalytic turnovers occurred before inactivation and the protection afforded by DMSO, mannitol, and tryptophan was advanced as evidence implicating hydroxyl radicals, OH•, as one possible cause.

The mode of action of the potent lathyrogen β-aminopropionitrile (BAPN) on lysyl oxidase is now much clearer and seems to belong in the category of suicide inhibition. It is known that BAPN is not oxidized to the cyanoaldehyde during the inactivation, and labeling studies have demonstrated that the intact carbon chain of BAPN is incorporated into the covalently labeled enzyme.[69] Pretreatment with borohydride or phenylhydrazine decreased the extent of label incorporation, suggesting the involvement of a Schiff base and it was established that the copper extraction was not involved. Not surprisingly, it was noted that BAPN was also a competitive inhibitor and, since lysyl oxidase is also suicide inhibited by β-bromoethylamine, a mechanism was suggested involving α-methylene proton abstraction, prototropic shift, ketenimine formation, and attack by a nucleophilic enzyme residue X. If Schiff base formations were involved the final structure produced would be:

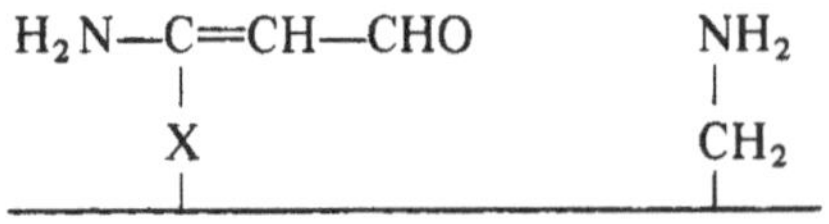

but in the absence of Schiff base formation the structure would be:

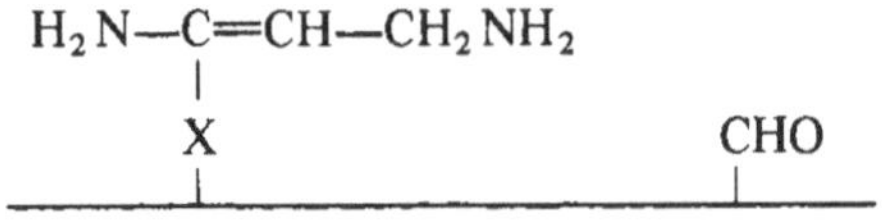

Suicide inhibition of the SKDAO has now been reported using allylamine, aminoacetonitrile, propargylamine, *cis*-imidazolylallylamine, and also 1,4-diamino-2-butyne,[70,71] and it seems that rapid developments can be expected in this area by design-directed synthesis of new compounds. It remains to be seen whether suicide inhibition by natural substrates or compounds incapable of elimination to give reactive species has been overlooked in the past. Also, it has yet to be established whether this phenomenon has a biological function in the case of copper AO or whether it is merely a chemical curiosity. Whatever the outcome of this will be, the demonstration of proton abstraction from the α-methylene group and associated possibilities for further reactions has opened up a fruitful new area in AO research.

VIII. IN VIVO EFFECTS OF SOME MISCELLANEOUS AO INHIBITORS

There have been numerous studies dealing with the in vivo effects of AO inhibitors. Usually, the aim has been to employ aminoguanidine as a supposedly specific inhibitor of kidney DAO in order to diminish diamine or histamine metabolism or to use lathyrogens such as aminoacetonitrile or BAPN to inhibit lysyl oxidase and impair collagen cross-linking. Incubation of polyamines with plasma AO leads to the formation of cytotoxic aminoaldehydes and acrolein and attempts are often made to block this using AO inhibitors. However, these views on the supposed specific mode of action of AO inhibition have often been reinforced by repetition rather than established by definitive experiments. In order to analyze the in vivo effects of an AO inhibitor it is necessary to know whether the agent prevents synthesis of AO or merely inhibits prexisting AO. Also, other parameters such as the permeability properties, elimination kinetics, metabolism, and effect on other enzyme, hormone, or transport systems should be defined. Further, it is now known that a large number of miscellaneous compounds can act as copper AO inhibitors and so it is important to first consider this subject.

The copper AO are inhibited by a number of substances that have not always been clearly fitted into any of the previous schemes. In many instances, it has not yet been established whether the inhibition is reversible, time-dependent, substrate protected, etc. and the experimental evidence necessary to make such decisions has not always been reported. The classical view of a clear distinction between MAO and DAO inhibitors requires qualification since it was shown that hog kidney and human placental AO were inhibited to some extent by tranylcypromine, harmine, isocarboxazid, nialamide, iproniazid, phenelzine, mebanazine, isoniazid, and pargyline.[72] Some of these MAO inhibitors are substrate analogs, some could be regarded as potential time-dependent inhibitors, and others are capable of more complex behavior, e.g., iproniazid, which can act instantaneously as a substrate analog but produces a carbonyl group reagent on hydrolysis or isoniazid, which can act as a substrate analog, carbonyl group reagent, and copper chelating reagent. Bulbocapnine[73] and numerous alkaloids[74] could be thought of as substrate analogs while hydroxybenzyloxamines[75] and glycine[76] are probably carbonyl group reagents. Inhibition of lysyl oxidase by estrone sulfate[77] or fatty acids[78] seem to be examples of nonspecific modifier action while antibodies against SKDAO gave competitive inhibition.[79] A distinctive PAO had been claimed in human pregnancy plasma since inhibition was produced by quinacrine,[80] a known inhibitor of the flavin AO of rat liver.[81] However, quinacrine is also a good inhibitor of copper AO[82,83] and it seems now that this PAO is just the well-characterized pregnancy plasma AO.[84] The sort of artifacts that can easily arise in complex systems are well illustrated by a study of the inactivation of PSAO that occurs in the presence of 1,5-diaminopentane and peroxidase.[85] It was shown that pyridoxal and unrelated compounds, including oxoglutarate, catechol, or *o*-aminobenzaldehyde, afforded protection from this inactivation and thus previous evidence for the supposed presence of pyridoxal phosphate was weakened by this result. Any ionizable substance can change the ionic strength of the assay medium at higher concentrations and it is known that the activation curve for SKDAO shows first a rise and then a steady decline as salt concentration increases,[61] an effect that depends on the nature of the substrate being used.[86] It has usually been supposed that 8-hydroxyquinoline inhibited AO by acting as a chelating agent but *Aspergillus niger* AO was reversibly inhibited without removal of copper.[87] The noncompetitive, instantaneous inhibition of the reaction was attributed to complex formation, possibly by ligand substitution at the copper, with the reduced form of the enzyme since the oxidative step was not affected. As a final example of the relatively nonspecific inhibition of AO, it should be pointed out that we studied 21 miscellaneous drugs used in pregnancy and all of them showed a measurable degree of inhibition of placental AO.[72]

There have been some good examples of in vivo plant studies. For instance, Macholán and Minar[88] performed a study of the effect of 1,4-diamino-2-butanone and 1,5-diamino-3-pentanone on pea seedlings. The compounds did not affect germination but they were absorbed and in relatively high concentrations delayed the growth of seedlings. It was shown that synthesis of AO was not depressed since most of the AO activity could be restored by dialysis. The results were interpreted as indicating that AO does not play an important part in auxin biosynthesis. The PAO of barley leaves was inhibited by polyamine analogs but methylglyoxal-*bis*(guanylhydrazone) had only a slight effect on spermine oxidation.[89] Aliphatic monoguanidines with a chain length of 8 to 10 produced the greatest inhibition of growth with cress, barley, and oat seedlings, but chain lengths of 10 to 14 were maximal in apple cell suspension cultures.[90] An investigation of the effect of an homologous series of mono- and diamines and guanidines on membrane permeability has been reported using the efflux of betacyanin from beet root discs and loss of total ions from beet root and swede discs, beet and spinach leaf discs, and apple cells in suspension culture.[91]

As typical examples of the use of AO inhibitors in mammalian systems, we could consider some studies by Sourkes et al.[92] using rat as the experimental animal. They showed that tranylcypromine, chlorgyline, and deprenyl inhibited the metabolism of *n*-pentylamine but not putrescine. Aminoguanidine inhibited the oxidation of putrescine only while iproniazid and isoniazid inhibited the metabolism of both mono- and diamine. Pargyline also inhibited putrescine oxidation. Subsequently, the kinetics of recovery from aminoguanidine inhibition was investigated,[93] and in vitro and in vivo inhibition of AO by the antimalarial drugs amodiaquine, quinacrine, and chloroquine was demonstrated.[82] Aminoguanidine has also been used to show that polyamine synthesis is increased in the rat during pregnancy but, due to the increase in placental AO, polyamine excretion does not rise.[94] The formation of amino acids from spermidine and spermine has also been studied using aminoguanidine.[95] Pargyline had in vivo but not in vitro effects and it was pointed out that this compound is an inhibitor of aldehyde dehydrogenase as well as flavin AO. Earlier it had been shown that pargyline can lead to abortion[96] but the mechanism of action was not clarified.

Interest in polyamines is very intense at present and AO inhibitors are often used in this field. Aminoguanidine, for example, has no effect on proliferation of human fibroblasts in culture but the inhibition brought about by polyamines was reversed, presumably because inhibitors of copper AO prevented formation of cytotoxic compounds by oxidation of polyamines.[97] In this context, an investigation reported by Pegg and McGill should be noted.[98] They showed that 1,1′-[(methylethanediylidene)-dinitrilo]-diguanidine (i.e., methylglyoxal *bis*-(guanylhydrazone) and 1,1′-[(methylethanediylidene)-dinitrilo]-*bis*(3-aminoguanide), two compounds currently used to inhibit spermidine formation, were more potent inhibitors of copper AO than polyamine synthesis. Not surprisingly, the compounds showed instantaneous effects as substrate analogs and also slower time-dependent inhibition.

Some other compounds that have recently been used as AO effectors include: folate antagonists as inhibitors of histamine-metabolizing enzymes,[99] phytic acid in germinating pea seeds,[100] methylguanide with kidney AO,[101] brocresine in plasma, liver, stomach, and intestine of rats,[102] an exogenous AO inhibitor in amniotic fluid,[103] an inhibitory factor in guinea pig skin,[104] alkaloids of Argemone mexicans-berberine and sanguinarine,[105] cycloheximide and heparin in rat tissue,[106] guanethidine and imidazole and histamine metabolism,[107] 3-hydrazinopyridazine derivatives with pyridoxal dependent enzymes and copper AO,[108,109] impromidine,[110] antihistaminic drugs,[111] a new H-2 receptor agonist dimaprit,[112] and nitrofuran derivatives.[113] It is known that spermatozoa are inhibited by the products of oxidation of polyamines by AO found in the semen and in the spermatozoa.[114] Levamisole, an antihelmintic drug, reduces sperm motility and it has been shown that semen AO is inhibited by this compound.[115] This result suggested an investigation of the effect of *bis*-onium compounds and other substrate analogs on sperm motility. Many of these substrate analogs have ganglion blocking properties but a novel result with sperm was that although many of these compounds simply decreased motility, the isothiuronium compounds induced a circular swimming motion.

Comparative studies of plant and mammalian AO using substrate analogs, imidazoles, carnosine, and anserine have led to the conclusion that the mammalian enzymes are more adapted to histamine oxidation and the plant enzymes for diamine oxidation.[116-118] The imidazoles inhibited mammalian more than plant AO and histamine receptor antagonists were powerful inhibitors of the mammalian enzymes. It was pointed out that diamine analogs and BAPN were more potent towards plant than mammalian AO but hydrazine and guanidine derivatives had comparable effects with both classes of enzymes.[30] Finazzi-Agrò et al. have confirmed that the pig kidney enzyme can be inhibited by a wide variety of compounds such as tryptamine, serotonin, phenformin, guanidine, amitryptiline, cimetidine, Br-ethylamine, 2-aminopyridine, benzophenonhydrazone, β-phenylethylhydrazine, and phenipyrazine.[111] The time-dependent behavior was not clarified but they found no inhibition at all with pargyline and no intercept effect with guanidine, thus contradicting earlier claims.[15,72] Some of the contradictions concerning inhibitor specificity reported in the literature could be due to the use of impure enzyme or reagents or different assay techniques. Another example of this is the oxidation of *p*-nitrobenzylamine by SKDAO,[55,119] which we could not repeat and which could not be demonstrated with the pea seedling enzyme.[71] Controversies such as these should be relatively easy to clear up with homogeneous reagents and proper statistical analysis of data. Now that a wide variety of histamine receptor antagonists are available and a number of promising results have been obtained, it would seem that a detailed study using pure reagents and careful statistical analysis could be very rewarding.

IX. CONCLUSIONS

The first phase in AO research was characterized by a rapid advance on a broad front leading to the identification of AO activity in a number of biological systems and a rough outline of cofactor requirement and substrate and inhibitor specificities.[1,2] Towards the end of this period we could identify important changes in emphasis with better purification procedures and the application of physical techniques. We are now in a second phase where physical techniques are becoming even more penetrating, affinity chromatography and monoclonal antibodies present new possibilities in the fields of protein identification, and purification and deeper understanding of complex kinetics is possible with the wider availability of computers for statistical analysis and modeling.

In this chapter I have sought to define the various ways that inhibitors of copper AO can bring about their effects and I have also attempted to identify those areas where success has recently been achieved and those where definitive experiments have yet to be performed. It now seems that, although the copper AO differ widely in substrate specificity, the inhibitor specificities do not vary so much. It remains true that they are all inhibited by copper-chelating and carbonyl group reagents, but also a large number of miscellaneous compounds are now known to be inhibitory. Sometimes, as with aminoguanidine, it is clear that rapid reversible and also slower time-dependent effects are operating simultaneously and suicide substrates and recent developments in affinity labeling of resistant active site groups promise a more detailed understanding of active site chemistry. In due course it should be possible to state precisely in molecular terms how any particular inhibitor brings about its effect: first in relatively simple conditions with isolated, purified enzymes, but ultimately in the more complicated context of intact biological systems.

ACKNOWLEDGMENTS

I would like to thank my research colleagues R. W. Lobley, C. M. Hill, J. S. Ashford, M. J. C. Crabbe, J. S. Shindler, R. E. Childs, R. D. Waight, and I. V. Scott for the stimulating times we shared together researching into amine oxidases.

REFERENCES

1. **Kapeller-Adler, R.,** *Amine Oxidases and Methods for Their Study*, John Wiley & Sons, New York, 1970.
2. **Costa, E. and Sandler, M., Eds.,** *Advances in Biochemical Psychopharmacology*, Vol. 5, Raven Press, New York, 1972.
3. **Yasunobu, K. T., Ishizaki, H., and Minamiura, N.,** The molecular, mechanistic and immunological properties of amine oxidases. *Mol. Cell. Biochem.*, 15, 3, 1976.
4. **Solano-Muñoz, F., Bardsley, W. G., and Indge, K. J.,** The probability that complex enzyme kinetic curves can be caused by activators or inhibitors, *Biochem. J.*, 195, 589, 1981.
5. **Massey, J. B. and Churchich, J. E.,** Effector binding sites in plasma amine oxidase, *Biochim. Biophys. Acta*, 480, 70, 1977.
6. **McEwen, C.M., Jr. and Sober, A.J.,** Rabbit serum monamine oxidase II determinants of substrate specificity, *J. Biol. Chem.*, 242, 3068, 1967.
7. **Quash, G., Calogero, H., Fossar, N., Ferdinand, A., and Taylor, D.,** Modification of diamine oxidase activity in vitro by metabolites of asparagine and differences in asparagine decarboxylation in normal and virus-transformed baby hamster kidney cells, *Biochem. J.*, 157, 599, 1976.
8. **Kelly, I. D., Knowles, P. F., Yadav, K. D. S., Bardsley, W. G., Leff, P., and Waight, R. D.,** Steady-state kinetic studies on benzylamine oxidase from pig plasma, *Eur. J. Biochem.*, 114, 133, 1981.
9. **Neufeld, E. and Chayen, R.,** An evaluation of three methods for the measurement of diamine oxidase (DAO) activity in amniotic fluid, *Anal. Biochem.*, 96, 411, 1979.
10. **Macholán, L.,** Selective and reversible inhibition of diamine oxidase by 1,5-diamino-3-pentanone, *Coll. Czech. Chem. Commun.*, 39, 653, 1974.
11. **Hill, C. M., Waight, R. D., and Bardsley, W. G.,** Does any enzyme follow the Michaelis-Menton equation?, *Mol. Cell. Biochem.*, 15, 173, 1977.
12. **Cleland, W. W.,** *The Enzymes*, 3rd ed., Boyer, P.D., Ed., Academic Press, New York, 1970, 2.
13. **Bardsley, W. G. and Ashford, J. S.,** Inhibition of pig kidney diamine oxidase by substrate analogues, *Biochem. J.*, 128, 253, 1972.
14. **Crabbe, M. J. C. and Bardsley, W. G.,** The inhibition of human placental diamine oxidase by substrate analogs, *Biochem. J.*, 139, 183, 1974.
15. **Shindler, J. S. and Bardsley, W. G.,** Human kidney diamine oxidase: inhibition studies, *Biochem. Pharmacol.*, 25, 2689, 1976.
16. **Bardsley, W. G., Hill, C. M., and Lobley, R. W.,** A reinvestigation of the substrate specificity of pig kidney diamine oxidase, *Biochem. J.*, 117, 169, 1970.
17. **Bardsley, W. G., Ashford, J. S., and Hill, C. M.,** Synthesis and oxidation of aminoalkyl-onium compounds by pig kidney diamine oxidase, *Biochem. J.*, 122, 557, 1971.
18. **Neumann, R., Hevey, R., and Abeles, R. H.,** The action of plasma amine oxidase on β-haloamines, *J. Biol. Chem.*, 250, 6362, 1975.
19. **Suva, R. H. and Abeles, R. H.,** Studies on the mechanism of action of plasma amine oxidase, *Biochemistry*, 17, 3538, 1978.
20. **Finazzi-Agrò, A., Guerrieri, P., Costa, M. T., and Mondoví, B.,** On the nature of chromophore in pig kidney diamine oxidase, *Eur. J. Biochem.*, 74, 435, 1977.
21. **Lindström, A. and Pettersson, G.,** The order of substrate addition and product release during the catalytic action of pig-plasma benzylamine oxidase, *Eur. J. Biochem.*, 84, 479, 1978.

22. **Lindström, A. and Pettersson, G.,** Active-site titration of pig-plasma benzylamine oxidase, *Eur. J. Biochem.*, 83, 131, 1978.
23. **Yamada, H. and Yasunobu, K. T.,** Monoamine oxidase. IV. Nature of the second prosthetic group of plasma monoamine oxidase, *J. Biol. Chem.*, 238, 2669, 1963.
24. **Rucker, R. B. and Goettlich-Riemann, W.,** Purification and properties of sheep plasma amine oxidase, *Enzymologia*, 43, 33, 1972.
25. **Nylén, U. and Szybeck, P.,** Kinetics and other characteristics of diamine oxidase of pea seedling, *Acta Chem. Scand*, B28, 1153, 1974.
26. **Suzuki, H., Ogura, Y., and Yamada, H.,** Stoichiometry of the reaction by amine oxidase from *Aspergillus niger*, *J. Biochem.*, 69, 1065, 1971.
27. **Suzuki, Y. and Yamasaki, K.,** Inhibition of pea diamine oxidase by cycloserine, *Enzymologia*, 35, 198, 1968.
28. **Childs, R. E. and Bardsley, W. G.,** Time dependent inhibition of enzymes by active-site directed reagents. A theoretical treatment of the kinetics of affinity labelling, *J. Theor. Biol.*, 53, 381, 1975.
29. **Crabbe, M. J. C., Childs, R. E., and Bardsley, W. G.,** Time-dependent inhibition of diamine oxidase by carbonyl-group reagents and urea, *Eur. J. Biochem.*, 60, 325, 1975.
30. **Biegański, T., Osińska, Z., and Maśliński, C.,** Inhibition of plant and mammalian diamine oxidases by hydrazine and guanidine compounds, *Int. J. Biochem.*, 14, 949, 1982.
31. **Pionetti, J. M.,** Analytical-band centrifugation of the active form of pig kidney diamine oxidase, *Biochem. Biophys. Res. Commun.*, 58, 495, 1974.
32. **Kagan, H. M., Sullivan, K. A., Olsson, T. A., III, and Cronlund, A. L.,** Purification and properties of four species of lysyl oxidase from bovine aorta, *Biochem. J.*, 177, 203, 1979.
33. **Falk, M. C.,** Stoichiometry of phenylhydrazine inactivation of pig plasma amine oxidase, *Biochemistry*, 22, 3740, 1983.
34. **Húcko-Haas, J. E. and Reed, D. J.,** Hydrazines as substrates for bovine plasma amine oxidase (PAO), *Biochem. Biophys. Res. Commun.*, 39, 396, 1970.
35. **Buffoni, F. and Ignesti, G.,** Active-site titration of pig plasma benzylamine oxidase with phenylhydrazine, *Biochem. J.*, 145, 369, 1975.
36. **Suzuki, H., Ogura, Y., Yamada, H., and Arima, K.,** Studies on sulfhydryl groups of *Aspergillus niger* amine oxidase, *Biochem. Biophys. Acta*, 403, 23, 1975.
37. **Zeidan, H., Watanabe, K., Piette, H., and Yasunobu, K. T.,** Electron spin resonance studies of bovine plasma amine oxidase. Probing the environment about the substrate-liberated sulfhydryl groups in the active site, *J. Biol. Chem.*, 255, 7621, 1980.
38. **Floris, G., Giartosio, A., and Rinaldi, A.,** Essential sulfhydryl groups in diamine oxidase from *Euphorbia characias* Latex, *Arch. Biochem. Biophys.*, 220, 623, 1983.
39. **Tsurushiin. S., Hiramatsu, A., Inamasu, M., and Yasunobu, K. T.,** The essential histidine residues of bovine plasma amine oxidase, *Biochim. Biophys. Acta*, 400, 451, 1975.
40. **Mondovì, B., Rotilio, G., Finazzi Agrò, A., and Antonini, E.,** Amine oxidases: a new class of copper oxidases, in *Magnetic Resonances in Biological Research*, Franconi, C., Ed., Gordon & Breach, 1971, 233.
41. **Bardsley, W. G. and Childs, R. E.,** Inhibition of enzymes by metal ion-chelating reagents. Theory and new graphical methods of study, *Biochem. J.*, 137, 55, 1974.
42. **Bardsley, W. G., Childs, R. E., and Crabbe, M. J. C.,** Inhibition of enzymes by metal ion-chelating reagents. The action of copper-chelating reagents on diamine oxidase, *Biochem. J.*, 137, 61, 1974.
43. **Bardsley, W. G., Crabbe, M. J. C., and Scott, I. V.,** The amine oxidases of human placenta and pregnancy plasma, *Biochem. J.*, 139, 169, 1974.
44. **Lindström, A., Olsson, B., and Pettersson, G.,** Effect of azide on some spectral and kinetic properties of pig plasma benzylamine oxidase, *Eur. J. Biochem.*, 48, 237, 1974.
45. **Barker, R., Boden, N., Cayley, G., Charlton, S. C., Henson, R., Holmes, M. C., Kelly, I. D., and Knowles, P. F.,** Properties of cupric ions in benzylamine oxidase from pig plasma as studied by magnetic-resonance and kinetic methods, *Biochem. J.*, 177, 289, 1979.
46. **Lindström, A. and Pettersson, G.,** The mechanism of inhibition of pig-plasma benzylamine oxidase by the copper-chelating reagent Cuprizone, *Eur. J. Biochem.*, 48, 229, 1974.
47. **Bardsley, W. G., Leff, P., Kavanagh, J., and Waight, R. D.,** Deviations from Michaelis-Menten kinetics. The possibility of complicated curves for simple kinetic schemes and the computer fitting of experimental data for acetylcholinesterase, acid phosphatase, adenosine deaminase, arylsulphatase, benzylamine oxidase, chymotrypsin, fumarase, galactose dehydrogenase, β-galactosidase, lactate dehydrogenase, peroxidase and xanthine oxidase, *Biochem. J.*, 187, 739, 1980.
48. **Burguillo, F. J., Wright, A. J., and Bardsley, W. G.,** Use of the F test for determining the degree of enzyme-kinetic and ligand-binding data. A Monte Carlo simulation study, *Biochem. J.*, 211, 23, 1983.
49. **Scott, I. V., Childs, R. E., Crabbe, M. J. C., Tindall, V. R., and Bardsley, W. G.,** *Clinical Enzymology Symposia*, Burlina, A. and Galzigna, I., Eds., Piccin Medical Books, Padua, 1977, 97.

50. **Hill, C. M. and Bardsley, W. G.,** Histamine and related compounds as substrates of diamine oxidase (histaminase), *Biochem. Pharmacol.*, 24, 253, 1975.
51. **Costa, M. T., Rotilio, G., Finazzi Agrò, A., Vallogini, M. P., and Mondovì, B.,** On the active site of diamine oxidase: kinetic studies, *Arch. Biochem. Biophys.*, 147, 8, 1971.
52. **McEwen, C. M., Jr.,** Human plasma monoamine oxidase. II. Kinetic studies, *J. Biol. Chem.*, 240, 2011, 1965.
53. **Bardsley, W. G. and Childs, R. E.,** Sigmoid curves, non-linear double-reciprocal plots and allosterism, *Biochem. J.*, 149, 313, 1975.
54. **Bardsley, W. G., Solano-Muñoz, F., Wright, A. J., and McGinlay, P. B.,** Inhibition of enzyme-catalysed reaction by excess substrate. A theoretical and Monte Carlo study of turning points in v(S) graphs, *J. Mol. Biol.*, 169, 597, 1983.
55. **Zeller, E. A.,** Diamine oxidase: specificity and mechanism, in *Advances in Biochemical Psychopharmacology*, Vol. 5, Costa, E. and Sandler, M., Eds., Raven Press, New York, 1972, 167.
56. **Kluetz, M. D. and Schmidt, P. G.,** Proton relaxation study of the hog kidney diamine oxidase active center, *Biochemistry*, 16, 5191, 1977.
57. **Shindler, J. S.,** An Investigation of some Enzymes from Human Tissues, *Ph.D. thesis*, University of Manchester, 1975.
58. **Oi, S., Inamasu, M., and Yasunobu, K. T.,** Mechanistic studies of beef plasma amine oxidase, *Biochemistry*, 9, 3378, 1970.
59. **Taylor, C. E., Taylor, R. S., Rasmussen, C., and Knowles, P. F.,** A catalytic mechanism for the enzyme benzylamine oxidase from pig plasma, *Biochem. J.*, 130, 713, 1972.
60. **Finazzi-Agrò, A., Rotilio, G., Costa, M. T., and Mondovì, B.,** Evidence for a ping-pong mechanism in the diamine oxidase reaction, *FEBS Lett.*, 4, 31, 1969.
61. **Bardsley, W. G., Crabbe, M. J. C., and Shindler, J. S.,** Kinetics of the diamine oxidase reaction, *Biochem. J.*, 131, 459, 1973.
62. **Bardsley, W. G., Crabbe, M. J. C., Shindler, J. S., and Ashford, J. S.,** Oxidation of *p*-dimethylaminomethylbenzylamine by pig kidney diamine oxidase. A new method for spectrophotometric assay, *Biochem. J.*, 127, 875, 1972.
63. **Bardsley, W. G., Crabbe, M. J. C., and Scott, I. V.,** A method for determining diamine oxidase (histaminase) in biological fluids, *Biochem. Med.*, 11, 138, 1974.
64. **Mondovì, B., Rotilio, G., Finazzi-Agrò, A., and Costa, M. T.,** Diamine oxidase inactivation by hydrogen peroxide, *Biochim. Biophys. Acta*, 132, 521, 1967.
65. **Rando, R. J. and De Mairena, J.,** Propargyl amine-induced irreversible inhibition of non-flavin-linked amine oxidases, *Biochem. Pharmacol.*, 23, 463, 1974.
66. **Hevey, R. C., Babson, J., Maycock, A. L., and Abeles, R. H.,** Highly specific enzyme inhibitors. Inhibition of plasma amine oxidase, *J. Am. Chem. Soc.*, 95(18), 6125, 1973.
67. **Maycock, A. L., Suva, R. H., and Abeles, R. H.,** Novel inactivators of plasma amine oxidase, *J. Am. Chem. Soc.*, 97(19), 5613, 1975.
68. **Kagan, H. M., Soucy, D. M., Zoski, C. G., Resnick, R. J., and Tang, S.-S.,** Multiple modes of catalysis-dependent inhibition and inactivation of aortic lysyl oxidase, *Arch. Biochem. Biophys.*, 221, 158, 1983.
69. **Tang, S.-S., Trackman, P. C., and Kagan, H. M.,** Reactions of aortic lysyl oxidase with β-aminopropionitrile, *J. Biol. Chem.*, 258, 4331, 1983.
70. **Tobes, M. C., Griffith, R. K., DiPietro, R. A., and Zoorob, H. H.,** Design and evaluation of mechanism-based inhibitors of diamine oxidase, *Fed. Proc.*, 42, 2042, 1983.
71. **Macholán, L.,** personal communication, 1983.
72. **Crabbe, M. J. C. and Bardsley, W. G.,** Monoamine oxidase inhibitors and other drugs as inhibitors of diamine oxidase from human placenta and pig kidney, *Biochem. Pharmacol.*, 23, 2983, 1974.
73. **Chapman, J. E. and Walaszek, E. J.,** Inhibition of diamine oxidase by bulbocapnine, *Biochem. Pharmacol.*, 11, 205, 1962.
74. **Peč, P. and Macholán, L.,** Inhibition of diamine oxidase by Cinchona alkaloids and lobeline, *Coll. Czech. Chem. Commun.*, 41, 3474, 1976.
75. **Gaugas, J. M.,** Potent inhibition of diamine oxidase with the hydroxybenzyloxamines NSD-1039, NSD-1531 and NSD-1024, *Biochem. Pharmacol.*, 30, 3121, 1981.
76. **Costa, M. T., Sabatini, S., Turini, P., and Mondovì, B.,** Inhibition of beef plasma amine oxidase activity by glycine, *FEBS Lett.*, 94, 112, 1978.
77. **Buffoni, F. and Della Corte, L.,** Pig plasma benzylamine oxidase, in *Advances in Biochemistry and Psychopharmacology*, Vol. 5, Raven Press, New York, 1972, 133.
78. **Kagan, H. M., Tseng, L., and Simpson, D. E.,** Control of elastin metabolism by elastin ligands, *J. Biol. Chem.*, 256, 5417, 1981.

79. **Oratore, A., Guerrieri, P., Ballini, A., Mondovì, B., and Finazzi-Agrò, A.,** Preparation and properties of antibody against pig kidney diamine oxidase, *FEBS Lett.*, 104, 154, 1979.
80. **Morgan, D. M. L.,** Polyamine oxidation and human pregnancy, in *Advances in Polyamine Research*, Vol. 4, Bachrach, U. and Chayen, R., Eds., Raven Press, New York, 1983, 155.
81. **Hölttä, E.,** Oxidation of spermidine and spermine in rat liver: purification and properties of polyamine oxidase, *Biochemistry*, 16, 91, 1977.
82. **Ma, K. and Sourkes, T. L.,** Inhibition of diamine oxidase by antimalarial drugs, *Agents Actions*, 10, 395, 1980.
83. **Chayen, R.,** personal communication.
84. **Gahl, W. A., Vale, A. M., and Pitot, H. C.,** Spermine oxidase in human pregnancy serum. Probable identity with diamine oxidase, *Biochem. J.*, 201, 161, 1982.
85. **Hill, J. M.,** The inactivation of pea-seedling diamine oxidase by peroxidase and 1,5-diaminopentane, *Biochem. J.*, 104, 1048, 1967.
86. **Guerrieri, P., Finazzi-Agrò, A., Costa, M. T., Rotilio, G., and Mondovì, B.,** Salt effect on diamine oxidase activity, *Ital. J. Biochem.*, 25, 160, 1976.
87. **Suzuki, H. and Ogura, Y.,** Mode of inhibition of the *Aspergillus* amine oxidase by 8-hydroxyquinoline, *J. Biochem.*, 72, 833, 1972.
88. **Macholán, L. and Minář, J.,** The depression of the synthesis of pea diamine oxidase due to light and the verification of its participation in growth processes using competitive inhibitors, *Biol. Plant. (Praha)*, 16, 86, 1974.
89. **Smith, T. A. and Bickley, D. A.,** Further properties of the polyamine oxidase of barley leaves, *Phytochemistry*, 13, 2437, 1974.
90. **Srivastava, S. K. and Smith, T. A.,** The inhibition of growth in higher plants by a homologous series of guanidines and its reversal by spermine, *Ann. Bot.*, 50, 265, 1982.
91. **Srivastava, S. K. and Smith, T. A.,** The effect of some oligo-amines and guanidines on membrane permeability in higher plants, *Phytochemistry*, 21, 997, 1982.
92. **Sourkes, T. L. and Missala, K.,** Action of inhibitors on monoamine and diamine metabolism in the rat, *Can. J. Biochem.*, 55, 56, 1976.
93. **Missala, K. and Sourkes, T. L.,** Putrescine catabolism in rats given heparin or aminoguanidine, *Eur. J. Pharmacol.*, 64, 307, 1980.
94. **Rojansky, N., Neufeld, E., and Chayen, R.,** Excretion of polyamines by the pregnant rat following inhibition of diamine oxidase, *Biochim. Biophys. Acta*, 586, 1, 1979.
95. **Seiler, N., Knödgen, B., Gittos, M. W., Chan, W. Y., Griesmann, G., and Rennert, O. M.,** On the formation of amino acids deriving from spermidine and spermine, *Biochem. J.*, 200, 123, 1981.
96. **Koren, Z., Pfeifer, Y., and Sulman, F. G.,** Induction of legal abortion by intra-uterine instillation of pargyline hydrochloride (Eutonyl), *J. Reprod. Fertil.*, 12, 75, 1966.
97. **Gahl, W. A. and Pitot, H. C.,** Reversal by aminoguanidine of the inhibition of proliferation of human fibroblasts by spermidine and spermine, *Chem. Biol. Interact.*, 22, 91, 1978.
98. **Pegg, A. E. and McGill, S. M.,** Inhibition of diamine oxidase by 1,1′-[(methylethanediylidene)-dinitrilo]-*bis*-(3-aminoguanidine) and 1,1′-[(methylethanediylidene)-dinitrilo]-diguanidine, *Biochem. Pharmacol.*, 27, 1625, 1978.
99. **Duch, D. S., Edelstein, M. P., and Nichol, C. A.,** Inhibition of histamine-metabolising enzymes and elevation of histamine levels in tissues by lipid-soluble anticancer folate antagonists, *Mol. Pharmacol.*, 18, 100, 1980.
100. **Prakash, V. and Sindhu, R. K.,** Effect of phytic acid on diamine oxidase activity in germinating pea seeds, *Experientia*, 35, 855, 1979.
101. **Gang, V., Berneburg, H., Hennemann, H., and Hevendehl, G.,** Diamine oxidase (histaminase) in chronic renal disease and its inhibition in vitro by methylguanidine, *Clin. Nephrol.*, 5, 171, 1976.
102. **Binder, B. and Seuring, K.,** The effect of brocresine (NSD-1055) on diamine oxidase activity in plasma, liver, stomach and upper small intestine of rats, *Naunyn. Schmiedebergs. Arch. Pharmacol.*, 278, 425, 1973.
103. **Törnqvist, A., Fredholm, A. M., Jonassen, F., and Samsioe, G.,** On the passage of an exogenous diamine oxidase inhibitor to the amniotic fluid, *Acta Obstet. Gynecol. Scand.*, 52, 25, 1973.
104. **Yamamoto, S.,** Histaminase and its inhibitory factor in guinea pig skin, *Hiroshima J. Med. Sci.*, 30, 251, 1981.
105. **Vaidya, A. B., Rajogopalan, T. G., Kale, A. G., and Levine, R. J.,** Inhibition of human pregnancy plasma diamine oxidase with alkaloids of *Argemone mexicana* — berberine and sanguinarine, *J. Postgrad. Med., (Bombay)*, 26, 28, 1980.
106. **Shaff, R. E. and Beaven, M. A.,** Turnover and synthesis of diamine oxidase (DAO) in rat tissues. Studies with heparin and cycloheximide, *Biochem. Pharmacol.*, 25, 1057, 1976.
107. **Dore, F., Leblanc-, J., and Cote-Jecote, J.,** Effects of guanaethidine and imidazole on histamine metabolism, *Can. J. Physiol., Pharmacol.*, 52, 978, 1974.

108. **Buffoni, F., Ignesti, G., and Pirisino, R.**, 3-Hydrazinopyridazine derivatives as inhibitors of the copper containing amine oxidases, *Farmaco (Sci)*, 32, 404, 1977.
109. **Buffoni, F., Carpi, C., Soldaini, G. B., and Raimondi, L.**, 3-Hydrazinopyridazine derivatives as inhibitors of pyridoxal phosphate dependent enzymes, *Farmaco (Sci)*, 35, 848, 1980.
110. **Beaven, M. A. and Roderick, N. B.**, Impromidine, a potent inhibitor of histamine methyltransferase (HMT) and diamine oxidase (DAO), *Biochem. Pharmacol.*, 29, 2897, 1980.
111. **Finazzi-Agrò, A., Floris, G., Fadda, M. B., and Crifo, C.**, Inhibition of diamine oxidase by antihistaminic agents and related drugs, *Agents Actions*, 9, 244, 1979.
112. **Shaff, R. E. and Beaven, M. A.**, Inhibition of histamine-*N*-methyltransferase and histaminase (diamine oxidase) by a new histamine H2-receptor agonist, Dimaprit, *Biochem. Pharmacol.*, 26, 2075, 1977.
113. **Quiring, K. and Palm, D.**, Inhibition of amine oxidases by nitrofuran derivatives: possible structure-activity relations and criteria of irreversibility, *Naunyn. Schmiedebergs. Arch. Pharmacol.*, 265, 397, 1970.
114. **Hölttä, E., Pulkkinen, P., Elfring, K., and Janne, J.**, Oxidation of polyamines by diamine oxidase from human seminal plasma, *Biochem. J.*, 145, 373, 1975.
115. **Kavanagh, J. P., Brightwell, R., Bardsley, W. G., and Schnieden, H.**, Levamisole and other diamine oxidase inhibitors as inhibitors of sperm motility, *Arch. Androl.*, 7, 51, 1981.
116. **Biegański, T., Kusche, J., Feussner, K.-D., Hesterberg, R., Richter, H., and Lorenz, W.**, Human intestinal diamine oxidase: substrate specificity and comparative inhibitor study, *Agents Actions*, 10, 108, 1980.
117. **Biegański, T., Kusche, J., Feussner, K.-D., Hesterberg, R., Richter, H., and Lorenz, W.**, The importance of human intestinal diamine oxidase in the oxidation of histamine and/or putrescine, *Arch. Immunol. Theor. Exp.*, 28, 901, 1980.
118. **Biegański, T., Osińska, Z., and Masliński, C.**, Inhibition of plant and mammalian diamine oxidase by substrate analogues, *Agents Actions*, 12, 41, 1982.
119. **Tourkov, M. I., Klimova, G. I., Davydova, G. A., Yermolaev, K. M., and Gorkin, V. Z.**, Colorimetric estimation of diamine oxidase activity, *Anal. Biochem.*, 64, 177, 1975.

Chapter 14

DIAMINE OXIDASE IN DEVELOPING TISSUES

Czeslaw Máslínski, Tadeusz Biegański, Wieslawa Agnieszka Fogel, and Mary Ellen Kitler

TABLE OF CONTENTS

I. INTRODUCTION

The view that homeostatic mechanisms control cell proliferation and differentiation is not new. But recently, when the polyamines were being implicated in the regulation of these events, the metabolic pathways and the patterns of changes in cellular content of these substances produced expectations of more substantial approaches to the problem.

Putrescine and aliphatic polyamines, spermidine, and spermine are normal cell constituents, which are important in cell proliferation and differentiation.[1-4] Histamine could be included in this group of growth-regulating substances, though the evidence for its role in cell proliferation and differentiation is circumstantial.[5-8]

Theoretically, homeostasis implies the existence of a self-regulatory system, which controls cell growth and division (i.e., proliferation) from one side, and cell differentiation (maturation) from the other. In this respect, diamine oxidase (DAO) may serve as a "sluice-gate", which adjusts inflow and outflow of putrescine, polyamines, and histamine, and which controls their levels as well as the levels of their metabolites in cells (Figure 1).

DAO is responsible for oxidative deamination of putrescine and histamine to aldehydes. Putrescine is transformed to γ-aminobutyraldehyde, which is further metabolized to γ-aminobutyric acid (GABA) and/or γ-aminobutanol. Histamine is transformed to imidazoleacetaldehyde, which is then converted to imidazoleacetic acid and/or histaminol. These substances may be involved in the process of cell proliferation. Therefore, it is conceivable that an enhancement of DAO activity may result in a decrease in putrescine levels with concomitant catabolites formation. In this situation, the amount of putrescine available to interact with aminopropyl group to form spermidine may decrease. In contrast, low DAO activity may result in an increase of polyamine biosynthesis.

Likewise, histamine may be a substrate for DAO. An enhancement of DAO activity should decrease histamine levels with a concomitant imidazoleacetic acid and/or histaminol increase, while the inhibition of DAO activity should increase tissue histamine level.

In reality, the situation is complex since many various factors influence the process. Unfortunately, the effect of many of the factors on the situation have never been thoroughly investigated. For example, histamine and putrescine, which are the main substrates for DAO, may possibly compete for DAO.

Other factors may also affect the enzymatic reaction. A few examples are: the presence of reaction products, the availability of cofactors, and the influence of pH and temperature.

The present chapter describes the involvement of DAO and the presumed role of its main substrates, putrescine and histamine (as well as their metabolites), in growing tissue in model situations, such as tissue culture, fetal growth, and normal tissue growth and development in an adult organism.

II. CELL CULTURE

The involvement of amine oxidases (AOs) in the regulation of cell proliferation has been extensively studied in different cell cultures. The importance of the AO-catalyzed reactions is due to both a breakdown of biologically active substances, such as putrescine, histamine, and polyamines, and a formation of their active metabolites.

The main activity of biogenic amines is the stimulation of growth processes. The main products of amine degradation, but not the exclusive products, are the corresponding aminoaldehydes. The role of these amine degradation products is the inhibition of cell proliferation. The function of putrescine and aliphatic polyamines in relation to cell proliferation has been extensively documented.[1,9,10]

Principally, all results support the theory that polyamines are essential in nucleic acid synthesis and behavior.[11] Thus, polyamines facilitate the progression through the cell cycle and lead to cell proliferation.[12] Ornithine decarboxylase (ODC) is the enzyme responsible

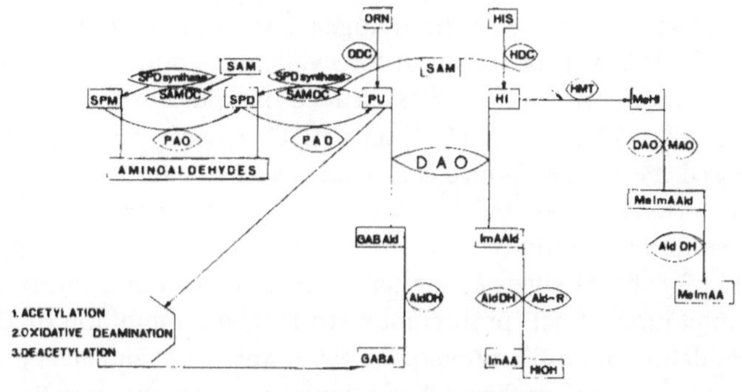

FIGURE 1. Metabolic pathways related to the DAO-catalyzed reaction. Abbreviations: ORN, ornithine; HIS, histidine; SAM, *S*-adenosylmethionine; HI, histamine; SPD, spermidine; SPM, spermine; PU, putrescine; MeHI, methylhistamine; GABAld, aminobutyraldehyde; GABA, aminobutyric acid; ImAAld, imidazoleacetaldehyde; ImAA, imidazoleacetic acid; HIOH, histaminol; MeImAAl, methylimidazoleacetaldehyde; MeImAA, methylimidazoleacetic acid; ODC, ornithine decarboxylase; HDC, histidine decarboxylase; HMT, histamine-*N*-methyltransferase; SAMDC, SAM decarboxylase; PAO, polyamine oxidase; DAO, diamine oxidase; MAO, monoamine oxidase; AldDH, aldehyde dehydrogenase; AldR, aldehyde reductase.

for the conversion of ornithine to putrescine, which is the key regulatory step in the synthesis of all aliphatic polyamines. Therefore, it is not surprising that ODC is directly involved in cell proliferation.[1]

The increase in ODC activity was observed in rat embryo fibroblasts,[13] in bone marrow cultures of erythroid cells,[14] in rat hepatoma cells,[15] in rat brain tumor cells,[16] and in leukemic cells.[17]

Likewise, in mouse neuroblastoma cells, induced to differentiation with cAMP analog, both ODC activity and spermidine levels were significantly lower than in growing nondifferentiating cells.[18] Using other types of cells as models, similar results were found. Difluoromethylornithine or vitamin A analogs inhibited ODC, which led to cell differentiation and inhibition of neoplastic growth.[19-22]

Kahlson et al. and others[5,23-25] have studied the involvement of histidine decarboxylase (HDC) and histamine forming capacity (HFC) in various growth processes, such as regeneration, fetal development, wound healing, tumor growth, and plant growth. However, the mechanisms by which histamine stimulates cell proliferation are unknown.

As was expected, the period of intensive formation of putrescine and aliphatic polyamines coincides with a concomitant change in DAO activity. When normal rat kidney and virally transformed cells were cultivated in vitro, the DAO activity increased rapidly during the early growth phase. The maximum activity in both cultures was seen 24 hr after seeding. Then, the DAO activity decreased during the exponential phase of growth.[26] Similar changes were observed in a culture of human fibroblasts.[27] These changes corresponded well to general patterns of enzymatic activities involved in polyamine metabolism, i.e., SAM-decarboxylase and spermidine and spermine synthases. Additional information is in a comprehensive review.[10] ODC seems to be the first rate-limiting trigger of all these events.

DAO activity may be located in the G_1 phase of the cell cycle, since during a period of exponential growth, DAO activities in different normal parental cell lines are several times higher than in corresponding transformed cells.[26] Transformed cells are known to have an impairment during the G_1 period, i.e., these cells lack a restriction point, which is sensitive to the deficiency of growth factors.[28-30] The growth factors, such as the epidermal growth factor[28] or serum growth factors,[30] or the mitogens, such as phorbol ester,[29] are able to induce ODC activity[32,34,35] by a mediation of the cyclic nucleotides.[31-33] HDC activity may also be induced by various mitogens.[36,37] By a feed-back mechanism, putrescine[38,39] and histamine[40-42] inhibit the induction of ODC and HDC, respectively, though the effect of putrescine is expressed less in transformed cells than in normal ones.[30,43] Thus, both histamine and putrescine are formed during the G_1 phase and these two amines exert, directly or indirectly, growth-stimulating activity. Therefore, it seems reasonable to suppose that DAO, which is responsible for the catalysis of the oxidative deamination of these amines, will be implicated in the inhibition of cell proliferation. In fact, the inhibition of DAO activity is responsible for stimulation of proliferation of human fibroblasts[44] and an application of DAO inhibitor, aminoguanidine, into erythroid cell colonies reverses the growth-inhibitory effect of difluoromethyl-ornithine.[14] Therefore, the biological significance of diamine and polyamine oxidases in cell growth and division may be related to their ability to destroy diamines and polyamines, respectively.

The other side of the problem appeared when experiments showed that the spermine-, spermidine-, and putrescine-derived aldehydes are able to inhibit thymidine uptake and its incorporation into DNA by rat thymic lymphocytes incubated with phytohemagglutinine.[45,46] Similar effects were observed in cultures of fibroblasts, human lymphocytes, melanoma cells, and Burkitt lymphoma cells.[45,47] The polyamine-oxidized products, derived from semen diamine oxidase reaction, have been found to influence the behavior and metabolism of spermatozoa. The high affinity of the seminal cells to these aldehydes,[48] the inhibitory effects of aminoaldehydes on glucose and fructose metabolism in spermatozoa,[49] and the loss of cell motility in the semen exhibiting high DAO activity[50,51] resembles the general effect of aminoaldehydes on cell proliferation and viability. It is worth noting that the aliphatic aldehydes[52] and benzaldehyde[53] were reported to exert carcinostatic effect. However, a growth inhibitory effect of benzaldehyde is indiscrimately expressed through the cell cycle,[53] which suggests that its mechanism of action differs from that of natural aminoaldehydes.

The inhibitory effect of aminoaldehydes on lymphocyte proliferation is expressed in G_1 phase or G_1/S interphase.[45,54] Thus, the same cell cycle period, which is influenced by a decrease in polyamines, may be inhibited or even arrested by its own catabolites. Such multiplication of inhibitory reactions might disturb homeostasis. However, the competition between polyamines and aldehydes in the regulation of proliferative processes is expected. Its existence may attenuate the final reaction intensity.

In most experiments evaluating growth suppression both the methodology used in the experiment and the conclusions drawn from the results are indirect. Aminoaldehydes are produced from the action of amine oxidase on their substrate amine. The growth experiment does not actually use the particular aminoaldehyde. Instead, the experiment uses various tissues, which are known to be rich sources of polyamine oxidase and/or diamine oxidase activity. Such sources include human pregnancy serum, bovine or fetal calf serum, or purified diamine oxidase from pig kidney. Thus, the methodology requires that the tissue be combined with the respective substrate, i.e., the particular amine. The aminoaldehyde under investigation is produced directly from the action of the amine oxidase on the particular amine. The rationale for this procedure is simple. Biogenic aldehydes are unstable.[55] Only coupled enzyme systems produce the conditions which permit the quantitative estimation of aldehyde formation. Supplementary experiments with selective inhibitors, e.g., aminoguanidine, may be useful to determine if observed effects are artifacts or if they result from the action of naturally formed aminoaldehydes.

III. PREGNANCY AND FETAL DEVELOPMENT

Pregnancy greatly modifies all metabolic patterns of histamine, putrescine, and polyamines. When histamine- and putrescine-synthesizing enzymes are stimulated,[5,23,56-59] the excretion of free amines and polyamines in urine is increased [5,23,60,61] and the oxidative deamination pathway is enhanced.[57-59,62-65]

During the 1930s human placenta and pregnancy plasma was shown to have high histamine degrading activity. Later, high histamine degrading activity associated with pregnancy was found in other species, such as rat, guinea pig, and rabbit.[5,64-68]

Histamine and putrescine are abundantly produced by the fetus. The evidence of the decidual origin of increased plasma DAO supported the theory of the protective role of DAO against histamine and putrescine (see Chapters 2 and 7). The evidence, though indirect, has been convincing: in deciduate placenta, DAO activity was high, while in non-deciduate ones, such as found in hog and horse, DAO activity could not be demonstrated;[64,65] after delivery or artificial abortion, plasma DAO activity returns within 24 hr to its nonpregnant value;[69] the ratio of maternal to fetal plasma DAO activity is high — 160 vs. 1 at the parturition;[70] an inhibitor of protein synthesis, cycloheximide, added to maternal circulation decreased DAO activity in in vitro perfused human placenta, but exerted no effect when added to the fetal circulation.[71] Direct proof has come from immunohistochemical studies. DAO is localized in the cytoplasm,[72,73] and in the intercellular space surrounding decidual cells.[73] Similarly, in rat placenta and in human placenta, most of the DAO activity is related to the maternal DAO activity. The reported values in the maternal placenta in the third trimester are several times higher than in fetal placenta.[57-59] Similarly, the guinea pig maternal placenta contains 20 to 100 times more DAO than the fetal placenta.[129] Transport of the enzyme from decidua does not involve the lymphatic system but probably occurs by diffusion into maternal plasma down a steep concentration gradient.[65,74]

By monitoring the changes in plasma DAO activity in female patients, it has been found that the increase in enzyme activity begins between 9 and 28 days following ovulation leading to pregnancy.[75] Usually, DAO activity in plasma can be demonstrated 5 to 6 weeks after the last menstrual period.[64,76,77] In the pregnant woman, DAO plasma activity increases rapidly. During the first 10 weeks of pregnancy, DAO enzyme activity doubles every 4 to 6 days.[75] From the first few weeks until the 18th to 20th week, the rate of increase in DAO activity is linear. After 20 weeks, the rate of increase drops and the DAO activity in plasma tends to plateau.[64,75-77] Near term, the enzyme activity in plasma is 500 to 1000 times that of a nonpregnant subject.[64,75] At delivery, DAO activity in the pregnant woman drops dramatically (see Chapter 18).

DAO diffuses from decidua through the membranes to amniotic fluid. The general pattern of changes of enzyme activity in amniotic fluid is similar to that in plasma. The distribution of DAO activity between these two fluids strongly depends on the length of gestation.[78] Vaginal secretion of pregnant women does not contain DAO, unless a rupture of fetal membranes has occurred. Thus, a measurement of DAO in vaginal fluid may be a reliable method for the definitive diagnosis.[79,80] Clinically, plasma DAO measurement may be useful in other diagnostic procedures[75-77,81] (see Chapter 18).

Diamine oxidase activity appears also in fetal tissues. In the guinea pig fetus, DAO can usually be detected in measurable quantities in liver in the middle of gestation. Thereafter, it increases quickly and between the 45th and 60th day of fetal development increases about three times to a level of more than twice the normal adult level. The peak value of DAO activity was found in 3-day-old neonates. In the following week, the enzyme tended to decrease its activity towards the adult level. During the entire study, the ability of developing guinea pig liver to oxidize putrescine and histamine progressed in a parallel manner.[68,82]

Our preliminary data from autopsied livers suggests that, in man, DAO activity also develops early; 3- to 6-month-old conceptuses showed hepatic DAO activity in a range 0.5 to 1.6 mU, somewhat higher than reported for adults — 0.21 mU.[83]

High DAO activity found in fetal tissues at the end of pregnancy raises questions concerning the role of the enzyme in developmental physiology. Fetal tissue produces large amounts of histamine, putrescine, and polyamines. The protective role of the enzyme DAO against these compounds in fetal tissue cannot be excluded. However, one should consider that oxidative deamination supplies the fetus also with biologically active substances. Aminoaldehydes may function as natural immunoregulatory agents, preventing the rejection of the fetoplacental unit;[84] they could also act on the DNA synthesis.[45,46,84] Also, GABA, a metabolite of putrescine, may stimulate protein synthesis.[85,86] Similarly, recent experiments described the effect of GABA on behavioral and developmental metamorphosis of planktonic molluscan larvae.[87] The theory that GABA may be a post-translational modifier of protein is strengthened by experiments showing that radioactive GABA is present in the acid-precipitable fraction of cultured human keratinocytes exposed to labeled putrescine.[88] No data are available concerning the importance of imidazoleacetic acid in growth processes. However, there is growing evidence that ImAA can mimic certain actions of GABA.[89-91]

IV. DAO AND INTESTINAL GROWTH

The intestine is convenient to use to study cell proliferation and its relation to DAO activity. There are two main reasons: (1) the intestine is a tissue with high DAO activity and (2) in the post-natal period, the intestine rapidly grows and develops. Likewise, in the adult the intestine has constant enterocyte proliferation and replacement.

A. Postnatal Development

In chick embryo (Figure 2) and in guinea pig (Table 1), relatively low DAO activities have been detected. Throughout the 21 days of the chick embryo development, the intestinal DAO activity on the 16th day and the 19th day is only about 4% of the adult value. The guinea pig pregnancy lasts approximately 68 to 70 days. At the beginning of the third trimester of pregnancy, the DAO activity in the guinea pig fetal intestine is about 33% of the adult value.

In both chick and guinea pig, there is a dramatic, though transient, rise in intestinal DAO activity immediately after birth. However, in rat, the activity in neonatal rat intestine is apparently lower than in adult rat intestine.[92-94] The variation in DAO activity may be due to different levels of food intake postnatally in the different species.

During the postnatal period, developmental changes last approximately 3 weeks in the pig[95] and rat.[96] During this time, cell addition rather than cell replacement occurs. Newly formed cells migrate onto the villi. The cell number in villus increases several times and villi become elongated.[95] During this early postnatal development activity in the rat intestinal mucosa, both ODC[93,97] and DAO[92,93] activity increase continuously. The stable adult level of both ODC and DAO activity is reached at approximately the 4th week.

The agreement of in vivo and in vitro data regarding the behavior of ODC and DAO activities during fast tissue growth is astonishing.

Difluoromethylornithine inhibits ODC activity irreversibly. Administration of difluoromethylornithine to rats during the postnatal development period produces an immediate decrease in putrescine level in gut mucosa and a decrease in both spermidine and spermine levels in gut mucosa. It should be noted that during ontogenesis in the rat, the gut mucosal content of spermidine and spermine is elevated. Concomitantly, DAO activity is suppressed. This suppression, however, does not appear to be related to the specific feed-back regulation (putrescine deprivation), but appears to be related to the general delay in enzyme maturation, since similar changes have been found in maltase activity.[93]

B. Permanent Enterocyte Proliferation and Replacement in Mature Gut

The rate of cellular proliferation (cell growth and division) in the ileal and jejunal villi is

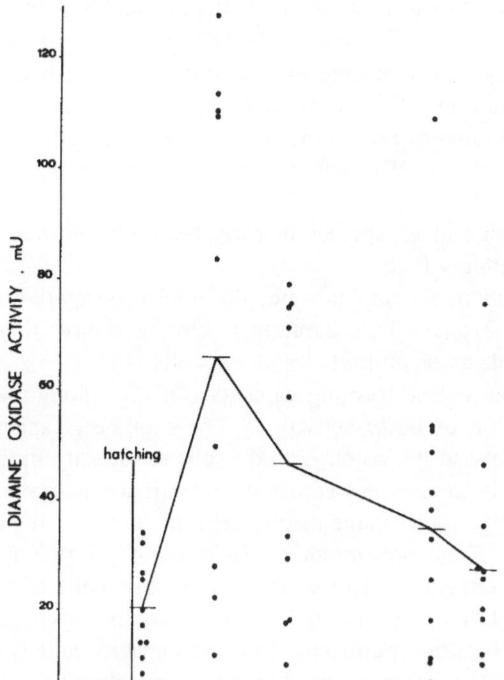

FIGURE 2. Changes in DAO activity in hen small intestine as a function of age. Points represent single animals. The average values of DAO activity expressed in mU/g wet tissue, are indicated. 1 mU corresponds to 1 nmol of putrescine deaminated per minute at 37°C. ^{14}C Putrescine method[128] was applied.

Table 1
DIAMINE OXIDASE ACTIVITY IN THE GUINEA PIG INTESTINE AT VARIOUS STAGES OF DEVELOPMENT

Age (days)	DAO Activity	(mU/g wet tissue)
Fetus		
45	4.3 ± 2.4	n = 5
60	12.9 ± 2.0	n = 3
After birth		
3	29.8 ± 3.8	n = 7
7	15.0 ± 3.3	n = 4
10	14.4 ± 0.9	n = 10
Adult	12.8 ± 4.9	n = 9

Note: 1 mU corresponds to 1 nmol of putrescine deaminated per min at 37°C. ^{14}C Putrescine method[128] was applied.

evidently about four times higher in weaned rats than in suckling animals.[98] It is known that food intake activates[99] and starvation lowers[100] ODC activity in intestinal wall. An epidermal growth factor (EGF) may be responsible for the continuous enhancement of intestinal ODC activity. In normal conditions, EGF is secreted by the salivary glands and by Brunner's glands in the gut.[101,102] Growth-promoting agent is also present in human milk.[103] EGF, given parenterally, stimulates ODC activity in the digestive tract[104] and enhances DNA synthesis in enterocytes.[105]

Intestine is unique since in all species investigated under normal conditions, ODC and DAO activities are extremely high.

In both adult man and in rodent intestine, the total proliferative cell cycle has a short duration of about 1 to 2 days. This duration is clearly shorter than the duration in fast proliferating leukemic blasts or multiple myeloma cells.[106]

The distribution of putrescine forming capacity (ODC)[97] along the GI tract corresponds exactly to the localization of DAO activity,[107-110] the highest capacity being in the small intestine, a lower capacity in the colon, and the lowest capacity in the stomach.

Also, localization of both enzymes across the intestinal mucosa is parallel. Activities of ODC and DAO are sharply higher in the mature cells of a villus tip region than in proliferating cells of the crypt areas. These proliferating cells of the crypt area apparently have approximately 15 to 20% less enzymatic activity then the mature cells of the villus tip region.[111]

However, this parallel distribution of the enzymes does not implicate the existence of any functional relationship between putrescine-forming capacity and DAO activity. Different half-life times for ODC (ca. 10 min) and DAO (several hours) suggest an involvement of different regulatory processes for these two enzyme proteins. Two phenomena occur simultaneously in the upper part of the villi. The first is that DNA synthesis stops, which leads to the cession of cell maturation.[106] The second is that DAO activity peaks. The fact that these two phenomena coexist in these cells may indicate that DAO is involved both in the retardation of growth and in the acceleration of the maturation process. This biological role of DAO could be expressed by its putrescine-degrading activity from one side, and by the formation of aminoaldehydes from another one. However, there is still no clear evidence to support this view.

Intestine is one of the organs richest in histamine,[112] the other biogenic substrate for DAO. Histamine stimulates fibroblast growth in vitro[24,25] and has a stimulatory effect on proliferation of jenunal epithelial cells.[113] Histamine H_2 receptor antagonists have been shown to produce suppression of xenografted human colorectal carcinomas.[114] The ability of intestinal DAO to destroy histamine and putrescine,[115-117] the potential growth stimulants, supports the theory that DAO may have a function as a growth retardant.

However, the ability of intestinal DAO to influence the histamine absorption from the gut lumen, and intestinal DAO involvement in the histamine methylation pathway, may be of primary importance in GI physiology. A recent review of the topic is available.[110]

Intestinal DAO is responsible for the oxidation of *N*-tele-methylhistamine. *N*-Methylhistamine is a potent inhibitor of histamine *N*-methyltransferase.[118] Thus, it can be seen that intestinal DAO accelerates histamine catabolism along the methylation pathway.

Putrescine and histamine are the two main substrates of DAO. The apparent preference of DAO for one of the substrates may possibly lead to a competition between them for DAO.

In many conditions, ODC as well as HDC activation takes place. This has been found after an application of phorbol esters,[35-37] after a provocation of nonspecific immune response,[119] during pregnancy,[5,57,58] and in other conditions.[120,121]

Investigations showed that if a mammalian enzyme was studied with a single substrate, i.e., either histamine or putrescine, the mammalian enzyme had a higher affinity for histamine than for putrescine. However, the mammalian enzyme had a higher reaction velocity with putrescine than with histamine.[115-117]

Table 2
PERCENT INHIBITION OF DAO BY IMIDAZOLE COMPOUNDS

Compound at 1×10^{-4} *M*	Human pregnancy plasma DAO	Pea seedling DAO
Histamine	84	15
2-Methylimidazole	84	No effect
N-Methylhistamine	69	No effect
1-Methylimidazole	65	No effect
N-Methylimidazole	48	No effect
Imidazole	44	No effect
Acetylhistamine	10	No effect
Histaminol	9	No effect
Imidazole-4-acetic acid	8	No effect

Note: The control values of enzymic activities without inhibitor were in the range of 80 to 110 dpm/min for pregnancy plasma AO and 200 to 480 dpm/min for pea seedling AO. ^{14}C putrescine method[128] was applied. Pea seedling AO was a 165-fold purified preparation; human pregnancy plasma unpurified, from late pregnancy. All values are means of 3 to 6 determinations.

In order to fully evaluate substrate preference by DAO, further detailed experiments are necessary. Experiments must be conducted with various in vitro and various in vivo conditions.

Substrate preference experiments should evaluate the preference of DAO for a particular substrate in the presence of both substrates.

V. PLANT AMINE OXIDASE

Also in plants, diamines and histamine are reported to be stimulators of growth and development.[5,122] The early report by Werle and Pechmann[123] concerns high DAO activity in leguminous seedling.

Pea seedling DAO is the enzyme involved in growth. The substrate properties of pea seedling DAO differ from the substrate properties of mammalian AOs connected with tissue proliferation, such as pregnancy plasma DAO or intestinal DAO.

For the plant enzyme, the ratio of reaction velocity with putrescine to that with histamine is about 20,[124,125] while the respective ratio value for the animal DAO is in a range two to three.[115-117] Thus, the plant enzyme prefers the aliphatic structures, whereas the mammalian one expresses a higher affinity to imidazole ring. The existence of a relationship between the enzyme and a structure of substrate became apparent in inhibitory studies with the substrates analogs. A compound with an aliphatic structure, 1,5 diamino-3-pentanone, is a more potent inhibitor of pea seedling AO ($K_i = 1.5 \times 10^{-8}$ *M*) than of pig kidney DAO ($K_i = 1.8 \times 10^{-6}$ *M*).[126] In contrast, the imidazole compounds inhibit human pregnancy DAO and have no effect on pea seedling enzyme (Table 2). Anserine and carnosine, which contain both an imidazole ring and aliphatic structure resembling putrescine, are equally potent inhibitors of either enzyme activity.[107,127] Thus, it is conceivable that mammalian enzymes are better adapted to the oxidative deamination of histamine, whereas plant enzymes prefer aliphatic diamines as their substrates (see Chapter 6).

REFERENCES

1. **Jänne, J., Pösö, H., and Raina, A.,** Polyamines in rapid growth and cancer, *Biochim. Biophys. Acta,* 473, 241, 1978.
2. **Russell, D. H. and Durie, B. G. M.,** Polyamines as biochemical markers of normal and malignant growth, in *Progress in Cancer Research and Therapy,* Vol.8, Raven Press, New York, 1978.
3. **Russell, D. H.,** Ornithine decarboxylase as a biological and pharmacological tool, *Pharmacology,* 20, 117, 1980.
4. **Heby, O.,** Role of polyamines in the control of cell proliferation and differentiation, *Differentiation,* 19, 1, 1981.
5. **Kahlson, G. and Rosengren, E.,** *Biogenesis and Physiology of Histamine,* Edward Arnold, London, 1971.
6. **Maśliński, Cz.,** Histamine and its metabolism in mammals. I. Chemistry and formation of histamine, *Agents Actions,* 5, 89, 1975.
7. **Maśliński, Cz.,** Histamine and its metabolism in mammals. II. Catabolism of histamine and histamine liberation, *Agents Actions,* 5, 183, 1975.
8. **Beaven, M. A.,** Histamine: its role in physiological and pathological processes, in *Monographs in Allergy,* Vol. 13, S. Karger, Basel, 1978.
9. **Anon.,** Polyamines as cellular regulators, *Medical Biology,* No. 5-6, Vol. 59, 1981.
10. **Bachrach, U.,** *Function of Naturally Occurring Polyamines,* Academic Press, New York, 1973.
11. **Kato, Y., Nishihara, H., Inoue, H., Gohda, E., and Takeda, Y.,** Polyamine metabolism and DNA synthesis in regenerating rat liver, *J. Biochem. (Tokyo),* 84, 1617, 1978.
12. **Duffy, P. E. and Kremzner, L. T.,** Ornithine decarboxylase activity and polyamines in relation to aging of human fibroblasts, *Exp. Cell Res.,* 108, 435, 1977.
13. **Rupniak, H. T. and Paul, D.,** Regulation of the cell cycle by polyamines in normal and transformed fibroblasts, in *Advances in Polyamine Research,* Vol. 1, Campbell, R. A., Morris, D. R., Bartos, D., Daves, G. D., and Bartos, F., Eds., Raven Press, New York, 1978, 117.
14. **Niskanen, E. O., Kalio, A., McCann, P. P., Pou, G., Lyda, S., and Thornhill, A.,** Divergent effect of ornithine decarboxylase inhibition on murine erythropoietic precursor cell proliferation and differentiation, *Cancer Res.,* 43, 1536, 1983.
15. **Mamont, P. S., Böhlen, P., McCann, P. P., Bey, P., Schuber, F., and Tardif, C.,** α-Methylornithine, a potent competitive inhibitor of ornithine decarboxylase, blocks proliferation of rat hepatoma cells in culture, *Proc. Natl. Acad. Sci. U.S.A.,* 73, 1626, 1976.
16. **Heby, O., Andersson, G., Gray, J. W., and Marton, L. J.,** Accumulation of rat brain tumor cells in G_1 and of Ehrlich ascites tumor cells in G_2 by partial deprivation of cellular polyamine content, in *Advances in Polyamine Research,* Vol.1, Campbell, R. A., Morris, D. R., Bartos, D., Daves, G. D., and Bartos, F., Eds., Raven Press, New York, 1978, 127.
17. **Luk, G. D., Civin, C. I., Weissman, R. M., and Baylin, S. B.,** Ornithine decarboxylase: essential in proliferation but not differentiation of human promyelocytic leukemia cells, *Science,* 216, 75, 1982.
18. **Chen, K. Y., Presepe, V., Parken, N., and Lin, A.Y.-C.,** Changes of ornithine decarboxylase activity and polyamine content upon differentiation of mouse NB-15 neuroblastoma cells, *J. Cell Physiol.,* 110, 285, 1982.
19. **Shindler, J., Kelly, M., and McCann, P. P.,** Inhibition of ornithine decarboxylase induces embryonal carcinoma cell differentiation, *Biochem. Biophys. Res. Commun.,* 114, 410, 1983.
20. **Haddox, M. K. and Russell, D. H.,** Cell cycle — specific locus of vitamin A inhibition of growth, *Cancer Res.,* 39, 2476, 1979.
21. **Haddox, M. K., Scott, K. F. F., and Russell, D. H.,** Retinol inhibition of ornithine decarboxylase induction and G_1 progression in Chinese hamster ovary cells, *Cancer Res.,* 39, 4930, 1979.
22. **Sporn, M. B. and Newton, D. L.,** Chemoprevention of cancer with retinoids, *Fed. Proc.,* 38, 2528, 1979.
23. **Kahlson, G. and Rosengren, E.,** Histamine: entering physiology, *Experientia,* 28, 993, 1972.
24. **Russell, J. D., Russell, S. B., and Trupin, K. M.,** The effect of histamine on the growth of cultured fibroblasts isolated from normal and keloid tissue, *J. Cell Physiol.,* 93, 389, 1977.
25. **Norrby, K.,** Effect of heparin, histamine and serotonin on the density-dependent inhibition of replication in two fibroblastic cell lines, *Virchows Arch. Abt. B Zellpath.,* 15, 75, 1973.
26. **Quash, G., Keolouangkhot, T., Gazzolo, L., Ripoll, H., and Saez, S.,** Diamine oxidase and polyamine oxidase activities in normal and transformed cells, *Biochem. J.,* 177, 275, 1979.
27. **Gahl, W. A. and Pitot, H.,** Diamine oxidase activity in fibroblasts from normal and cystic fibrosis patients, *Pediatr. Res.,* 14, 118, 1980.
28. **Pardee, A. B., Dubrow, R., Hamlin, J. L., and Kletzien, R. F.,** Animal cell cycle, *Ann. Rev. Biochem.,* 47, 715, 1978.

29. **Moses, H. L., and Robinson, R. A.,** Growth factors, growth factor receptors, and cell cycle control mechanisms in chemically transformed cells, *Fed. Proc.*, 41, 3008, 1982.
30. **Haddox, M. K., Magun, B. E., and Russell, D. H.,** Ornithine decarboxylase induction during G_1 progression of normal and Rous sarcoma virus-transformed cells, *Cancer Res.*, 40, 604, 1980.
31. **Perchellet, J. P. and Boutwell, R. K.,** Enhancement by 3-isobutyl-1-methylxanthine and cholera toxin of 12-*o*-tetradecanoylphorbol-13-acetate-stimulated cyclic nucleotide levels and ornithine decarboxylase activity in isolated epidermal cells, *Cancer Res.*, 40, 2653, 1980.
32. **Guroff, G., Dickens, G., and End, D.,** The induction of ornithine decarboxylase by nerve growth factor and epidermal growth factor in PC12 cells, *J. Neurochem.*, 37, 342, 1981.
33. **Beck, W. T., Bellantone, R. A., and Canellakis, E. S.,** The in vivo stimulation of rat liver ornithine decarboxylase activity by dibutyryl cyclic adenosine 3′ 5′ monophosphate, theophylline and dexamethasone, *Biochem. Biophys. Res. Commun.*, 48, 1649, 1972.
34. **O'Brien, T. G., Simsiman, R. C., and Boutwell, R. K.,** Induction of the polyamine-biosynthetic enzymes in mouse epidermis by tumor-promoting agents, *Cancer Res.*, 35, 1662, 1975.
35. **Yuspa, S. H., Lichti, V., Ben, T., Patterson, E., Hennings, H., Slaga, T. J., Colburn, H., and Kelsey, W.,** Phorbol esters stimulate DNA synthesis and ornithine decarboxylase activity in mouse epidermal cell cultures, *Nature (London)*, 262, 402, 1976.
36. **Watanabe, T., Taguchi, Y., Sasaki, K., Tsuyama, K., and Kitamura, Y.,** Increase in histidine decarboxylase activity in mouse skin after application of the tumor promoter tetra-decanoylphorbol acetate, *Biochem. Biophys. Res. Commun.*, 100, 427, 1981.
37. **Endo, Y.,** Induction of histidine decarboxylase by mitogens and the effects of actinomycin D and anti-inflammatory agents, *J. Pharm. Dyn.*, 6, 5, 1983.
38. **Jänne, J. and Hölttä, E.,** Regulation of ornithine decarboxylase activity by putrescine and spermidine in rat liver, *Biochem. Biophys. Res. Commun.*, 61, 399, 1974.
39. **Clark, J. L. and Fuller, J. L.,** Regulation of ornithine decarboxylase in 3T3 cells by putrescine and spermidine: indirect evidence for translational control, *Biochemistry*, 14, 4403, 1975.
40. **Johansson, I., Lundell, L., Rosengren, E., and Svensson, S. E.,** Gastric secretion and its facilitation as related to gastric mucosal histamine, *J. Physiol.*, 226, 431, 1972.
41. **Garbarg, M. and Halpern, B.,** The regulation of gastric histidine decarboxylase activity. I. Effect of pyridoxine deficiency, *Life Sci.*, 10, 1211, 1971.
42. **Garbarg, M. and Halpern, B.,** The regulation of gastric histidine decarboxylase activity. II. Effects of physiological stimuli during pyridoxine deficiency, *Life Sci.*, 10, 1219, 1971.
43. **Bethell, D. R. and Pegg, A. E.,** Effect of diamines on ornithine decarboxylase activity in control and virally transformed mouse fibroblasts, *Biochem. J.*, 180, 87, 1979.
44. **Duffy, P. E. and Kremzner, L. T.,** Ornithine decarboxylase activity and polyamines in relation to aging of human fibroblasts, *Exp. Cell Res.*, 108, 435, 1977.
45. **Allen, J. C., Smith, J., Hussain, J. I., Thomas, J. M., and Gaugas, J. M.,** Inhibition of lymphocyte proliferation by polyamines requires ruminant-plasma polyamine oxidase, *Eur. J. Biochem.*, 102, 153, 1979.
46. **Gaugas, J. M. and Curzen, P.,** Polyamine interaction with pregnancy serum in suppression of lymphocyte transformation, *Lancet*, i, 18, 1978.
47. **Gaugas, J. M. and Dewey, D. L.,** Hog kidney diamine oxidase conversion of biogenic diamines to inhibitors of cell proliferation, *J. Pathology*, 134, 243, 1981.
48. **Pulkkinen, P., Kanerva, S., Elfving, K., and Jänne, J.,** Association of spermine and diamine oxidase activity with human spermatozoa, *J. Reprod. Fertil.*, 43, 49, 1975.
49. **Pulkkinen, P., Piik, K., Koso, P., and Jänne, J.,** Effect of polyamines and their oxidized derivatives on the metabolism of rat epidydymal spermatozoa, *Acta Endocrinol.*, 87, 845, 1978.
50. **Jänne, J., Hölttä, E., Haaranen, P., and Elfving, K.,** Polyamines and polyamine metabolizing enzymes in human semen, *Clin. Chim. Acta*, 48, 393, 1973.
51. **Crabbe, M. J. C.,** The development of a qualitative assay for male infertility from a study of enzymes in human semen, *J. Reprod. Fertil.*, 51, 73, 1977.
52. **Perin, A., Sessa, A., and Ciaranfi, E.,** Carcinostatic effect of aliphatic aldehydes and aldehyde dehydrogenase activity in Ehrlich carcinoma, sarcoma 180, and Yoshida AH 130 hepatoma, *Cancer Res.*, 38, 2180, 1978.
53. **Pattersen, E. O., Nome, O., Ronning, O. W., and Oftebro, R.,** Effects of benzaldehyde on survival and cell-cycle kinetics of human cells cultivated in vitro, *Eur. J. Cancer Clin. Oncol.*, 19, 507, 1983.
54. **Gaugas, J. M. and Dewey, D. L.,** Evidence for serum binding of oxidized spermine and its potent G_1-phase inhibition of cell proliferation, *Br. J. Cancer*, 39, 548, 1979.
55. **Kivits, G. A. A. and Hora, J.,** A derivative of (4-imidazolyl) acetaldehyde of proven structure, *J. R. Nether. Chem. Soc.*, 94, 245, 1975.

56. **Maslinski, C. and Niedzielski, A.,** Histidine decarboxylase in mouse fetal tissues. II. Physiological role of the enzyme in delivery, *Eur. J. Pharmacol.*, 5, 196, 1969.
57. **Kim, K. S. and Harris, M. E.,** Mechanisms of increased diamine oxidase activity in pregnancy and the possible roles of diamine oxidase in the uterus and the ovary, in *Histamine*, Maslinski, C., Ed., Dowden, Hutchinson & Ross, Stroudsburg, 1973, 312.
58. **Maudsley, D. V. and Kobayashi, Y.,** Biosynthesis and metabolism of putrescine in the rat placenta, *Biochem. Pharmacol.*, 26, 121, 1977.
59. **Guha, S. K., and Jänne, J.,** The synthesis and accumulation of polyamines in reproductive organs of the rat during pregnancy, *Biochim. Biophys. Acta*, 437, 244, 1976.
60. **Russell, D. H., Giles, H. R., Christian, C. D., and Campbell, J. L.,** Polyamines in amniotic fluid, plasma, and urine during normal pregnancy, *Am. J. Obstet. Gynecol.*, 132, 649, 1978.
61. **Andersson, A-Ch., Henningsson, S., and Rosengren, E.,** Increased formation of diamines and polyamines in the pregnant rat, *J. Physiol.*, 285, 311, 1978.
62. **Danforth, D. N. and Gorham, F.,** Plasma histaminase, *Am. J. Physiol.*, 119, 294, 1937.
63. **Marcou, I., Athanasiu-Vergu, E., Chiriceanu, D., Cosma, G., Gingold, N., and Parhon, C. G.,** Metabolism of l'histaminase, *Presse Med.*, 46, 371, 1938.
64. **Ahlmark, A.,** Studies on the histaminolytic power of plasma with special reference to pregnancy, *Acta Physiol. Scand.*, 9(28), 1, 1944.
65. **Swanberg, H.,** Histaminase in pregnancy. With special reference to its origin and formation, *Acta Physiol. Scand.*, 23(79), 1, 1950.
66. **Kobayashi, Y.,** Plasma diamine oxidase titres of normal and pregnant rats, *Nature (London)*, 203, 146, 1964.
67. **Maslinski, C. and Niedzielski, A.,** The serum histaminase activity in guinea pig pregnancy, *Experientia*, 20, 372, 1964.
68. **Fogel, W. A., Bieganski, T., and Maslinski, C.,** Gamma-aminobutyric acid (GABA) formation from putrescine in guinea pig liver during ontogenesis, *Comp. Biochem. Physiol.*, 73C, 431, 1982.
69. **Maslinski, C., Niedzielski, A., and Rędziejowska, B.,** The histaminase activity in pregnant women and after the artificial abortion, *Gynaecologia*, 157, 43, 1964.
70. **Southren, A. L., Kobayashi, Y., Brenner, P., and Weingold, A. B.,** Diamine oxidase activity in human maternal and fetal plasma and tissues at parturition, *J. Appl. Physiol.*, 20, 1048, 1965.
71. **Southren, A. L., Galanog, A. M., Mishr, S., and Weingold, A. B.,** Effect of cycloheximide on production of diamine oxidase by the human placenta, *J. Appl. Physiol.*, 29, 684, 1970.
72. **Weisburger, W. R., Mendelsohn, G., Eggleston, J. C., and Baylin, S. B.,** Immunohistochemical localization of histaminase (diamine oxidase) in decidual cells of human placenta, *Lab. Invest.*, 38, 703, 1978.
73. **Lin, C. W., Chapman, C. M., DeLellis, R. A., and Kirley, S.,** Immunofluorescent staining of histaminase (diamine oxidase) in human placenta, *J. Histochem. Cytochem.*, 26, 1021, 1978.
74. **Hansson, R., Tryding, N., and Törnqvist, A.,** Diamine oxidase (histaminase) in human pregnancy, *Acta Obstet. Gynec. Scand.*, 48, 8, 1969.
75. **Beaven, M. A., Marshall, J. R., Baylin, S. B., and Sjoerdsma, A.,** Changes in plasma histaminase activity during normal early human pregnancy and pregnancy disorders, *Am. J. Obstet. Gynecol.*, 123, 605, 1975.
76. **Kobayashi, Y.,** Blood levels of diamine oxidase as a test for pregnancy, *J. Lab. Clin. Med.*, 62, 699, 1963.
77. **Elmfors, B. and Tryding, N.,** Date of confinement prediction from serum diamine oxidase determination in early pregnancy, *Br. J. Obstet. Gynaecol.*, 83, 6, 1976.
78. **Ward, H., Whyley, G. A., and Millar, M. D.,** A comparative study of diamine oxidase in amniotic fluid and serum in normal pregnancy and in pregnancy complicated by rhesus iso-immunization, *J. Obstet. Gynaecol.*, 80, 525, 1973.
79. **Elmfors, B., Tryding, N., and Tufvesson, G.,** The diagnosis of ruptured fetal membranes by measurement of the diamine oxidase (DAO) activity in vaginal fluid, *J. Obstet. Gynaecol.*, 81, 361, 1974.
80. **Gahl, W. A., Kozina, T. J., Fuhrmann, D. D., and Vale, A. M.,** Diamine oxidase in the diagnosis of ruptured fetal membranes, *Obstet. Gynecol.*, 60, 297, 1982.
81. **Southren, A. L., Kobayashi, Y., Sherman, D. H., Levine, L., Gordon, G., and Weingold, A. B.,** Plasma diamine oxidase in human pregnancy: plasma diamine oxidase in non-pregnant subjects and in normal pregnancy, *Am. J. Obstet. Gynecol.*, 89, 199, 1964.
82. **Fogel, W. A., Bieganski, T., Bechtold, E., and Maslinski, C.,** Diamine oxidase activity and imidazoleacetic acid formation in the fetal and maternal guinea pig liver, *Agents Actions*, 12, 49, 1982.
83. **Kusche, J., Bieganski, T., Hesterberg, R., Stahlknecht, C.-D., Feussner, K.-D., Stahlenberg, I., and Lorenz, W.,** The influence of carcinoma growth on diamine oxidase activity in human gastrointestinal tract, *Agents Actions*, 10, 110, 1980.

84. **Morgan, D. M. L. and Illei, G.,** Polyamine- polyamine oxidase interaction: part of maternal protective mechanism against fetal rejection, *Br. Med. J.*, 280, 1295, 1980.
85. **Tewari, S. and Baxter, C. F.,** Stimulatory effect of γ-aminobutyric acid upon amino acid incorporation into protein by a ribosomal system from immature rat brain, *J. Neurochem.*, 16, 171, 1969.
86. **Sandoval, M. E. and Tapia, R.,** GABA metabolism and cerebral protein metabolism, *Brain Res.*, 96, 279, 1975.
87. **Morse, D. E., Hooker, N., Duncan, H., and Jensen, L.,** γ-Aminobutyric acid, a neurotransmitter, induces planktonic abalone larvae to settle and begin metamorphosis, *Science*, 204, 407, 1979.
88. **Canellakis, Z. N., Milstone, L. M., Marsh, L. L., Young, P. R., and Bondy, P. K.,** GABA from putrescine is bound in macromolecular form in keratinocytes, *Life Sci.*, 33, 599, 1983.
89. **Godfraind, J. M., Krnjevic, K., Maretic, H., and Pumain, R.,** Inhibition of cortical neurones by imidazole and some derivatives, *Can. J. Physiol. Pharmac.*, 51, 790, 1973.
90. **Barker, J. L.,, Nicoll, R. N., and Padjen, P.,** Studies on convulsants in the isolated frog spinal cord. I. Antagonism of amino acid responses, *J. Physiol.*, 245, 521, 1975.
91. **Bowery, N. G. and Jones, G. P.,** A comparison of aminobutyric acid and semi-rigid analogues 4-aminotetrolic acid, 4-aminocrotonic acid and imidazole-4-acetic acid on isolated superiol cervical ganglion of the cat, *Br. J. Pharmac.*, 56, 323, 1976.
92. **Luk, G. D., Bayless, T. M., and Baylin, S. B.,** Diamine oxidase (histaminase), a circulating marker for rat intestinal mucosal maturation and integrity, *J. Clin. Invest.*, 66, 66, 1980.
93. **Luk, G. D., Marton, L. J. M., and Baylin, S. B.,** Ornithine decarboxylase is important in intestinal mucosal maturation and recovery from injury in rats, *Science*, 210, 195, 1980.
94. **Jänne, J. and Hölttä, E.,** Putrescine metabolizing enzyme activities in some rat tissues during postnatal development, *Acta Chem. Scand.*, 27, 2399, 1973.
95. **Smith, M. W. and Jarvis, L. G.,** Growth and cell replacement in the newborn pig intestine, *Proc. R. Soc. London B.*, 203, 69, 1978.
96. **Buts, J.-P., Meyer, R., and Kolanowski, J.,** Ontogeny of cell proliferation and DNA synthesis in rat colon: role of glucocorticoids, *Am. J. Physiol.*, 244, G469, 1983.
97. **Ball, W. J. and Balis, M. E.,** Changes in ornithine decarboxylase activity in rat intestines during aging, *Cancer Res.*, 36, 3312, 1976.
98. **Koldovsky, O., Sunshine, P., and Kretchmer, N.,** Cellular migration of intestinal epithelia in suckling and weaned rats, *Nature (London)*, 212, 1389, 1966.
99. **Maudsley, D. V., Leif, J., and Kobayashi, Y.,** Ornithine decarboxylase in rat small intestine: stimulation with food or insulin, *Am. J. Physiol.*, 231, 1557, 1976.
100. **Stanley, B. A. and Kazarinoff, M. N.,** Induction of ornithine decarboxylase in colon and liver by starvation and refeeding: a comparison of effects on total and holoenzyme, *Biochem. Biophys. Res. Commun.*, 105, 773, 1982.
101. **Elder, J. B., Williams, G., Lacey, E., and Gregory, H.,** Cellular localization of human urogastrone epidermal growth factor, *Nature (London)*, 271, 466, 1978.
102. **Gregory, H., Walsh, S., and Hopkins, C. R.,** The identification of urogastrone in serum, saliva and gastric juice, *Gastroenterology*, 77, 313, 1979.
103. **Carpenter, G.,** Epidermal growth factor is a major growth promoting agent in human milk, *Science*, 210, 198, 1980.
104. **Feldman, E. J., Aures, D., and Grossman, M. I.,** Epidermal growth factor stimulates ornithine decarboxylase activity in digestive tract of mouse, *Proc. Soc. Exp. Biol. Med.*, 159, 400, 1978.
105. **Dembinski A., Gregory, H., Konturek, S. J., and Polanski, M.,** Trophic action of epidermal growth factor on the pancreas and gastroduodenal mucosa in rat, *J. Physiol.*, 325, 35, 1982.
106. **Lipkin, M. and Bell, B.,** Cell proliferation, in *Handbook of Physiology* Vol. 5, Sec. 6, *Alimentary Canal*, Code, C. F. and Heidel, W., Eds., American Physiological Society, Washington, D.C., 1968, 2861.
107. **Bieganski, T. and Ulatowska, M.,** Diamine oxidase in the hen, *Agents Actions*, 13, 257, 1983.
108. **Kusche, J., Lorenz, W., Stahlknecht, C.-D., Friedrich, A., Schmidt, A., Boo, K., and Reichert, G.,** Diamine oxidase activity in gastric and duodenal mucosa of man and other mammals with special reference to the pyloric junction, *Agents Actions*, 8, 366, 1978.
109. **Kim, K. S., Backus, B., Harris, M., and Rourke, P.,** Distribution of diamine oxidase and imidazole-N-methyltransferase along the gastrointestinal tract, *Comp. Biochem. Physiol.*, 31, 137, 1969.
110. **Bieganski, T.,** Biochemical, physiological and pathophysiological aspects of intestinal diamine oxidase, *Acta Physiol. Pol.*, 34, 139, 1983.
111. **Baylin, S. B., Stevens, S. A., and Shakir, K. M. M.,** Association of diamine oxidase and ornithine decarboxylase with maturing cells in rapidly proliferating epithelium, *Biochim. Biophys. Acta*, 541, 415, 1978.
112. **Neugebauer, E. and Lorenz, W.,** Histamine in health and disease, *Behring Inst. Mitt.*, 68, 102, 1981.
113. **Tutton, P. J. M.,** The influence of histamine on epithelial cell proliferation in the jejunum of the rat, *Clin. Exp. Pharmacol. Physiol.*, 3, 369, 1976.

114. **Tutton, P. J. M. and Steel, G. G.,** Influence of biogenic amines on the growth of xenografted human colorectal carcinomas, *Br. J. Cancer*, 40, 743, 1979.
115. **Bieganski, T., Kusche, J., Lorenz, W., Hesterberg, R., Stahlknecht, C.-D., and Feussner, K.-D.,** Distribution and properties of human intestinal diamine oxidase and its relevance for the histamine catabolism, *Biochim. Biophys. Acta*, 756, 196, 1983.
116. **Kusche, J., Lorenz, W., and Schmidt, J.,** Oxidative deamination of biogenic amines by intestinal amine oxidases: histamine is specifically inactivated by diamine oxidase, *Hoppe-Seyler's Z. Physiol. Chem.*, 356, 1485, 1975.
117. **Kusche, J., Schmidt, J., Schmidt, A., and Lorenz, W.,** Diamine oxidase in the small intestine of rabbits, dogs and pigs, separation from soluble monoamine oxidases and substrate specificity of the enzymes, *Agents Actions*, 5, 440, 1975.
118. **Barth, H., Lorenz, W., and Niemeyer, I.,** Inhibition and activation of histamine methyltransferase by methylated histamine, *Hoppe-Seyler's Z. Physiol. Chem.*, 354, 1021, 1973.
119. **Endo, Y.,** Simultaneous induction of histidine and ornithine decarboxylases and changes in their product amines following the injection of *Escherichia coli* lipopolysaccharide into mice, *Biochem. Pharmacol.*, 31, 1643, 1982.
120. **Graham, P., Harris, J. M., and Salter, M. R.,** The effect of ultraviolet light on skin histidine decarboxylase activity, *Br. J. Dermatol.*, 106, 732, 1982.
121. **Lowe, N. J., Verma, A. K., and Boutwell, R. K.,** Ultraviolet light induces epidermal ornithine decarboxylase activity, *J. Invest. Dermatol.*, 71, 417, 1978.
122. **Galston, A. W.,** Polyamines as modulators of plant development, *BioScience*, 33, 382, 1983.
123. **Werle, E. and Pechmann, E.,** Über die Diamin-oxydase der Pflanzen und ihre adaptive Bildung durch Bakterien, *Liebig's Ann. Chem.*, 562, 44, 1949.
124. **Hill, J. M. and Mann, P. J. G.,** Further properties of the diamine oxidase of pea seedling, *Biochem. J.*, 91, 171, 1964.
125. **Suzuki, Y.,** Some properties of the amine oxidase of soybean seedlings, *Plant Cell Physiol.*, 14, 413, 1973.
126. **Macholan, L.,** Selective and reversible inhibition of diamine oxidase by 1,5-diamino-3-pentanone, *Collection Czechoslov. Chem. Commun.*, 39, 653, 1974.
127. **Bieganski, T., Osinska, Z., and Maslinski, C.,** Inhibition of plant and mammalian diamine oxidase by substrate analogues, *Agents Actions*, 12, 41, 1982.
128. **Okuyama, T. and Kobayashi, Y.,** Determination of diamine oxidase activity by liquid scintillation counting, *Arch. Biochem. Biophys.*, 95, 242, 1961.
129. **Fogel, W. A.,** personal communication.

Chapter 15

PATHOPHYSIOLOGICAL FUNCTIONS OF DIAMINE OXIDASE: AN EVALUATION IN ANIMAL STUDIES USING A PROBABILISTIC MODEL WITH SEVERAL TYPES OF CAUSAL RELATIONSHIPS

Jörg Sattler, Rudolf Hesterberg, Wilfried Lorenz, Madeleine Ennis, Carl-Detmar Stahlknecht, and Jürgen Kusche

TABLE OF CONTENTS

I. INTRODUCTION

Enzymes existing in man and animals are shown or are thought to have specific functions. To elucidate these functions usually both the substrates and the products of the enzyme-catalyzed reaction must be investigated, which themselves possess dynamic and/or stationary (structural) properties.

Diamine oxidase (EC. 1.4.3.6) (DAO) catalyzes the oxidative deamination of the potent pharmacological agent histamine and of polyamines such as putrescine, cadaverine, and, to a lesser extent, spermidine and spermine. The products of the catalytic action of DAO on histamine and putrescine are imidazole acetaldehyde and γ-aminobutyric aldehyde, respectively. These have no obvious biological functions. Hence, more attention has been directed to the substrates histamine and putrescine.

Although the *physiological* functions of histamine have not been clarified in general, it is at least an important messenger in gastric acid secretion.[1] Diamine oxidase activity is low in the normal oxyntic mucosa and only appears to have a minor function in healthy stomachs. However, DAO seems to serve as a protective factor where free histamine levels are endogenously or exogenously raised in normal conditions, such as feeding.[2-6] For putrescine the situation is similar. Its synthesis from ornithine in vivo[7,8] with the subsequent formation of spermidine and spermine is well documented. However, the contribution of the oxidative deamination catalyzed by DAO as a regulatory principle in this metabolic pathway is not clear. In certain growing processes, such as wound healing, both spermidine synthesis and DAO activity are enhanced, and blocking of DAO results in further elevated spermidine levels. Hence DAO is supposed to be a regulator of these growing processes,[9,10] but further work is needed to establish this physiological function of DAO.

A *pathological* role for the two major interesting substrates of DAO is easier to find. Histamine as a ligand at receptors has an important function in allergic and pseudoallergic reactions,[11-13] in inflammatory diseases[14] including chronic peptic ulcer,[1] in intestinal ischemia,[5] and in drug-induced diseases.[13,13a] Putrescine turnover in association with spermidine formation has been implicated in rapid cell proliferation of carcinomas and similar disease states;[9,10,16,17] however, more work is necessary to establish this pathological function.

Most of the studies on a pathological role of DAO with respect to histamine and putrescine have been conducted in animals. Hence this review will consider only animal models and will offer possibilities to demonstrate pathophysiological functions of DAO.

II. DAO ACTIVITY AS A PATHOLOGICAL FACTOR: APPLICATION OF A PROBABILISTIC MODEL ON CAUSAL RELATIONS WITH NECESSARY AND SUFFICIENT DETERMINANTS AND CONTRIBUTORY CONDITIONS

In the last 2 years a method has been developed to demonstrate histamine as a pathological factor even in the presence of a battery of other mediators[18] (Figure 1). It is based on the concept of several kinds of causal relationship, including necessary and sufficient determinants and contributory conditions.[19] When a drug is administered in anesthesia and surgery, when histamine release is measured and the clinical syndrome of an adverse reaction is observed, then only one prerequisite has to be fullfilled in all three types of causal relationship: the preceding step has to be histamine release before the adverse reaction is detectable. Koch-Dale's[20] criteria for a pathological factor are included in this model.

Not only small molecules, such as histamine, but also proteins and enzymes[21,22] can serve as mediators in pathological reactions. Thus, a similar concept as shown for histamine can be used for analyzing their pathophysiological role. The situation is more complex, however, for demonstrating a similar function for DAO. This is illustrated in Figure 2. To characterize several states of enzymic activity and of health and disease we define:

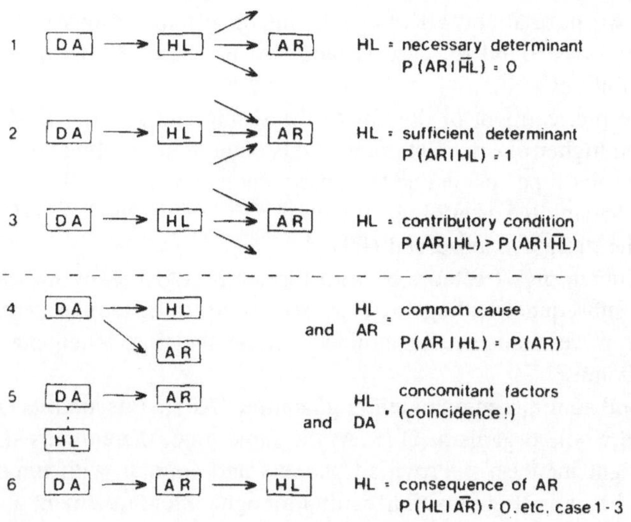

FIGURE 1. A probabilistic model for describing several types of causal relationship between drug administration, histamine release, and adverse reactions. Vector of time indicated by arrows. DA = drug administration; HL = histamine liberation; AR = allergic and pseudoallergic response; P = probability; P(ARIHL) = population probability for the event AR given that the event HL took place. The statistical model in type 5 is omitted. It may be similar to 4 but the situation is more complex. For example, the presence of HL might make that of DA more probable and hence increase the probability of AR. Flow diagrams constructed according to Wulff.[19] For further explanation and detailed examples, see Lorenz et al.[18]

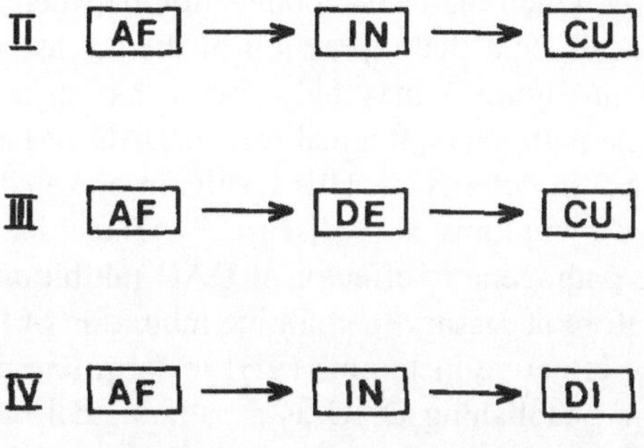

FIGURE 2. Four possibilities (categories) of causal relationships between an etiological factor (AF), changes in DAO activity (decrease [DE] or increase [IN]), and disease presentation (DI) or prevention and cure of a disease (CU). Vector of time indicated by arrows.

- An etiological factor of a disease as AF
- The increase of DAO activity as IN
- The decrease (either inhibition or reduction of DAO activity) as DE
- The appearance of a disease or further enhancement of clinical symptoms of a present disease as DI
- The disappearance of a disease or lowering of some symptoms (cure, change for the better) as CU

Using these definitions we expect four major possibilities to show a causality chain between the etiological factors, altered DAO activities, and the states of disease (Figure 2). Indeed, in the literature we have found all of them, but most publications reflect cases I and II. Only one example was found for case III and two for Case IV. They are given below for each of the possibilities

Case I — The pretreatment of dogs with the strong and specific DAO inhibitor aminoguanidine leads to higher plasma histamine levels compared to saline controls after intestinal ischemia and vascular reperfusion and in consequence to a drastically reduced survival time. In this case aminoguanidine as etiological factor (AF) inhibits the DAO (DE) and as a result, aggravation of the disease is observed (DI).[23]

Case II — Guinea pigs treated s.c. with turpentine (AF) show increased plasma DAO levels (IN). On subsequent testing for their sensitivity to histamine aerosol there is a significant increase in resistance to histamine (CU) at the time when plasma DAO activity reaches its maximum.[24]

Case III — Oral administration of aminoguanidine (AF) to rats inhibits DAO and decreases its activity in the whole organism (DE). At the same time, the animals show stimulation of appetite and weight increase.[25] Provided patients and animals with tumors suffering from anorexia and cachexia will gain weight with aminoguanidine treatment and decreased DAO activity, this effect of the drug could be interpreted as an improvement of the disease (CU).[25]

Case IV/1 — In many patients with gastric carcinoma (AF) increased DAO activities are found in the noninfiltrated parts of the gastric mucosa (IN) which, in addition to macroscopic inspection, was shown to be normal by microscopical examination. However, in these patients low acid secretion or even achlorhydria is observed that counts as disease (DI).[1]

Case IV/2 — Age-related rise in blood pressure that gradually develops in rats (AF) is followed by a significant increase in putrescine concentration and DAO activity (IN). The authors[10] speculate that the increase in putrescine turnover could play a role in maintaining the cardiomegaly in hypertensive rats (DI).

The question of whether or not DAO is a pathological factor cannot be answered in real life by drawing arrows as vectors of time (Figure 2). The four possibilities remain speculative if no method is available by which the causal connection between the postulated etiological factor, the pathogenetic factor, and the expression of the disease can be tested. The probabilistic model illustrated in Figure 1 may help, but it has to be slightly extended to be suitable for investigating the pathophysiological role of DAO. It has been demonstrated that inhibition of DAO (decrease in activity = DE) results in elevated plasma histamine levels which aggravate histamine symptoms as a disease.[25a] Thus, increased plasma histamine levels is thought to be the pathogenetic effector of DAO inhibition as already postulated in the introduction. It is therefore necessary to combine inhibition of DAO (DE) with elevation of free, pharmacologically effective histamine (EH) to form *one* pathogenetic factor. Then the probabilistic model for establishing DAO as a pathological factor can be developed in the same way as for histamine release (Figure 3). Augmented putrescine levels in connection with DAO inhibition can be handled in the same way.

The literature is full of examples of case I in Figure 2. In association with etiological factors (drugs, environment, genetic factors, operations), a decrease in DAO activity is observed which is claimed to cause a certain clinical syndrome or a disease. For this reason we have formulated the application of the probabilistic model in Figure 3 only for the decrease in DAO activity. However, the same 6 possibilities as shown in Figure 3 for the decrease in DAO activity can also be shown for the cases II to IV in Figure 2.

III. ETIOLOGICAL FACTORS, DECREASE IN DAO ACTIVITY AND APPEARANCE OF DISEASE, COMPLICATION, OR AGGRAVATION OF SYMPTOMS: AN EXAMPLE FOR SEVERAL POSSIBILITIES OF CAUSAL RELATIONSHIP IN THE NEW MODEL

In many reports in the literature experimental studies are not designed with the aim to

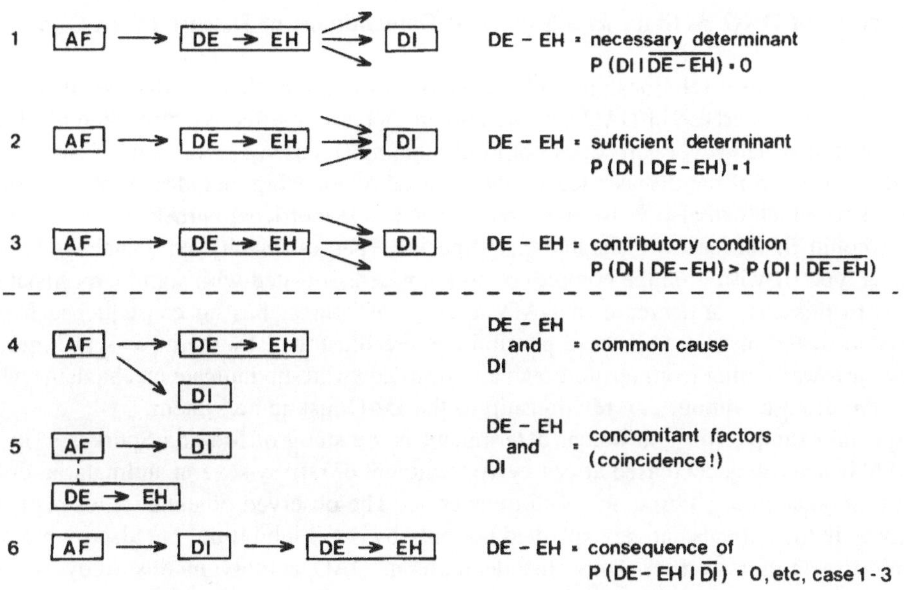

FIGURE 3. A probabilistic model for denoting several types of causal relationship between etiological factors, decrease in DAO activity followed by elevated histamine levels and presentation of a disease. AF = etiological factor, DE = decrease or inhibition of DAO activity, EH = elevated level of free, pharmacologically active histamine, DI = occurrence of symptoms, complications, or aggravation of a disease. Vector of time indicated by arrows. P(DI|DE-EH) = population probability for the occurrence of disease given that a decrease in DAO activity takes place. According to Lorenz et al.,[18] the statistical model in type 5 is omitted since the situation is fairly complex. Flow diagrams constructed according to Wulff.[19] For further explanations, see text.

demonstrate the full causality chain from etiological factors via pathogenetical factors to the disease. However, in some cases, several studies have to be combined to fill the gaps or some connections have to be discussed only theoretically.

A. Decrease of DAO Activity as a Necessary Determinant of Disease (Type 1 in Figure 3)

If the decrease of DAO activity (DE) and, consecutively, the elevated effective histamine level (EH) is supposed to be a necessary determinant of disease (DI), DE *always* has to precede DI. In addition, no expression of disease or worsening of disease can be expected to occur if the decrease of DAO activity and the increase in effective histamine levels cannot be demonstrated. However, the disease does not necessarily succeed the decrease in DAO activity (arrows with up and down direction). For instance, minor changes in DAO activity and histamine levels are not necessarily followed by clinical symptoms, or blocking histamine receptors by the corresponding antagonists may prevent expression of the disease despite considerable changes in DAO activity and histamine levels.

In our opinion, the experiments on blocking DAO by aminoguanidine and intestinal ischemia[5,23] are examples for the causal relationship type I. However, only the aggravation of the disease, the shortening of survival time after superior mesenteric artery occlusion, and vascular reperfusion can be considered as type 1 relationship, not intestinal ischemia per se, which is highly multifactorial. The finding that inhibition of DAO and elevation of effective plasma histamine levels is a necessary determinant for the expression of disease is supported by the complete abolishing of the effects of DAO blockade by histamine H_1 + H_2-receptor antagonists.[5]

B. Decrease of DAO Activity as a Sufficient Determinant of Disease (Type 2 in Figure 3)

This type of causal relationship is characterized by the fact that the disease always has to be present if a decrease in DAO activity and an increase in effective histamine levels can be demonstrated. However, this does not imply that a decrease of DAO activity always has to be demonstrated if the disease occurs. This causal relationship includes all other possible mechanisms which can also cause the disease, such as leukotrienes, prostaglandins, kinins, and serotonin in the case of anaphylaxis. Special types of anaphylaxis, such as allergic reactions type III with immune complexes may not be associated with significant histamine release. In this case, a decrease in DAO activity, of course, has no consequence for the expression of the disease. All these possibilities are illustrated by the arrows in Figure 3. The two arrows coming from upsite-down and from downsite-up indicate mechanisms which induce the disease without any relationship to the DAO/histamine system.

A special example for a sufficient determinant is the study of Ma and Sourkes.[26] In this study DAO activity is decreased after in vivo treatment of rats by several antimalarial drugs, such as amodiaquine, chloroquine, and quinacrine. The observed potentiation of histamine sensitivity in the animals can be explained not only by DAO inhibition, but also by blocking histamine methyltransferase. Thus, the decrease in DAO activity in this study is not a necessary determinant for elevated histamine levels, but only a sufficient determinant.

C. Decrease of DAO Activity as a Contributory Condition of Disease (Type 3 in Figure 3)

In type 3 of the probabilistic model, decrease of DAO activity and elevated levels of free histamine do not *always* lead to symptoms, aggravation, or complication of a disease, but there exists an *increased probability* of the disease if a decrease in DAO activity is observed (conditional probability). In other words, in cases in which an etiological factor and DE-EH are demonstrated for a contributory condition, the incidence of the disease is expected to be higher than in those cases in which the etiological factor and the disease are present, but no decrease in DAO activity and no increase in free histamine can be measured ($P[DI|DE\text{-}EH] > P[DI|\overline{DE\text{-}EH}]$). The ratio $P(DI|DE\text{-}EH)$: $P(DI|\overline{DE\text{-}EH})$ is called the *likelihood ratio* and has to be remarkably greater than 1 or better (after several repetitions of the experiments) *significantly* greater than 1. Assessment of the likelihood ratio is a fairly sensitive method to establish a causal relationship in those cases in which a connection between two events cannot be recognized at first sight.[18] It still works when in some individuals the disease is present without a decrease in DAO activity and in some others a decrease in DAO activity is observed without the eruption of clinical symptoms of a disease. As in other fields of experimental and clinical medicine, most of the causal relationships belong to that of a contributory condition. Symptoms of a disease in which DAO activity may be involved are often caused by a variety of chemical factors, such as allergic and inflammatory diseases, ischemia, tumors, and other complicated disorders. In addition, contributory conditions are more easily observed or suspected than sufficient and necessary determinants, but it is a difficult way to find, in the complicated sequence of the events, exactly that particular step which fulfills the criteria of a sufficient determinant or even that of a necessary determinant. In the field of DAO and its substrates it seems useful first to recognize any kind of causal relationship and then to start analyzing the complex in vivo-conditions with a more specialized design of an experimental study.

The probabilistic model also works in the case of putrescine when in the likelihood ratio EH is substituted by EP, i.e., by elevated levels or decreased turnover of putrescine. Many examples for contributory conditions will be provided in the tables of Section IV.

D. Types of Coordination Other than Causal Relationship Between Decrease of DAO Activity and Disease (Types 4 to 6 in Figure 3)

Along with the three causal connections, other types of coordination are possible. The decrease of DAO activity and the appearance of disease may have a common cause (type 4) or the decrease in DAO activity and disease are just coincidental (type 5). These coordinations, however, can be identified usually if the likelihood ratio $P(DI|DE\text{-}EH: P(DI|\overline{DE\text{-}EH})$ tends to be 1 (equal). The last interesting type of coordination is the possibility that a decrease in DAO activity is not the cause, but the consequence of the disease. This causal relationship can be identified by exchanging event and conditional event in the formulae, such as: $P(DE\text{-}EH|\overline{DI}) = 0$, $P(DE\text{-}EH|DI) = 1$, etc. The advantage of a more formal mathematical analysis of coordination between several events is found in much more differentiated answers to the question whether changes in DAO activity in a given set of data are of pathophysiological significance than in the classical method, including Koch-Dale's criteria alone, which only works satisfactorily in the case of a necessary determinant.

IV. ANALYSIS OF ANIMAL STUDIES WITH THE PROBABILISTIC MODEL: FACTS AND HYPOTHESES ON PATHOPHYSIOLOGICAL FUNCTIONS OF DAO

Although many authors have examined alterations of DAO activity in animal studies, very few have investigated the pathophysiogical significance of these changes. Complete studies, with determination of all relevant parameters for a causality chain are especially thin on the ground. To analyze, therefore, a pathophysiological function of DAO with our probabilistic model, we are forced to fill the gaps in the causality chains by hypotheses which are more or less evident in medical practice.

Most of the pathological states in which changes of DAO activity have been reported can be put into category I or II in Figure 2. These are, therefore, especially itemized in Tables 1 and 2 with respect to their inferred type of causal relationship as described in Section III. The procedure of classification as a necessary and sufficient determinant or as a contributory condition is described in the text for some especially interesting examples. Disagreement may exist between the authors and several readers, but an extensive and fruitful discussion is expected from that.

Intestinal ischemia is one example in which the factors of the causality chain have been extensively studied in several animal species. Superior mesenteric artery occlusion and subsequent vascular reperfusion induces death of the animals within several hours. Aminoguanidine, as a potent inhibitor of DAO, decreases DAO activity and aggravates the clinical syndrome. The animals die ten times faster with aminoguanidine pretreatment than with saline treatment.[5,23,28] In addition, it has been shown that plasma histamine levels are elevated only in the animals treated with aminoguanidine. This supports a type 1 casual relationship between aminoguanidine treatment and shortening of survival times, $P(DI|\overline{DE\text{-}EH}) = 0$, where DI is the aggravation of death from mesenteric artery occlusion by aminoguanidine. Administration of H_1- and H_2-receptor antagonists completely prevents the reduction in survival times caused by aminoguanidine, providing further support. However, for the disease intestinal ischemia alone, many other mediators such as prostaglandins and oxygen radicals are also thought to be pathogenetical factors. Until further studies are carried out, these mediators including histamine can at best only be considered as sufficient determinants: $P(DI|DE\text{-}EH = 1)$, casual type 2 in Table 1. Other examples for a type 2 causal relationship are constructed in Table 1.

However, in most of the pathological states in which a pathophysiological function of DAO is considered only a type 3 causal relationship can be postulated (Table 1 and 2). Rabbits fed a cholesterol diet have decreased plasma DAO levels.[38] Cholesterol diet is the

Table 1
ANALYSIS OF CAUSAL RELATIONS BETWEEN ETIOLOGICAL FACTORS, DECREASE IN DAO ACTIVITY, AND EXPRESSION OF DISEASE (CATEGORY I IN FIGURE 2)

	Animal species	Etiological factor	Decrease of DAO (measured in)	Disease or complication	Inferred type of relationship	Ref.
1.	Dog, pig, rabbit	Mesenteric infarction	Intestine, plasma + aminoguanidine	Aggravation of circulatory shock	Causal, type 1	5, 23, 28
2.	Sheep	Aminoguanidine feeding	(Intestine)	Pseudoallergic reactions	Causal, type 2	27
3.	Cat	DAO inhibitors intravenously	(Intestine, kidney)	Increased histamine sensitivity	Causal, type 2	29
4.	Guinea pig	Histamine challenge	Liver (+ aminoguanidine)	Aggravation of anaphylactic shock	Causal, type 2	30, 31
5.	Sheep	Intrauterine device (plastic)	Uterine horn	Inflammation, cycle disturbance	Causal, type 3	32
6.	Dog	Mesenteric infarction	Intestine (ischemic)	Circulatory shock	Causal, type 3	33
7.	Dog	Treatment with iproniazid or aminoguanidine	(Intestine)	(Acute peptic ulcers)	Causal, type 3	34
8.	Cat, rat	Adrenalectomy	Liver, kidney, intestine	Pseudoallergic reactions	Causal, type 3	35, 36
9.	Cat	Reserpine intravenously	Liver	Hyperchlorhydria	Causal, type 3	37
10.	Rabbit	Cholesterol diet	Plasma	Atherosclerosis	Causal, type 3	38
11.	Guinea pig	Tuberculosis	Blood	Allergic reactions	Causal, type 3	39
12.	Guinea pig, rat	Food poisoning (Cadaverine)	(Intestine)	Pseudoallergic shock	Causal, type 3	40, 41
13.	Rat	Antimalarial drugs	Intestine	(Catalepsy), potentiation of histamine effects	Causal, type 3	26
14.	Mouse	Schistosomal infection or schistosomal drugs	Intestine	(Hypersensitivity, allergic reactions)	Causal, type 3	42
15.	Rat	Pyridoxine deficiency	Skin, intestine	(Avitaminosis)	Not causal	43
16.	Mouse	Inoculation of Ehrlich ascites tumor	Plasma (postheparin)	(Increased sensitivity to allergic reactions)	Not causal	43
17.	Mouse	Thyroxine treatment	Intestine	(Hyperthyreosis)	Not causal	45

Note: 17 examples for the first category of causal relationship (case 1 in Figure 1). A decrease (DE) of DAO activity is considered partly responsible for a disease or complication of a disease. Terms in parentheses indicate that no data were found in the literature and the gaps had to be filled by hypothetical factors.

Table 2
ANALYSIS OF CAUSAL RELATIONS BETWEEN ETIOLOGICAL FACTORS, INCREASE IN DAO ACTIVITY, AND EXPRESSION OF DISEASE (CATEGORY II IN FIGURE 2)

Animal species	Etiological factor	Increase of DAO (measured in)	Cure of a disease	Inferred type of relationship	Ref.
1. Guinea pig	Treatment with exogen DAO in shock	(Whole body)	Inhibition of bronchoconstriction	Causal, type 2	46
2. Guinea pig, rabbit	Heparin in shock	Plasma	Protection against shock	Causal, type 2	31, 47
3. Guinea pig	Turpentine inflammation	Plasma	Resistance to histamine	Causal, type 2	24
4. Guinea pig	Wound incision	Skin	Wound healing	Causal, type 3	9
5. Rat	Heparin in liver disease	Plasma	(Inhibition of liver necrosis)	Causal, type 3	48, 49
6. Rat	Olive oil feeding	Intestinal lymph	Antihistaminic effect on blood vessels	Causal, type 3	50
7. Mice	Heparin intrapulmonary (+ surgery)	Plasma	(Antihistaminic action)	Causal, Type 3	51

Note: 7 examples for the second category of causal relationship (case II in Figure 1). An increase (IN) in DAO activity induces an improvement of disease or attenuates the severity of clinical symptoms. Parentheses indicate events which are likely to occur, but which have not been investigated in the particular study. These events are suggested by the authors as possible hypotheses.

etiological factor. The relationship between atherosclerosis as the disease and the decrease of DAO activity as the pathogenetic factor is only of type 3 since many other factors can produce atherosclerosis and not every decrease in DAO activity induces atherosclerosis. However, it is important to note that the incidence of atherosclerosis in the animals in the presence of a decrease of DAO activity is higher than in its absence; thus, $P(DI|DE) > P(DI|\overline{DE})$. Hence the study of Balakrishna and Zaidi[38] is an interesting example to illustrate how the probabilistic model works.

V. CONCLUSIONS

A probabilistic model is used to investigate the pathophysiological functions of DAO in animal studies. It replaces the traditional criteria for establishing a chemical factor and distinguishes between several types of causal relationship: necessary and sufficient determinants and contributory condition.

Studying the connections between etiological factors, pathogenetic factors, and expression of disease in the field of DAO and its substrates histamine and putrescine, it becomes evident that the first type of causal relationship is very rare, the second type is more common (but not yet sufficiently established, and the third type, the contributory condition, can be attributed to most of the pathophysiological states investigated.

In many animal studies, unfortunately, biochemical and pharmacological parameters are studied extensively and precisely, but no comments are given on clinical symptoms observed during the experiments. It is necessary to speculate about their occurrence extrapolating clinical data in man to animals. It is recommended to design animal studies in such a manner that etiological factors, pathogenetic factors, and expression of the disease are defined or registered with an equivalent reliability.

The recognition of the pathophysiological role of DAO in various disease states in animals is directed towards clinical studies and trials for therapy. However, such experimental findings in animals are only pointers, and clinical trials are necessary to establish the pathophysiological role of DAO in human diseases.

ACKNOWLEDGMENT

This work was supported by a Grant of the Deutsche Forschungsgemeinschaft Lo 199/12-5. The excellent technical assistance of Marlies Crombach and Doris Weber is highly appreciated.

REFERENCES

1. **Lorenz, W., Thon, K., Barth, H., Neugebauer, E., Reimann, H.-J., and Kusche, J.,** Metabolism and function of gastric histamine in health and disease, *J. Clin. Gastroenterol.*, 5(1), 37, 1983.
2. **Best, C. H. and McHenry, E. W.,** The inactivation of histamine, *J. Physiol. (London)*, 70, 349, 1930.
3. **Buffoni, F.,** Histamine and related amine oxidases, *Pharmacol. Rev.*, 18, 1163, 1966.
4. **Kahlson, G. and Rosengren, E.,** New approaches to the physiology of histamine, *Physiol. Rev.*, 48, 155, 1968.
5. **Kusche, J., Lorenz, W., Stahlknecht, C.-D., Richter, H., Hesterberg, R., Schmal, A., Hinterlang, E., Weber, D., and Ohmann, Ch.,** Intestinal diamine oxidase and histamine release in rabbit mesenteric ischaemia, *Gastroenterology*, 80, 980, 1981.
6. **Wollin, A. and Navert, H.,** Release of intestinal diamine oxidase by histamine in rats, *Can. J. Physiol. Pharmacol.*, 61, 349, 1983.

7. **Jänne, J. and Raina, A.,** Stimulation of spermidine synthesis in the regenerating rat liver: relation to increased ornithine decarboxylase activity, *Acta Chem. Scand.*, 22, 1349, 1968.
8. **Pegg, A. E. and Williams-Ashman, H. G.,** Biosynthesis of putrescine in the prostate gland of the rat, *Biochem. J.*, 108, 533, 1968.
9. **Woźniak, J., BIegański, T., and Maśliński, Cz.,** Diamine oxidase activity during wound healing in guinea pig skin, *Agents Actions*, 9(1), 45, 1979.
10. **Perin, A., Sessa, A., and Desiderio, M. A.,** Polyamine levels and diamine oxidase activity in hypertrophic heart of spontaneously hypertensive rats and of rats treated with isoproterenol, *Biochim. Biophys. Acta*, 755, 344, 1983.
11. **Coombs, R. R. A. and Gell, P. G. H.,** Classification of allergic reactions responsible for hypersensitivity and disease, in *Clinical Aspects of Immunology*, Gell, P. G. H., Coombs, R. R. A., and Lachmann, P. J., Eds., Blackwell Scientific, Oxford, 1975, 761.
12. **Kalloś, P. and Kalloś, L.,** Histamine and some other mediators of pseudo-allergic reactions, in *PAR. Pseudo-Allergic Reactions. Involvement of Drugs and Chemicals*, Vol. 1, Dukor, P., Kalloś, P., Schlumberger, H. D., and West, G. B., Eds., Karger, Basel, 1982, 28.
13. **Watkins, J. and Thornton, J. A.,** Immunological and non-immunological mechanisms involved in adverse reactions to drugs, *Klin. Wochenschr.*, 60, 958, 1982.
13a. **Lorenz, W., Doenicke, A., Schöning, B., and Neugebaver, E.,** The role of histamine in adverse reactions to intravenous agents, in *Adverse Reactions of Anesthetic Drugs*, Thornton, A., Ed., Elsevier/North-Holland, Amsterdam, 1981, 169.
14. **Plaut, M. and Lichtenstein, L. M.,** Inflammatory mediators released from cells, in *Inflammation*, Vane, J. R. and Reweira, S. H., Eds., Springer, Berlin, 1978, chap. 11.
15. **Russell, D. and Snyder, S. H.,** Amine synthesis in rapidly growing tissues: ornithine decarboxylase activity in regenerating rat liver, chick embryo and various tumors, *Proc. Natl. Acad. Sci. U.S.A.*, 60, 1420, 1968.
16. **Snyder, S. H., Kreuz, D. S., Medina, V. J., and Russell, D. H.,** Polyamine synthesis and turnover in rapidly growing tissues, *Ann. N.Y. Acad. Sci.*, 171, 749, 1970.
17. **Matsuzaki, S., Suzuki, M., Hamana, K., and Itoh, K.,** Elevated levels of polyamines and histamine in adenocarcinomas of the thyroid, *J. Clin. Endocrinol. Metab.*, 47, 1038, 1978.
18. **Lorenz, W., Röher, H. D., Doenicke, A., and Ohmann, Ch.,** Histamine release in anaesthesia and surgery: a new method to evaluate its clinical significance with several types of causal relationship, *Clin. Anesthesiol.*, 2, 403, 1984.
19. **Wulff, H. R.,** *Rational Diagnosis and Treatment*, Blackwell Scientific, Oxford, 1976, 182.
20. **Dale, H. H.,** Croonian lectures on some chemical factors in the control of the circulation, *Lancet*, 1, 1285, 1929.
21. **Altura, B. M.,** Reticuloendothelial cells and host defense, *Adv. Microcirc.*, 9, 252, 1980.
22. **Riches, D. W. H. and Stanworth, D. R.,** Studies on the possible involvement of complement component C3 in the initiation of acid hydrolase secretion by macrophages, *Immunology*, 44, 29, 1981.
23. **Kusche, J., Stahlknecht, C.-D, Lorenz, W., Reichert, G., and Richter, H.,** Diamine oxidase activity and histamine release in dogs following acute mesenteric artery occlusion, *Agents Actions*, 7, 81, 1977.
24. **Biegański, T., and Maśliński, Cz.,** The histaminase activity of guinea-pig's plasma during experimental inflammation. II. Some properties of the enzyme, *Agents Actions*, 4, 69, 1974.
25. **Baylin, S., Horakova, Z., and Beaven, M. A.,** Increase in food consumption and growth after treatment with aminoguanidine, *Experientia*, 31, 562, 1975.
25a. **Lorenz, W., Thermann, M., Messmer, K., Schmal, A., Dormann, P., Kusche, J., Barth, H., Tauber, R., Hutzel, M., Mann, G., and Uhlig, R.,** Evaluation of histamine elimination curves in plasma and whole blood of several circulatory regions: a method for studying kinetics of histamine release in the whole animal, *Agents Actions*, 4, 188, 1974.
26. **Ma, K. and Sourkes, T. L.,** Inhibition of diamine oxidase by antimalarial drugs, *Agents Actions*, 10, 395, 1980.
27. **Sjaastadt, O. V.,** Potentiation of aminoguanidine of the sensitivity of sheep to histamine given by mouth. Effect of aminoguanidine in urinary excretion of endogenous histamine, *Q. J. Exp. Physiol.*, 52, 319, 1967.
28. **Kusche, J. Stahlknecht, C.-D., Lorenz, W., Reichert, G., and Dietz, W.,** Comparison of alterations in the histamine-diamine oxidase system during acute intestinal ischaemia in pigs, dogs and rabbits: evidence for a uniform pathophysiological mechanism?, *Agents Actions*, 9, 49, 1979.
29. **Lindell, S.-E. and Westling, H.,** Potentiation by histaminase inhibitors of the blood pressure responses to histamine in cats, *Acta Physiol. Scand.*, 37, 307, 1956.
30. **Hahn, F., Pröhle, F., Mitze, R., and Degand, L.,** Changes in the histaminolytic activity of the guinea-pig liver and plasma during anaphylaxis and after heparin injection, *Int. Arch. Allergy Appl. Immunol.*, 40, 340, 1971.

31. **Giertz, H., Hahn, F., Seseke, G., and Schmutzler, W.,** Über die Wirkung von Heparin und Aminoguanidin auf den anaphylaktischen, Anaphylatoxin- und Histaminschock des Meerschweinchens, *Naunyn-Schmiedebergs Arch. Pharmakol. Exp. Pathol.*, 256, 26, 1967.
32. **Cooper, B. S., Hawk, H. W., and Scommegna, A.,** Influence of ovarian status and intrauterine devices on the diamine oxidase and histamine content of sheep endometrium, *J. Anim. Sci.*, 31, 1191, 1970.
33. **Kusche, J., Richter, H., Schuman, K. T., Schmidt, J., and Lorenz, W.,** Diamine oxidase, histamine and histidine in the small intestine after experimental ischaemia, *J. Pharmacol. (Paris)*, 3, 526, 1972.
34. **Liepins, K. W., Ivy, A. C., and Suzuki, R.,** Effect of certain histaminase inhibitors on gastric secretion, *Am. J. Physiol.*, 195, 209, 1958.
35. **Carlsten, A.,** On the sources of the histaminase present in thoracic duct lymph, *Acta Physiol. Scand.*, 20(70), 5, 1950.
36. **Haeger, K., Jacobsohn, D., and Kahlson, G.,** The levels of histaminase and histamine in the gastrointestinal mucosa and kidney of the cat deprived of the hypophysis or the adrenal glands, *Acta Physiol. Scand.*, 25, 243, 1952.
37. **Udwadia, B. P., Sachdev, K. S., and Joshi, D. J.,** Effect of reserpine on the histaminolytic activity of guinea-pig liver, *J. Pharm. Pharmacol.*, 18, 321, 1966.
38. **Balakrishna, I. and Zaidi, S. H.,** Serum histaminase activity in experimental atherosclerosis, *J. Sci. Ind. Res. (India)*, 20C, 65, 1961.
39. **Werle, E. and Koch, O.,** Die Aktivität der Bluthistaminase bei der experimentellen Tuberkulose des Meerschweinchens, sowie bei Scharlach und Diptherie, *Beitr. Klin. Tuberk.*, 101, 151, 1948.
40. **Bjeldanes, L. F., Schultz, D. E., and Morris, M. M.,** On the aetiology of scombroid poisoning: cadaverine potentiation of histamine toxicity in the guinea-pig, *Food Cosmet. Toxicol.*, 16, 157, 1978.
41. **Lyons, D. E., Beery, J. T., Lyons, S. A., and Taylor S. L.,** Cadaverine and aminoguanidine potentiate the uptake of histamine in vitro in perfused intestinal segments of rats, *Toxicol. Appl. Pharmacol.*, 70, 445, 1983.
42. **Mikhail, M. M., Ghoneim, M. Th., Abdel-Maguid, R., and Nasser, S. H.,** Effect of schistosomal infection and of schistosomicidal drugs on histamine forming capacity and diamine oxidase activity in the mice, *Bull. Alex. Fac. Med.*, 17, 319, 1981.
43. **Sinclair, H. M.,** Pyridoxal phosphate as coenzyme of histaminase, *Biochem. J.*, 51, 10, 1952.
44. **Gäng, V.,** Die Postheparin-Diaminoxidase (Histaminase) bei Entzündungen, Habilitation thesis, University of Würzburg, 1978, 99.
45. **Spencer, P. S. J. and West, G. B.,** Further observations on the effect of the thyroid gland on the sensitivity of mice to histamine, *Arch. Int. Pharmacodyn. Ther.*, 148, 31, 1964.
46. **Kitao, T. and Hattori, K.,** Purification of pig kidney diamine oxidase (histaminase) and in vivo effects on passive cutaneous anaphylaxis and histamine induced bronchoconstriction of guinea-pigs, *Experientia*, 38, 906, 1982.
47. **Dhar, H. L., Mukherjee, B., and Sanyal, R. K.,** The effect of heparin on the heart in anaphylaxis, *Am. Heart J.*, 74, 489, 1967.
48. **Bartsokas, S. K., Papadimitrion, D. G., Papacharalampous, N. X., Katapoti, Z., and Kannos, D.,** Heparinoids — a new inhibitor of experimental liver necrosis in rats, *Acta Hepato-Gastroenterol.*, 21, 253, 1974.
49. **Schütterle, G. and Platt, D.,** Die Beeinflussung der experimentellen Glomerulonephritis durch Heparin und Prednisolon, *Klin. Wochenschr.*, 48, 179, 1970.
50. **Wollin, A. and Jaques, L. B.,** Increased diamine oxidase activity in the intestinal lymph of rats on fat ingestion, *Can. J. Physiol. Pharmacol.*, 52, 760, 1974.
51. **Mahadoo, J., Wright, C. J., and Jaques, L. B.,** Effect of intrapulmonary heparin on plasma diamine oxidase (histaminase) activity in mice, *Agents Actions*, 11, 335, 1981.

Chapter 16

DIAMINE OXIDASE IN REGENERATING AND HYPERTROPHIC TISSUES

Antonio Perin, Angela Sessa, and M. Alfonsina Desiderio

TABLE OF CONTENTS

I. INTRODUCTION

To contribute to the knowledge of the biology of cellular growth, in recent years we have studied diamine oxidase (EC 1.4.3.6), which, besides its role in histamine and cadaverine catabolism,[1-3] is rate limiting in the oxidative metabolism of putrescine in vivo.[4,5] Since putrescine rather than spermidine and spermine is considered a critical regulatory molecule for cell proliferation and growth, it has been suggested that DAO is implicated in the regulation of these processes.[6]

Although the occurrence, purification, and mechanisms of action of DAO have been object of numerous of reports, the behavior and function of this enzyme in normal and neoplastic growth need further investigation.

An increase in DAO activity has been described in some forms of rapid growth, such as in the uterus during pregnancy,[7,8] in human[9] and rat placenta,[7] as well as in the wound-healing process.[10] Elevated DAO activity has also been reported in cultures of normal and transformed rat kidney cells during the lag phase of growth[11] and of C_6-BU_1 glioma cells which have been stimulated to grow,[12] as well as in experimental rat tumors (4-dimethylaminoazobenzene-induced and Yoshida ascites hepatomas[13]), human neoplasms (medullary thyroid carcinoma, small cell carcinoma of the lung[14-16]), and human melanoma cell lines[17] (see Chapter 17).

This chapter is concerned with the behavior of DAO activity in some hyperplastic and hypertrophic tissues, and considers some aspects of its control and possible functional significance in growing cells.

II. DIAMINE OXIDASE ACTIVITY IN ORGANS UNDERGOING HYPERPLASIA AND HYPERTROPHY

Although DAO in the rat, as in other mammals, is chiefly a soluble cytoplasmatic protein,[1] the total DAO activity has been studied on homogenates because part of the enzyme activity appears to be bound to mitochondrial and microsomal membranes, probably as an artifact due to homogenization,[18] and because it is not known whether the growth process influences its subcellular distribution.

Of the various methods available for the determination of DAO activity, the radiochemical assay of Okuyama and Kobayashi[19] as described by Kusche et al.,[20] has been employed, being considered the most specific and sensitive. This method is based on the measurement of [^{14}C]Δ^1-pyrroline and its polymers derived from spontaneous cyclization of [^{14}C]γ-aminobutyraldehyde, the first product of the oxidative deamination of [^{14}C]putrescine, which is one of the most preferred substrates of DAO.[1,18] However, it should be noted that in tissue homogenates a fraction of γ-aminobutyraldehyde, instead of cyclizing, may be further oxidized to GABA[4] or reduced to alcohol[21] by aldehyde-metabolizing enzymes (Figure 1). For this reason, a better approximation of the actual values of DAO activity in tissues can be obtained in the presence of inhibitors of aldehyde-metabolizing enzymes.[13,22-24] As shown in Table 1, acetaldehyde, a competitive inhibitor of these enzymes,[25] enhanced Δ^1-pyrroline formation in rat liver and kidney without modifying DAO activity, and did not affect the activity of a commercial preparation of SKDAO. On the other hand, other inhibitors enhanced or diminished the activity of tissue DAO as well as of the enzymatic preparation. Consequently, the values of enzyme activity reported here were obtained by the addition of 10 m*M* acetaldehyde to the incubation medium, which increases the yield in Δ^1-pyrroline almost twofold.

The specificity of the assay was also proven by the measurement of an aminoguanidine-sensitive activity, which corresponded to the amount of Δ^1-pyrroline obtained by the difference between the values measured in vitro in the absence and in the presence of aminoguanidine, a specific inhibitor of DAO.[19]

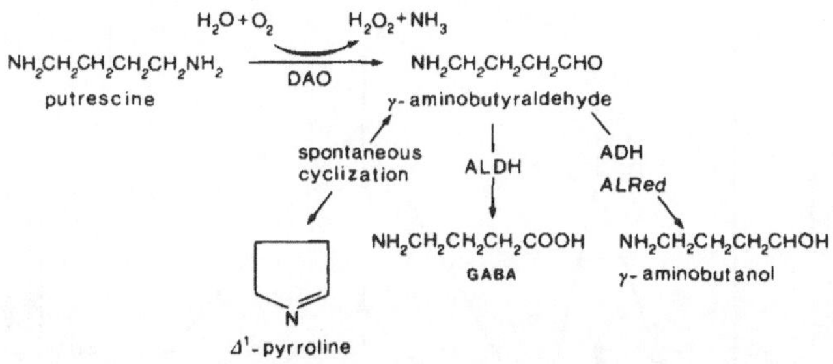

FIGURE 1. Catabolism of putrescine by DAO. ALDH, aldehyde dehydrogenase; ADH, alcohol dehydrogenase; ALRed, aldehyde reductase; GABA, γ-aminobutyric acid.

Table 1
EFFECT OF INHIBITORS OF ALDEHYDE-METABOLIZING ENZYMES ON Δ¹-PYRROLINE FORMATION FROM PUTRESCINE BY RAT LIVER AND KIDNEY HOMOGENATES AS WELL AS BY HOG KIDNEY DAO[13,23]

		Δ¹-Pyrroline formation (%)[a]		
Addition	Concentration (M)	Liver homogenate	Kidney homogenate	Hog kidney DAO
Acetaldehyde	1×10^{-3}	181 ± 13	160 ± 13	99 ± 6
	1×10^{-2}	261 ± 19	212 ± 19	98 ± 7
	5×10^{-2}	280 ± 20	220 ± 21	97 ± 8
Chloral hydrate	1×10^{-3}	179 ± 18	153 ± 14	130 ± 8
	1×10^{-2}	207 ± 20	194 ± 19	132 ± 8
Disulfiram	1×10^{-4}	57 ± 6	59 ± 5	47 ± 3
	2×10^{-4}	31 ± 4	65 ± 4	36 ± 2
Pyrazole	2×10^{-3}		83 ± 7	46 ± 3
	1×10^{-2}		38 ± 4	9 ± 2
Phenobarbital	1×10^{-3}		165 ± 15	96 ± 9
	1×10^{-2}		53 ± 5	38 ± 4

[a] Compared to control values obtained without any addition.

DAO activity was shown to increase in regenerating rat liver, with a peak between 16 and 48 hr after partial hepatectomy.[13] The values of this peak, which varied from one rat colony to another, were about 3 to 15 times higher than those observed in normal liver, where this enzyme activity is very low (about 1.5 nmol/hr) $\cdot g^{-1}$, wet weight, and 7.8 pmol/hr $\cdot mg^{-1}$, homogenate protein).[13,26] At 48 hr after hepatectomy, DAO activity had decreased and approached normal values 7 days after the operation.[13]

In the rat kidney undergoing compensatory hypertrophy after unilateral nephrectomy, a biphasic increase in DAO activity was observed, with peak values three to four times higher than those observed in sham-nephrectomized rats on days 2 and 6 (Figure 2).[23]

In rat heart during the development of cardiac hypertrophy sympathetically induced by isoproterenol (5 mg/kg body weight, twice daily), DAO activity showed two- to fourfold increases. With the regression of cardiac hypertrophy after suspension of the drug, DAO activity returned to basal values[27] (Figure 3). In the hypertrophic heart of spontaneously hypertensive rats, DAO activity was about 2.5 times higher than that of normotensive rats.[27]

All these findings give evidence that a variety of stimuli, which can determine in vivo cell proliferation and/or increased protein synthesis, cause an enhancement in DAO activity.

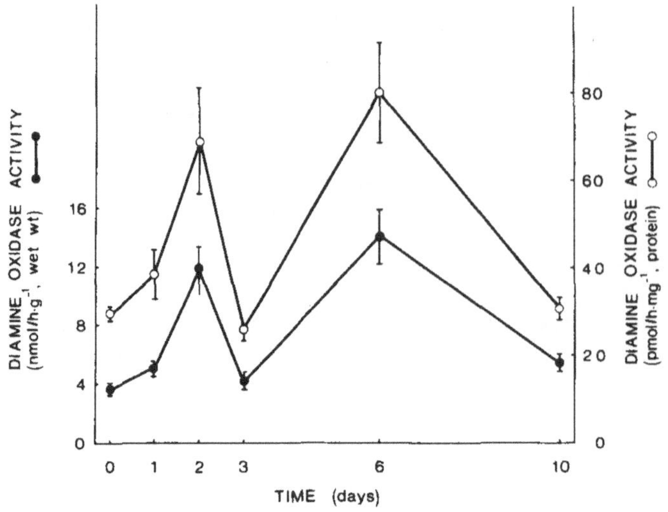

FIGURE 2. Time course of changes in DAO activity in kidney during compensatory hypertrophy. (From Desiderio, M. A., Sessa, A., and Perin, A., *Biochim. Biophys. Acta*, 714, 243, 1982. With permission.)

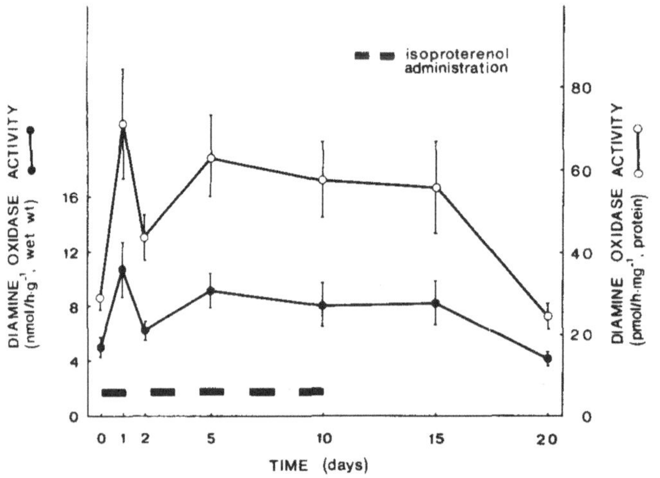

FIGURE 3. Time course of changes in DAO activity during the development of cardiac hypertrophy induced by isoproterenol and its regression. (From Perin, A., Sessa, A., and Desiderio, M. A., *Biochim. Biophys. Acta*, 755, 344, 1983. With permission.)

III. CONTROL OF DIAMINE OXIDASE ACTIVITY IN GROWING TISSUES

Changes in enzyme turnover or in the amount of enzyme protein for *de novo* synthesis and/or post-translational modifications could be hypothesized to explain the increase in DAO activity in growing tissues.

The in vivo turnover of DAO, studied as degradation of enzyme activity after cycloheximide injection, was not modified by growth stimuli. Half-life values of about 15 hr in normal and regenerating liver and about 14 hr in normal and hypertrophic kidney of rat have been found.[23,26]

The increase in DAO activity in the growing tissues that were examined was the consequence of a new synthesis of the enzyme protein with synthesis of mRNAs that code for DAO because the administration of cycloheximide or actinomycin D (inhibitors of protein or DNA-dependent RNA synthesis) prevented the enhancement of this enzyme activity.[23,26,27] Other mechanisms, such as the release of an inactive enzyme from subcellular cryptic structures and/or post-translational modifications of the enzyme, if they occur, probably are less important in determining the increase in DAO activity in hypertrophic tissues.

The in vivo induction of DAO may be mediated by the action of plasmatic catecholamines on cellular β receptors. In fact, adrenalectomy or the administration of β antagonist to nephrectomized rats prevented the enzyme increase during the first stages of kidney hypertrophy.[28] Further investigation is required to better understand whether such control mechanism also occur in other growth processes.

IV. BIOLOGICAL SIGNIFICANCE OF DIAMINE OXIDASE INDUCTION IN GROWING TISSUES

Regardless of the type of stimulus, DAO induction in hypertrophic and hyperplastic tissues occurs in the early phases of growth, even if its significant increment follows the stimulation of other enzymes, such as ornithine decarboxylase and spermidine N^1-acetyltransferase, involved in putrescine formation. Since, as pointed out by Schimke and Doile[29] and Tabor and Tabor,[30] a general enhancement in protein synthesis manifests itself first in those enzymes with a short half-life, the delay in DAO response can be explained in terms of a slower turnover rate (14 to 15 hr[23,26]) than those of ODC (10 to 60 min[31,32]) and spermidine N^1-acetyltransferase (40 min[33]).

The significance of the increase in DAO activity in growth processes is not completely clear. However, some data provide evidence for its role in the regulation of putrescine levels. In all the growth processes in which an increase in DAO activity has been observed, an enhancement in putrescine and other polyamine levels also occurs.[34] Moreover, a correlation between the increase in DAO activity and that of the enzymes involved in putrescine formation may be found in several growth conditions. In fact, in proliferating epithelium of rat small intestine,[35] and in cultured rat glioma cells stimulated to growth,[12] high levels of DAO activity associated with an increase in ODC activity have been observed. An enhancement in DAO activity has been found in the hypertrophic kidney[23] and heart,[27] where an increase in ODC activity was described,[36,37] as well as in regenerating liver,[13,26] where the induction of ODC [38,39] and spermidine N^1-acetyltransferase[40] contributes to the increase in putrescine content.

Further proof of the influence of DAO on putrescine levels in growing tissues is given by the results of in vivo experiments with aminoguanidine. In the hypertrophic kidney and heart of rats treated with this inhibitor, an increase in putrescine levels proportional to the inhibition of DAO activity has been found (Table 2).[27,41] Moreover, the administration of aminoguanidine has been reported to increase the putrescine levels in regenerating rat liver.[42]

DAO activity is therefore involved in putrescine catabolism of hyperplastic and hypertrophic tissues, and it may be considered part of a cellular system that also includes the acetylation/oxidation pathway of polyamines for the removal of these substances when a diminution of their levels in the cell is required. In addition, an increase in DAO activity might also represent, through the modulation of putrescine levels, a subtle regulation mechanism of polyamine and rRNA syntheses, as this substrate exerts a physiological control of ODC[43-46] and together they form a complex that could be an unstable subunit of RNA polymerase I.[47]

In contrast to the large number of investigations on putrescine levels and metabolism in growing tissues, very few studies on the behavior of DAO on other physiological substrates such as histamine and cadaverine[1] have been carried out. A substantial histamine formation

Table 2
EFFECT OF AMINOGUANIDINE IN VIVO ON DAO ACTIVITY AND PUTRESCINE LEVELS IN RAT HEART FOLLOWING ISOPROTERENOL TREATMENT AND IN RAT KIDNEY AFTER UNILATERAL NEPHRECTOMY

Organ	Treatment	DAO activity (nmol/hr· g^{-1}, wet wt)	Putrescine (nmol/g, wet wt)
Heart	Saline	5.11 ± 0.31	35 ± 4
	Isoproterenol	8.17 ± 0.53	50 ± 4
	Isoproterenol + aminoguanidine	1.98 ± 0.17	82 ± 8
Kidney	Sham-operation	4.17 ± 0.27	60 ± 4
	Nephrectomy	11.66 ± 1.55	64 ± 5
	Nephrectomy + aminoguanidine	4.39 ± 0.41	100 ± 9

From Perin, A., Sessa, A., and Desiderio, M. A., *Biochim. Biophys. Acta*, 755, 344, 1983. With permission.

in tissues undergoing rapid growth and repair has been described.[48] The biosynthesis of cadaverine in rat placenta[49] and in mouse kidney stimulated to grow by an anabolic steroid[50] has been reported, even though the accumulation of cadaverine in cultured animal cells may indicate a mycoplasma infection.[51] The role of histamine and cadaverine as regulatory growth factors and the importance of DAO in the control of their cellular levels must be further investigated (see Chapters 14 and 17).

V. CONCLUSIONS

In the previous sections it has been shown that in hypertrophic and hyperplastic processes an induction of DAO takes place and is associated with an increased activity of the enzymes involved in putrescine formation. DAO, due to its inducibility, is probably implicated in the regulation of putrescine levels as part of a multienzymatic system for the removal of excess polyamines in the cell. Our findings, in addition to the results of other authors,[9,10] suggest that the catabolic pathway catalyzed by DAO is enhanced in the pathophysiological states with elevated levels of polyamines and that this enzyme could play a prominent role in such conditions or help to monitor them. However, on the basis of our conclusions, it cannot be stated whether this enzyme may have other functions besides the regulation of putrescine levels in cells exposed to growth stimuli.

Further investigations on the inductions of DAO could help to better understand the possible mechanism(s) involved in the modulation of enzyme synthesis. Whether DAO is a critical enzyme and thus a possible target for the regulation of normal and neoplastic growth, or whether it plays a role in cell differentiation, are unanswered questions. Moreover, it is still unknown whether the specific oxidation products of putrescine via DAO may influence development, differentiation, the normal and neoplastic growth, as well as the course of some human illnesses.

ACKNOWLEDGMENTS

Research in the author's laboratory was supported by Grants 80,1479.96, Target Project ''Control of Tumor Growth'', from the Italian National Research Council, and Grants 1801.150 L and 1801.124 E from the Italian Ministry of Public Education, Rome.

REFERENCES

1. **Zeller, E. A.,** Diamine oxidases, in *The Enzymes,* Vol. 8, 2nd ed., Boyer, P. D., Lardy, H., and Myrbäck, K., Eds., Academic Press, New York, 1963, 313.
2. **Bardsley, W. G. and Ashford, J. S.,** Inhibition of pig kidney diamine oxidase by substrate analogues, *Biochem. J.,* 128, 253, 1972.
3. **Kapeller-Adler, R.,** *Amine Oxidases and Methods for their Study,* Wiley-Interscience, New York, 1970, 73.
4. **Seiler, N. and Eichentopf, B.,** 4-Aminobutyrate in mammalian putrescine catabolism, *Biochem. J.,* 152, 201, 1975.
5. **Missala, K. and Sourkes, T. L.,** Putrescine catabolism in rats given heparin or aminoguanidine, *Eur. J. Pharmacol.,* 64, 307, 1980.
6. **Heby, O.,** Role of polyamines in the control of cell proliferation and differentiation, *Differentiation,* 19, 1, 1981.
7. **Guha, S. K. and Jänne, J.,** The synthesis and accumulation of polyamines in reproductive organs of the rat during pregnancy, *Biochim. Biophys. Acta,* 437, 244, 1976.
8. **Maudsley, D. V. and Kobayashi, Y.,** Biosynthesis and metabolism of putrescine in the rat placenta, *Biochem. Pharmacol.,* 26, 121, 1977.
9. **Baylin, S. B. and Gordon, D. L.,** Diamine oxidase (histaminase) activity in human disease, in *The Biochemistry of Disease,* Vol. 8, Morris, D. R. and Marton, L. J., Eds., Marcel Dekker, New York, 1981, 377.
10. **Woźniak, J., Biegański, T., and Maśliński, Cz.,** Diamine oxidase activity during wound-healing in guinea-pig skin, *Agents Actions,* 9, 45, 1979.
11. **Quash, G., Keolouangkhot, T., Gazzolo, L., Ripoll, H., and Saez, S.,** Diamine oxidase and polyamine oxidase activities in normal and transformed cells, *Biochem. J.,* 177, 275, 1979.
12. **Bachrach, U.,** Metabolism of polyamines by cultured glioma cells. Effect of asparagine on γ-aminobutyric acid concentrations, *Biochem. J.,* 188, 387, 1980.
13. **Sessa, A., Desiderio, M. A., Baizini, M., and Perin, A.,** Diamine oxidase activity in regenerating rat liver and in 4-dimethylaminoazobenzene-induced and Yoshida AH 130 hepatomas, *Cancer Res.,* 41, 1929, 1981.
14. **Baylin, S. B., Beaven, M. A., Engelman, K., and Sjoerdsma, A.,** Elevated histaminase activity in medullary carcinoma of the thyroid gland, *N. Engl. J. Med.,* 283, 1239, 1970.
15. **Baylin, S. B., Beaven, M. A., Keiser, H. R., Tashjian, A. H., Jr., and Melvin, K. E. W.,** Serum histaminase and calcitonin levels in medullary carcinoma of the thyroid, *Lancet,* 1, 455, 1972.
16. **Baylin, S. B., Weisburger, W. R., Eggleston, I. C., Mendelsohn, G., Beaven, M. A., Abeloff, M. D., and Ettinger, D. S.,** Variable content of histaminase, L-dopa decarboxylase and calcitonin in small-cell carcinoma of the lung, *N. Engl. J. Med.,* 299, 105, 1978.
17. **Thomasset, N., Quash, G., and Doré, J. F.,** Diamine oxidase activity in human melanoma cell lines with different tumorigenicity in nude mice, *Br. J. Cancer,* 46, 58, 1982.
18. **Biegański, T., Kusche, J., Lorenz, W., Hesterberg, R., Stahlknecht, C.-D., and Feussner, K.-D.,** Distribution and properties of human intestinal diamine oxidase and its relevance for the histamine catabolism, *Biochim. Biophys. Acta,* 756, 196, 1983.
19. **Okuyama, T. and Kobayashi, Y.,** Determination of diamine oxidase activity by liquid scintillation counting, *Arch. Biochem. Biophys.,* 95, 242, 1961.
20. **Kusche, J., Richter H., Hesterberg, R., Schmidt, J., and Lorenz, W.,** Comparison of the ^{14}C-putrescine assay with the NADH test for the determination of diamine oxidase: description of a standard procedure with a high precision and an improved accuracy, *Agents Actions,* 3, 148, 1973.
21. **Dupré, S., Granata, F., Federici, G., and Cavallini, D.,** Hydroxyamino compounds produced by the oxidation of diamines with diamine oxidase in the presence of alcohol dehydrogenase, *Physiol. Chem. Phys.,* 7, 517, 1975.
22. **Fogel, W. A. Biegański, T., Woźniak, J., and Maśliński, C.,** Interference of aldehyde metabolizing enzymes with diamine oxidase (histaminase) activity as determined by ^{14}C putrescine method, *Biochem. Pharmacol.,* 27, 1159, 1978.
23. **Desiderio, M. A., Sessa, A., and Perin, A.,** Induction of diamine oxidase activity in rat kidney during compensatory hypertrophy. *Biochem. Biophys. Acta,* 714, 243, 1982.
24. **Perin, A., Sessa, A., and Desiderio, M. A.,** Induction of diamine oxidase activity in some processes of growth, *Adv. Polyamine Res.,* 4, 175, 1983.
25. **Deitrich, R. A. and Erwin, V. G.,** Involvement of biogenic amine metabolism in ethanol addiction, *Fed. Proc.,* 34, 1962, 1975.
26. **Sessa, A., Desiderio, M. A., and Perin, A.,** Diamine oxidase activity induction in regenerating rat liver, *Biochim. Biophys. Acta,* 698, 11, 1982.

27. **Perin, A., Sessa, A., and Desiderio, M. A.,** Polyamine levels and diamine oxidase activity in hypertrophic heart of spontaneously hypertensive rats and of rats treated with isoproterenol, *Biochim. Biophys. Acta*, 755, 344, 1983.
28. **Sessa, A., Desiderio, M. A., and Perin, A.,** β-Recettori e induzione di diammino ossidasi nell'ipertrofia renale compensatoria, presented at the 29th Congress of the Italian Biochemical Society, Saint Vincent, September 26—28, 1983, 340.
29. **Schimke, R. T. and Doyle, D.,** Control of enzyme levels in animal tissues, *Annu. Rev. Biochem.*, 39, 929, 1970.
30. **Tabor, C. W. and Tabor, H.,** 1,4-Diaminobutane (putrescine), spermidine, and spermine, *Annu. Rev. Biochem.*, 45, 285, 1976.
31. **Russell, D. H. and Snyder, S. H.,** Amine synthesis in regenerating rat liver: extremely rapid turnover of ornithine decarboxylase, *Mol. Pharmacol.*, 5, 253, 1969.
32. **Canellakis, E. S., Heller, J. S., Kyriakidis, D., and Chen, K. Y.,** Intracellular levels of ornithine decarboxylase, its half-life, and a hypothesis relating polyamine-sensitive membrane receptors to growth, *Adv. Polyamine Res.*, 1, 17, 1978.
33. **Matsui, I. and Pegg, A. E.,** Effect of inhibitors of protein synthesis on rat liver spermidine N^1-acetyltransferase, *Biochim. Biophys. Acta*, 675, 373, 1981.
34. **Jänne, J., Pösö, H., and Raina, A.,** Polyamines in rapid growth and cancer, *Biochim. Biophys. Acta*, 473, 241, 1978.
35. **Baylin, S. B., Stevens, S. A., and Shakir, K. M. M.,** Association of diamine oxidase and ornithine decarboxylase with maturing cells in rapidly proliferating epithelium, *Biochim. Biophys. Acta*, 541, 415, 1978.
36. **Brandt, J. T., Pierce, D. A., and Fausto, N.,** Ornithine decarboxylase activity and polyamine synthesis during kidney hypertrophy, *Biochim. Biophys. Acta*, 279, 184, 1972.
37. **Bartolome, J., Huguenard, J., and Slotkin, T. A.,** Role of ornithine decarboxylase in cardiac growth and hypertrophy, *Science*, 210, 793, 1980.
38. **Russell, D. H. and Snyder, S. H.,** Amine synthesis in rapidly growing tissues: ornithine decarboxylase activity in regenerating rat liver, chick embryo, and various tumors, *Proc. Natl. Acad. Sci. U.S.A.*, 60, 1420, 1968.
39. **Jänne, J. and Raina, A.,** Stimulation of spermidine synthesis in the regenerating rat liver: relation to increased ornithine decarboxylase activity, *Acta Chem. Scand.*, 22, 1349, 1968.
40. **Matsui, I. and Pegg, A. E.,** Effect of thioacetamide, growth hormone or partial hepatectomy on spermidine acetylase activity of rat liver cytosol, *Biochim. Biophys. Acta*, 633, 87, 1980.
41. **Perin, A., Desiderio, M. A., and Sessa, A.,** Role of diamine oxidase in the regulation of putrescine levels in growing tissues, *Ital. J. Biochem.*, 33, 50A, 1984.
42. **Moulinoux, J.-P, Quemener, V., and Chambon, Y.,** Action du sulfate d'aminoguanidine sur les taux hépatiques de polyamines après hépatectomie partielle chez le rat, *C. R. Soc. Biol.*, 175, 828, 1981.
43. **Kay, J. E. and Lindsay, V. J.,** Control of ornithine decarboxylase activity in stimulated human lymphocytes by putrescine and spermidine, *Biochem. J.*, 132, 791, 1973.
44. **McCann, P. P., Tardif, C., Mamont, P. S., and Schuber, F.,** Biphasic induction of ornithine decarboxylase and putrescine levels in growing HTC cells, *Biochem. Biophys. Res. Commun.*, 64, 336, 1975.
45. **Heller, J. S., Chen, K. Y., Kyriakidis, D. A., Fong, W. F., and Cannellakis, E. S.,** The modulation of the induction of ornithine decarboxylase by spermine, spermidine and diamines, *J. Cell Physiol.*, 96, 225, 1978.
46. **Jänne, J. and Hölttä, E.,** Regulation of ornithine decarboxylase activity by putrescine and spermidine in rat liver, *Biochem. Biophys. Res. Commun.*, 61, 449, 1974.
47. **Russell, D. H.,** Ornithine decarboxylase: a key regulatory protein, *Med. Biol.*, 59, 286, 1981.
48. **Kahlson, G. and Rosengren, E.,** Biogenesis and physiology of histamine, *Monogr. Physiol. Soc.*, Edward Arnold, London, 1971.
49. **Andersson, A.-CH., Henningsson, S., and Rosengren, E.,** Formation of cadaverine in the pregnant rat, *Acta Physiol. Scand.*, 105, 508, 1979.
50. **Andersson, A.-CH., Hammar, L., Henningsson, S., and Löwendahl, G. O.,** Changes in cadaverine and putrescine metabolism in the mouse kidney induced to grow by an anabolic steroid, *Acta Physiol. Scand.*, 114, 225, 1982.
51. **Alhonen-Hongisto, L., Viejalainen, P., EK-Kommonen, C., and Jänne, J.,** Polyamines in mycoplasmas and in mycoplasma-infected tumour cells, *Biochem. J.*, 202, 267, 1982.

Chapter 17

DIAMINE OXIDASE ACTIVITY IN HUMAN TUMORS: CLINICAL AND BIOLOGIC SIGNIFICANCE

Stephen B. Baylin and Gordon D. Luk

TABLE OF CONTENTS

I. INTRODUCTION

In the present review, we concentrate on the findings that the activity of the amine-metabolizing enzyme, diamine oxidase (DAO, EC 1.4.3.6-histaminase) is high in several types of human cancers. Investigators have been interested in the role of the synthesis and metabolism of amines in neoplastic tissues for many years. Although the specific functions of individual amines for mediation of tumor growth remain unknown, these agents could well play an important role in the proliferation and differentiation of neoplastic cells. For example, putrescine, the preferred in vivo substrate for DAO, is an amine whose synthesis is closely regulated during growth and differentiation.[1,2] Pools of this amine provide the substrate for the biosynthesis of the polyamines, spermidine and spermine, which may be essential for the rapid growth of normal and neoplastic cells.[1,2] Thus, the metabolism of putrescine could play an obvious potential role in regulation of the quantities of these molecules, and could then be important to the growth of cancer cells. Similarly, levels of metabolizing enzymes for different amines could be clinically useful as tumor biomarkers.

In this article, we will discuss the relationship of DAO activity to the cell populations present in certain human tumors, to the normal cell types from which the tumors may arise, and to the clinical monitoring of such neoplasms. Similarly, characterization of the DAO molecule in human tumor tissues vs. that found in normal tissues is considered. Finally, the interaction of the potential substrates for the tumor DAO with the growth and differentiation of the neoplastic cells is briefly considered.

II. BRIEF HISTORY OF DETECTION OF DAO IN HUMAN NEOPLASMS

The interest in DAO in tumor tissues probably can be traced back to original findings of Ahlmark[3] and Danforth and Gorham[4] concerning high activities of this enzyme in the circulation and placenta of humans and other mammalian species. Although initial biochemistry studies by Swanberg[5] and later, immunohistochemistry studies from our own group[6] and from Lin and colleagues[7] suggested that the mature maternal (i.e., the decidual cells) rather than fetal cells of the placenta were responsible for the high activity, the finding of elevated levels of DAO in this rapidly growing tissue logically suggested that investigations of tumor tissues should ensue. Indeed, using fluorometric assays, several investigators began to associate increased tissue and/or serum levels of DAO with a number of human malignancies, including those related to the uterus, such as endometrial carcinoma.[8]

In the early 1970s, using specific radioassays with either histamine and/or putrescine as substrates, high DAO activity was found to be a consistent property of the human endocrine tumor, medullary thyroid carcinoma.[9,10] The enzyme activity was increased not only in the tissues, but in the circulation of some patients with this cancer, and predominantly in those patients known to have metastatic disease.[9,10] Subsequently, increased DAO activity was associated with another human tumor which often has endocrine cell properties, small cell carcinoma of the lung,[11,12] but later also with lung cancers of other histologic types.[13,14] Simultaneously with these observations, Lin and colleagues also found increased DAO activity in ascitic and pleural fluid from patients with various types of cancer, including ovarian carcinoma,[15,16] and we verified that the tumor tissue was the source of the increased ascitic fluid DAO in patients with ovarian carcinoma.[17] In all of these situations where the enzyme was measured in tumor tissues, activity was greater than in normal tissue surrounding the primary lesions. In medullary thyroid carcinoma[9,10] and lung cancers,[11,12] increased DAO activity was also found in metastases as compared to surrounding normal tissue.

III. USE OF DAO AS A CLINICAL TUMOR MARKER

All of the above findings concerning the presence of increased DAO activity in human tumors raised the possibility that measurements of the enzyme activity could prove clinically useful for monitoring patients with the types of malignancies involved. However, this has not proven to be the case, and there is currently no defined use for measurements of DAO as a means for the diagnosis and/or monitoring of neoplastic disease. Particularly instructive in this regard is the history of the association of the enzyme activity with medullary thyroid carcinoma. Although the finding of increased circulating DAO activity certainly correlates with the presence of metastatic disease in some patients, many individuals with metastatic tumor, containing high DAO levels, have normal circulating levels of DAO activity. The factors underlying this discrepancy between circulating levels of DAO and tumor content of the enzyme are not all delineated, but relationships of the cellular populations within the tumor to the presence of DAO activity are now being appreciated, and these relationships, at least partially, explain the inability to use circulating levels of DAO as a tumor marker.

In medullary thyroid carcinoma,[18] ovarian carcinoma,[17] and small cell lung carcinoma,[12] immunohistochemistry studies of DAO localization show that the cell populations containing different amounts of the enzyme are heterogeneously distributed. Thus, in typical situations for these tumors, some populations of cells may stain heavily for DAO while a surrounding population of cells demonstrates no stain at all. The tumor populations of cells obviously may shift in content during the life history of the neoplasm and those cells making high amounts of DAO may not increase as tumor mass increases in patients with medullary thyroid carcinoma and other types of DAO-containing malignancies. The reasons for this heterogeneity of cellular distribution of DAO are not known, but may be intrically involved with the biological function of DAO within both normal and neoplastic cells. This fact emphasizes the need to study, in much more detail, the biology of DAO in self-renewing cell systems, as is discussed in a section to follow concerning the relationship of the amines and their metabolism to the growth and differentiation of epithelial cell systems (see Chapter 18).

IV. STRUCTURAL CHARACTERIZATION OF DAO IN HUMAN TUMOR TISSUES

It has been emphasized above that the cellular distribution of DAO in human tumors is generally an heterogeneous one. Also, within any one category of tumor type, the levels of activity of DAO can differ dramatically.[12-14] These findings raise questions concerning the specific forms of DAO present in the tumors, and whether the structural characteristics of the protein(s) responsible for the enzyme activity may differ between tumor types and/or between tumor cells in the same tumor, and account for some of the conflicting results found in the literature concerning the presence of DAO in human malignancies. While many more studies must be carried out to fully define the nature of the DAO present in human tumors, some structural characterizations of the enzyme performed in recent years are very pertinent to the points under consideration in this review. Recent purifications and immunologic characterization of human DAO will thus be discussed briefly.

At least four laboratories over the past several years have purified, to a high degree, DAO from human placenta (see Chapter 7). The studies indicate that human DAO is a complex, multisubunit glycoprotein[20] which exists in multiple forms in both normal and neoplastic tissues. The enzyme has been purified by Bardsley et al.[19] using classic column procedures and by an affinity chromatography method developed in our laboratory[21,22] and modified by others.[23,24] All of these studies indicate that the monomeric subunit size of the placental DAO is approximately 90 kD.

The native size of the polymeric DAO molecule has been more difficult to establish. Tufvesson has reported[23] that multiple charge forms of DAO exist including a major form having a Stokes radius size of 500 kD and a minor form of 240 kD. We would agree with these findings, and our own characterization of human placental DAO, purified to homogeneity, indicates by electron microscopy that the largest form of human placental DAO exists as an aggregate of 6 of the 90 kD subunits. Multiple other forms, which differ in size and charge, were detected in our purification studies and probably represent partial aggregates of the subunits. Some of these different forms, on native polyacrylamide gels, can be converted from one to another via proteolysis, indicating that products of protease activity in the placental homogenates probably account for some of the DAO forms acquired during purification. The fact that the human DAO is glycosylated may also account for some of the differences in charged species. It must be emphasized that, in our own recent studies, antibodies to human placental DAO recognize all of the different charged and size species which we can identify in human placenta.

The complex structure of the human enzyme plus the different forms of the 90-kD subunit structure which can be purified from human placenta may well explain some of the differences in enzyme activity and forms of the enzyme reported in human tumors. Thus, differences between tumor tissues in the kM for substrate utilization reported by Thomasset et al.[25] could well be due to the fact that the different forms of the enzyme have somewhat different affinities in the assay systems used. Importantly, however, studies by our own laboratory[22] and those of Lin and colleagues[24] indicate that the major antigenic determinants of human DAO in multiple normal tissues and in all the tumor tissues examined are very similar or identical. Thus, one major gene product or closely related gene products seem to be responsible for the expression of DAO in all of the tumor types examined to date (see Chapter 7).

While structural heterogeneity of the human DAO molecule could well explain some of the heterogeneous distribution of the enzyme between tumor cells and between tumors of the same type from different patients, the precise explanations for these findings would best be resolved by examination of the gene or genes responsible for coding of human DAO mRNA species. Direct examination of the mRNA, particularly in different tumor types and different normal tissues, might clarify the number of enzymes which exist which express classic DAO activity. It is probable that investigators interested in DAO, through the use of recombinant DNA technology, will be providing answers to such questions.

V. RELATIONSHIPS BETWEEN DAO ACTIVITY TO CELL GROWTH AND DIFFERENTIATION

One important area concerning the findings of high DAO activity in certain tumors is the relationship of this enzyme to states of cell proliferation and differentiation. It is becoming increasingly apparent that many tumors represent clonal expansions of a given state of differentiation within the normal cell lineage from which the neoplasms arise. Consequently, the presence of DAO in some tumor cell populations could reflect relationships between this enzyme activity and events in normal cell renewal and differentiation. These cellular events, with respect to DAO, have not been studied in great detail, but some investigations have been carried out and are worthy of comment in this review.

From the standpoint of normal cell renewal and differentiation, our own laboratory has studied the presence of DAO activity in the rat intestinal mucosa as a function of the differentiation pattern. This epithelial system represents an important model for such studies of DAO since these cells contain the highest levels of the enzyme in normal tissues.[10,26] We have found that the DAO activity increases in parallel with the degree of cell maturation, being highest in the differentiated villus tip cells and lowest in the proliferative compartment of cells in the crypt regions.[27,28] An additional piece of data linking high DAO activity to mature intestinal epithelial components of rat intestinal mucosa has come from our studies of maturation of the intestinal epithelium in the newborn rat.[29,30] As mature elements of the rat intestinal mucosa appear approximately 3 weeks after birth, DAO activities first appear in high amounts and rapidly reach, by week 4, levels seen in adult animals.[29,30] These data suggest, at least in renewing intestinal epithelium, that DAO activities bear a close relationship to the state of cell differentiation. They also stress that the appearance of the enzyme in certain cell populations within human tumors may reflect the metabolic status of the cells as a function of similar stages of maturation. It is clear that more work needs to be done to decipher the processes by which DAO activities are regulated as a function of maturation status in epithelial cell systems such as the intestinal mucosa. It is likely that an understanding of the presence of the enzyme in such cells will also shed light, on the relationships between DAO activity and neoplastic cells (see Chapter 14).

The precise role of DAO in cell renewing systems such as the intestinal epithelium is not known. However, our data[28,30] suggest that the enzyme could play an important regulatory role in the polyamine biosynthesis pathway. The polyamines, of course, have long been implicated in the process of cell proliferation.[1,2] We have used the specific inhibitor, α-difluoromethylornithine (DFMO),[31] to block ornithine decarboxylase activity (ODC) (the first and rate-limiting step in the polyamine biosynthesis pathway) to show that polyamine biosynthesis is a necessary component of both cell renewal and maturation in the intestinal mucosa.[30,32] Also, our own work and that of Porter et al.[33] has shown that the putrescine-synthesizing enzyme, ODC, has a similar relationship to the pattern of cellular differentiation in the rat intestinal mucosa as that outlined above for DAO. This close relationship between the enzyme responsible for putrescine synthesis, and the enzyme, DAO, for which putrescine is an excellent substrate, suggests that further work in the intestinal mucosa could provide insight into relationships between the regulatory role of DAO and renewing cell systems.

Some studies have been done which are addressed to the question of a specific relationship between DAO and cell proliferation. Quash and colleagues have found, in studies of tumor cells in culture, that DAO activity is highest at the outset of growth and falls during the logarithmic growth phases of the cells.[25,34] They suggest that the association of DAO with the most proliferative component of cells in some tumors could explain the high activity seen in some instances of neoplasia. It is not possible, at present, to reconcile these results with our work as detailed above, which associated high DAO activity with mature rather than proliferative cells in intestinal mucosa. It is clear that much further work will be necessary to document the precise relationships between levels of DAO activity and events in DNA synthesis, cell replication, and cell differentiation to decipher further the role of the enzyme in renewing cell populations. The studies cited, however, suggest that this area of investigation may prove to be a fertile one for understanding the presence of high DAO activity in some neoplastic tissues (see Chapter 14 to 16 and 18).

VI. THE USE OF DAO AS A MARKER OF INTESTINAL MUCOSAL INTEGRITY IN PATIENTS TREATED FOR NEOPLASTIC DISORDERS

One area which does not relate directly to the presence of DAO activity in tumor tissue but which could be important for the application of DAO measurements to the therapy of patients with cancer regards the use of the enzyme as a marker for the status of intestinal epithelial cell integrity. Many of the agents used in chemotherapy have a profound toxic effect on the rapidly dividing cells of the intestinal mucosa. In some instances, this effect of the treatment agents may be the major limiting factor in their continued use in patients with some cancers.[35] The close association of DAO with maturation events in the intestinal mucosa, as discussed in the previous section, have led us to question whether the enzyme activity in blood might be a monitor for the integrity of the intestinal mucosa. Since the intestinal epithelium is the major source of the circulating DAO in rats[29,36,37] and probably in man,[38] circulating levels of the enzyme might be used as a monitor to gauge the extent of intestinal epithelial damage. Indeed, our own studies in the rat[36,37] and initial studies in humans[36] suggest that this is the case.

In our rodent model, circulating levels of DAO directly reflect the extent of induced damage to the mature intestinal epithelial cells.[29,37] The ability to detect these types of changes in the circulating levels of the enzyme is enhanced with the use of heparin to mobilize intestinal epithelial DAO into the circulation.[37] The fact that heparin increases circulating DAO in many species, including man,[38-40] and that the intestine is the prime source for this increased blood DAO has been known from the work of several laboratories.[27,29,37,41,42] The rapid increase in blood DAO following heparin seems to occur because this highly charged polyanion displaces DAO directly from sites on the intestinal vascular endothelium.[27]

In humans receiving the cell cycle specific agent, ara-C, which is known to have profound toxic effects on the renewal of the intestinal epithelial mucosa, levels of DAO activity in the blood appear to act as a monitor of the integrity of the mucosa as they do in the rat.[36] Further clinical studies are warranted to see if the measurement of circulating levels might allow more accurate assessment of the toxic effects of agents such as ara-C and other chemotherapeutic drugs. The availability of such a monitoring mechanism might allow for reduction of the toxic effects of such treatment protocols so that optimal therapeutic schedules might be followed (see Chapters 15 and 18).

VII. SUMMARY

In the present review, we have focused on recent data concerning the presence of DAO in some human neoplasms and the implications for these findings for use of the enzyme as a clinical marker and for the biological events which might explain relationships between high DAO activity and neoplastic cells. We have stressed that the distribution of DAO in those tumors found to have high activity appears to be an heterogeneous one and that some cell populations are associated with high DAO, while others in the same neoplastic tissues are not. We have detailed how this heterogeneous cellular distribution probably accounts for the fact that circulating levels of DAO have proven to be a disappointing marker for tumor diagnosis and the monitoring of tumor burden in individual patients. We have also stressed, however, that is it important to decipher what cellular events determine the presence of high DAO in neoplastic tissues and have suggested that: (1) an understanding of the relationships of DAO to normal cell renewal and maturation may be important for such understanding — investigations of the intestinal mucosal epithelium as a model system for the dissection of these relationships has been discussed and (2) relationships between polyamine biosynthesis and DAO activity may be linked to the role of the enzyme in both normal and neoplastic cells. Finally, we have discussed findings which suggest that monitoring of blood DAO levels may be a tool for assessing the integrity of the intestinal mucosa. This may be clinically useful in patients undergoing therapy with agents known to have toxic effects on the normal intestinal epithelium. Thus, the presence of DAO activity in certain populations of human neoplastic cells and in normal cell renewing systems remains an important area of cell biology investigation, both for understanding basic aspects of tumor growth and for developing the use of DAO as a biomarker in delineated clinical areas.

REFERENCES

1. **Tabor, C. W. and Tabor, H.,** 1,4-Diaminobutane (putrescine), spermidine and spermine, *Ann. Rev. Biochem.*, 45, 285, 1976.
2. **Pegg, A. E. and McCann, P. P.,** Polyamine metabolism and function, *Am. J. Physiol.*, 243, C212, 1982.

3. **Ahlmark, A.,** Studies on the histaminolytic power of plasma with special reference to pregnancy, *Acta Physiol. Scand.* Suppl. 28, 1, 1944.
4. **Danforth, D. N. and Gorham, F.,** Placental histaminase, *Am. J. Physiol.*, 119, 294, 1937.
5. **Swanberg, H.,** Histaminase in pregnancy with special reference to its origin and formation, *Acta Physiol. Scand.*, 23(79), 1, 1950.
6. **Weisburger, W. R., Mendelsohn, G., Eggleston, J. C., and Baylin, S. B.,** Immunohistochemical localization of histaminase (diamine oxidase) in decidual cells of human placenta, *Lab. Invest.*, 38, 703, 1978.
7. **Lin, C.-W., Chapman, C. M., DeLellis, R. A., and Kirley, S. D.,** *J. Histochem. Cytochem.*, 26, 1021, 1978.
8. **Borglin, N. E. and Willert, B.,** Increased histaminolytic power of plasma in endometrial carcinoma, *Cancer*, 15, 271, 1962.
9. **Baylin, S. B., Beaven, M. A., Engelman, K., and Sjoerdsma, A.,** Elevated histaminase activity in medullary carcinoma of the thyroid gland, *N. Engl. J. Med.*, 283, 1239, 1970.
10. **Baylin, S. B., Beaven, M. A., Buja, L. M., and Keiser, H. R.,** Histaminase activity: a biochemical marker for medullary carcinoma of the thyroid, *Am. J. Med.*, 53, 723, 1972.
11. **Baylin, S. B., Abeloff, M. D., Tomford, J. W., Wieman, K. C., and Ettinger, D. S.,** Elevated histaminase (diamine oxidase) activity in small cell carcinoma of the lung, *N. Engl. J. Med.*, 293, 1286, 1975.
12. **Baylin, S. B., Weisburger, W. R., Eggleston, J. C., Mendelsohn, G., Beaven, M. A., Abeloff, M. D., and Ettinger, D. S.,** Variable content of histaminase L–dopa decarboxylase and calcitonin in small cell carcinoma of the lung, *N. Engl. J. Med.*, 299, 105, 1978.
13. **Baylin, S. B., Abeloff, M. D., Goodwin, G., Carney, D. N., and Gazdar, A. F.,** Activities of L-dopa decarboxylase and diamine oxidase (histaminase) in human lung cancers and decarboxylase as a marker for small (oat) cell cancer in cell culture, *Cancer Res.*, 40, 1990, 1980.
14. **Berger, C. L., Goodwin, G., Mendelsohn, G., Eggleston, J. C., Abeloff, M. D., Aisner, S., and Baylin, S. B.,** Endocrine related biochemistry in the spectrum of human lung carcinoma, *J. Clin. Endocrinol. Metab.*, 53, 422, 1981.
15. **Lin, C.-W., Orcutt, M. L., and Stohlbach, L. L.,** Elevation of histaminase and its concurrence with Regan isoenzyme in ovarian cancer, *Cancer Res.*, 35, 2762, 1975.
16. **Lin, C.-W., Inglis, N. R., Rule, A. H., Turksoy, R. N., Chapman, L. M., Kirley, S. D., and Stolbach, L. L.,** Histaminase and other tumor markers in malignant effusion fluids, *Cancer Res.*, 39, 4894, 1979.
17. **Ettinger, D. S., Rosenshein, N. B., Parmley, T. H., Mendelsohn, G., and Baylin, S. B.,** Tumor cell origin of histaminase activity in ascites fluid from patients with ovarian carcinoma, *Cancer*, 45, 2568, 1980.
18. **Mendelsohn, G., Eggleston, J. C., Weisburger, W. R., Gann, D. S., and Baylin, S. B.,** Calcitonin and histaminase in C-cell hyperplasia and medullary thyroid carcinoma. A light microscopic and immunohistochemical study, *Am. J. Pathol.*, 92, 35, 1978.
19. **Bardsley, W. G., Crabbe, M. J. C., and Scott, I. V.,** The amine oxidases of human placenta and pregnancy plasma, *Biochem. J.*, 139, 169, 1974.
20. **Baylin, S. B., Wieman, K. C., O'Neil, J. A., and Roos, B. A.,** Multiple forms of human tumor calcitonin demonstrated by denaturing polyacrylamide gel electrophoresis and lectin affinity chromatography, *J. Clin. Endocrinol. Metab.*, 53, 489, 1981.
21. **Baylin, S. B. and Margolis, S.,** Purification of histaminase (diamine oxidase) from human pregnancy plasma by affinity chromatography, *Biochim. Biophys. Acta*, 397, 294, 1975.
22. **Baylin, S. B.,** Histaminase (diamine oxidase) activity in human tumors: an expression of a mature genome, *Proc. Natl. Acad. Sci. U.S.A.*, 74, 883, 1977.
23. **Tufvesson, G.,** Purification and properties of human amniotic fluid diamine oxidase, *Scand. J. Clin. Lab. Invest.*, 38, 463, 1978.
24. **Lin, C.-W., Kirley, S. D., and St. Pierre, M.,** Tumor and placental histaminase. I. Affinity chromatography purification and characterization of the placental enzyme, *Oncodev. Biol. Med.*, 2, 1981.
25. **Thomasset, N., Quash, G., and Dore, J. F.,** Diamine oxidase activity in human melanoma cell lines with different tumorigenicity in nude mice, *Br. J. Cancer*, 46, 58, 1982.
26. **Buffoni, F.,** Histaminase and related amine oxidases, *Pharmacol. Rev.*, 18, 1163, 1966.
27. **Shakir, K. M. M., Margolis, S., and Baylin, S. B.,** Localization of histaminase (diamine oxidase) in rat small intestinal mucosa: Site of release by heparin, *Biochem. Pharm.*, 26, 2343, 1977.
28. **Baylin, S. B., Stevens, S. A., and Shakir, K. M. M.,** Association of diamine oxidase and ornithine decarboxylase with maturing cells in rapidly proliferating epithelium, *Biochim. Biophys. Acta*, 541, 415, 1978.
29. **Luk, G. D., Bayless, T. M., and Baylin, S. B.,** Diamine oxidase (histaminase). A circulating marker for intestinal mucosal maturation and integrity, *J. Clin. Invest.*, 66, 66, 1980.

30. **Luk, G. D., Marton, L. J., and Baylin, S. B.,** Ornithine decarboxylase is important in intestinal mucosal maturation and recovery from injury in rats, *Science*, 210, 195, 1980.
31. **Metcalf, B. W., Bey, P., Danzin, C., Jung, M. J., Casara, P., and Vevert, J. P.,** Catalytic irreversible inhibition of mammalian ornithine decarboxylase (EC 4.1.1.17) by substrate and product analogues, *J. Am. Chem. Soc.*, 100, 2551, 1978.
32. **Luk, G. D. and Baylin, S. B.,** Inhibition of intestinal epithelial DNA synthesis and adaptive hyperplasia after jejunectomy in the rat by suppression of polyamine biosynthesis, *J. Clin. Invest.*, 74, 698, 1984.
33. **Porter, C. W., Dworaczyk, D., Ganis, B., and Weiser, M. M.,** Polyamines and biosynthetic enzymes in the rat intestinal mucosa and the influence of methyl-gly-oxal-bis(guanylhydrazone), *Cancer Res.*, 40, 2330, 1980.
34. **Quash, G., Keoloungkhot, T., Gazzolo, L., Ripoll, H., and Saez, S.,** Diamine oxidase and polyamine oxidase activities in normal and transformed cells, *Biochem. J.*, 177, 275, 1979.
35. **Young, R. C., Goldberg, D., and Schein, P. S.,** Enhanced antitumor effect of cytosine arabinoside given in a schedule dictated by kinetic studies in vivo, *Biochem. Pharm.*, 22, 277, 1973.
36. **Luk, G. D., Vaughan, W. P., Burke, P. J., and Baylin, S. B.,** Diamine oxidase as a plasma marker of rat intestinal mucosal injury and regeneration after administration of 1-β-D-arabinofuranosylcytosine, *Cancer Res.*, 41, 2334, 1981.
37. **Luk, G. D., Bayless, T. M., and Baylin, S. B.,** Plasma post heparin diamine oxidase. Sensitive provocative test for quantitating length of acute intestinal mucosal injury in the rat, *J. Clin. Invest.*, 71, 1308, 1983.
38. **Ettinger, D. S., Baylin, S. B., Minaberry, D., Abeloff, M. D., and Mellits, E. D.,** Response of plasma histaminase activity to heparin in normal subjects and patients with small cell carcinoma of the lung, *J. Natl. Cancer Inst.*, 60, 1239, 1978.
39. **Baylin, S. B., Beaven, M. A., Krauss, R. M., and Keiser, H. R.,** Response of plasma histaminase activity to small doses of heparin in normal subjects and patients with hyperlipoproteinemia, *J. Clin. Invest.*, 52, 1985, 1973.
40. **Hansson, R., Holmberg, C. G., Tibbling, G., Tryding, N., Westling, H., and Wetterqvist, H.,** Heparin-induced diamine oxidase increase in human blood plasma, *Act. Med. Scand.*, 180, 533, 1966.
41. **Shaff, R. E. and Beaven, M. A.,** Turnover and synthesis of diamine oxidase (DAO) in rat tissues. Studies with heparin and cyclohexamide, *Biochem. Pharmacol.*, 25, 1057, 1976.
42. **Kobayashi, Y., Kupelian, J., and Maudsley, D. V.,** Release of diamine oxidase by heparin in the rat, *Biochem. Pharmacol.*, 18, 1585, 1969.

Chapter 18

COPPER-DEPENDENT AMINE OXIDASES; CLINICAL ASPECTS

G. Bombardieri, A. Milani, and L. Rossi

TABLE OF CONTENTS

I. GENERAL ASPECTS

In the wide and heterogeneous group of amine oxidases (AOs) the copper-dependent AOs are of special clinical interest.

Human serum amine oxidase (HSAO) and diamine oxidase (DAO) are commonly considered the copper-dependent AOs in the human body. However, the presence of copper as a cofactor has been proved only in purified human DAO;[1] in HSAO the presence of copper in the active site of the enzyme is strongly suggested by indirect evidence[2,3] and by analogy with other copper AOs purified from animal plasma,[4,5] but no direct demonstration has been provided, since the enzyme has not yet been fully purified. In humans HSAO was detected only in serum, and its origin is still unknown.

Because of its activity on primary amines (such as benzylamine, tyramine, tryptamine and dopamine) which are also oxidized by mitochondrial monoamine oxidases (MAO), HSAO has been also referred to as human plasma MAO.[2,3] However, the evidence of substantial differences between partially purified HSAO[2,6] and the purified mitochondrial MAOs (see Chapter 23), concerning molecular and biochemical features and behavior with substrates and inhibitors, has been definitely achieved.

Four different HSAO isoenzymes have been separated by means of chromatographic and electrophoretic techniques,[7-9] though it is known that the separation of the isoenzymes may result from a technical artifact.[10,11] Lin and colleagues[8,9] demonstrated that the HSAO isoenzyme pattern in patients with liver fibrosis was quite different from a normal one, with an increase of isoenzymes whose activity is negligible in normal serum. Since only healthy subject HSAO has been extensively characterized after partial purification,[2,6] we do not know if the above-mentioned consideration on differences between HSAO and intracellular MAO may concern all the HSAO isoenzymes.

DAO activity in humans has been found either in serum, and in normal and pathological tissues and fluids; it was purified from placenta,[12] intestinal mucosa,[13] plasma during pregnancy or after heparin administration,[14] and amniotic fluid.[14,15]

All these enzymes have been demonstrated to be different, but common antigenic properties and similar behaviors with different substrates and inhibitors characterize all human DAOs.

The physiological role of DAO is related to the biological function of its natural substrates, namely histamine[16] and polyamines (see Chapter 14).

Histamine is chiefly involved in gastric acid output and in inflammatory, allergic, and ischemic phoenomena, while polyamines play an important role in the active cellular proliferation, both in normal and in neoplastic tissues.

The preeminent DAO activity on histamine or on polyamines in individual organs is still under discussion. However, in human plasma and tissues the oxidizing activities on histamine and polyamines could never be separated, so that the identity of histaminase and DAO is postulated[17] (see Chapter 7).

II. HUMAN SERUM AMINE OXIDASE IN PHYSIOLOGICAL AND PATHOLOGICAL CONDITIONS

Several methods (spectrophotometric, fluorimetric, and radioenzymatic) have been proposed for HSAO assay[18-24] employing different substrates, but the majority of studies concerning HSAO in different clinical conditions were carried out with benzylamine as substrate.[18,22] Unless specified differently, even here reported data refer to HSAO as serum benzylamine oxidase.

Physiological age-related HSAO modifications have been reported. The enzyme activity is increased in childhood, and falls during adolescence to reach a minimum in young adults.[25] A slow increase of HSAO with aging was found,[25-28] but some authors did not confirm these data.[3] Also, sex-related small differences in HSAO (higher levels in women) are the object of conflicting reports.[3,25,26,29,30]

In pregnancy, HSAO was found within normal limits,[3,18,30,31] while serum DAO is increased several-fold.[32,33]

However, Gilmore and colleagues demonstrated that normal HSAO activity before delivery becomes significantly lower in post-partum. Since, in the same period, plasma DAO dramatically falls from the highly increased pregnancy levels to the low normal ones, the authors added to HSAO assays the selective DAO inhibitor aminoguanidine, reporting an unaffected post-partum HSAO activity and an ante-partum HSAO decreased to the post-partum levels. They suggested that normally negligible benzylamine oxidation by plasma DAO becomes significant when DAO reaches the increased pregnancy levels, concluding that the real HSAO activity is lowered in late pregnancy. However, this conclusion clashes with the observation that purified pregnancy plasma DAO is inactive on benzylamine.[12]

Modifications of HSAO during the menstrual cycle, with low values before ovulation and higher levels in the lutheinic phase, were reported only when tyramine was used as substrate,[34] instead of benzylamine.[30]

HSAO levels in various diseases have been widely investigated. They were reported to be significantly increased in chronic liver diseases with fibrosis (including Wilson's disease),[8,23,24,35-43] congestive heart failure,[44,45] hyperthyroidism,[45,46] diabetes mellitus,[25,45] acromegaly,[40] Addison's diseases,[105] carcinoid syndrome,[44] and progressive systemic sclerosis.[40,47]

HSAO was unaffected in acute hepatitis[3,23] and myocardial infarction (unless heart failure is present),[45] indicating that cytolisis does not affect the enzyme activity.

In malignancies,[45,48,49] severely burned patients,[49] and rheumatic fever,[50] low HSAO levels are reported. Alterations of HSAO were also detected in mental disorders with depression.[51] Drugs may affect HSAO: corticosteroid treatment lowers it,[45] while dopamine administration enhances the enzyme activity.[52]

Sex hormone-related changes in HSAO were observed only when tyramine was used as substrate: Klaiber and colleagues[34] found a decrease in HSAO activity when estrogen was administered to post-menopausal or amenorrhoeic women, followed by an increase of the enzyme when progestin was added to the estrogen treatment. Moreover, HSAO decreased during chemical contraception.[25,53]

The effect of some anti-MAO drugs and of isoniazid on HSAO has been also described: the enzyme activity is completely and irreversibly inhibited, and after drug withdrawal it takes months to return to the normal levels.[22,54-56]

No significant HSAO inhibition has been observed during treatment with the copper-chelating agent D-penicillamine, employed in Wilson's disease.[39]

The effort to understand the physiopathological mechanisms of HSAO modifications in health and disease help to explain the biological role of this enzyme.

It has been observed that in chronic liver disease HSAO increase is correlated with the extent of fibrous tissue deposition in the liver,[40] and with portal hypertension;[3,36] furthermore, several clinical conditions with high HSAO levels are associated with activated connective tissue metabolism (childhood and senescence, hyperthyroidism, acromegaly, progressive systemic sclerosis, congestive heart failure).

Also, the effect of corticosteroid hormones on HSAO may be explained by the drug influence on connective tissue metabolism, and the long time required for HSAO recovery after isoniazid and MAO inhibitors indicates that the enzyme resynthesis rate or the enzyme-producing cell multiplication are very slow, and that fits with fibroblast characteristics.

All these considerations support the hypothesis of HSAO origin from fibroblasts.

Fibroblasts contain both A- and B-MAO,[57] and lysil oxidase,[3] whose biological role seems to be related to the cross-linkage of collagen fibers, stabilizing their structure.[3]

Despite these suggestions, no fibroblast enzyme has been identified with HSAO, but information on the various HSAO isoenzymes in pathological conditions are still too few to draw a conclusion.

Also, sex hormone-induced HSAO alterations during the menstrual cycle or after estrogen and progestin administration (observed only with tyramine as substrate) may indicate a selective influence on individual HSAO isoenzymes.

The low HSAO levels in malignancies may be explained by the generally depressed biosynthetic functions, including those involving connective tissue. Low HSAO levels in rheumatic fever, and the absence of HSAO increase in other connective tissue diseases, may be explained by the usual corticosteroid therapy, but unfortunately the studies on this field do not report details on patient's therapy.

In conclusion, the clinical employment of HSAO assays appears to be restricted to the evaluation of liver fibrosis, and, possibly, to the detection of neoplasms.

Liver fibrosis is an unfavorable evolution of many chronic liver diseases, and fibrosis reduction and prevention by medical treatment is an important goal of modern hepatology. A serious difficulty in this field is the lack of reliability of clinical, histological, and biochemical parameters of liver fibroplasia.

HSAO activity seems to be well related, in comparison to other biological markers, to the actual presence of fibrous tissue in the liver, unless a large overlapping of HSAO activity in the different stages of fibrosis occurs.

Thus, HSAO cannot substitute the invasive approach to the diagnosis of liver fibrosis, but it may be employed in the follow up of individual patients.

Nevertheless, we were unable to correlate HSAO levels and clinical evolution of cirrhotic patients under antifibrotic treatment with colchicine[105] as suggested by Kershenobich and colleagues.[58]

We also found that normal or lowered HSAO levels, unusual in liver cirrhosis, were significantly more frequent when cirrhosis was associated with hepatocarcinoma.[105]

HSAO decrease in malignancies is not constant, and usually it does not appear in the early stages of the disease; however, in 9 out of 162 cancer patients in our study,[48] the diagnosis of cancer was first suggested by low HSAO levels and further confirmed by clinical investigation.

III. HUMAN DAO IN PHYSIOLOGICAL AND PATHOLOGICAL CONDITIONS

Several reports on DAO assays in human tissues, serum, and fluids are available. Different assay methods for human DAO have been described and their reliability has been extensively discussed.[59,60] A method based on indygo disulfate oxidation[61] is unspecific, because histamine itself may oxidize indygo disulfate.[17]

DAO assay employing *O*-dianisidine as oxygen acceptor[62] may provide false results when hemoglobin, which is also an oxygen acceptor, is present in homogenate: moreover, *O*-dianisidine may also be directly oxidized by ceruplasmin.[63]

The aldehyde produced by DAO activity on amines may undergo further oxidation in tissue homogenates containing aldehyde dehydrogenase,[60] causing incorrect results: the addition of aldehyde dehydrogenase inhibitors to the assay medium,[60] or the employment of dimethylaminomethilbenzylamine as substrate[64] has been proposed to overcome this problem (see Chapter 16).

In humans, DAO has been detected in intestinal mucosa,[13,65] kidney, spleen and liver,[66] placenta,[1,12] eosinophil and neutrophil granulocytes,[67] semen,[68] lymph,[69] plasma,[32] amniotic fluid,[70] and urine.[71]. In the intestinal tract, DAO activity increases from stomach downward to the bowel.[65] Placenta is an especially DAO-rich organ, and the enzyme is located on the decidual portion.[72]

Serum DAO has a low activity in all physiological conditions except pregnancy, and its values show a large individual dispersion. An inherited familial plasma DAO hyperactivity has been described, without any associated disease.[73]

No age- and sex-related changes have been reported using putrescine as substrate, but employing a nonspecific assay method with histamine as substrate[61] DAO was found to be significantly increased in children and adult women, in comparison with men.[74]

The source of the nonpregnancy plasma enzyme is uncertain, but the finding of an enzyme activity in thoracic duct lymph higher than in plasma suggested the hypothesis of the intestinal origin of plasma DAO in humans.

During normal pregnancy, because of placental enzyme production, serum DAO rises 200- to 1000-fold, beginning from the first weeks and reaching a maximum in the third trimester.[75,76] This physiological increase has been assessed as a favorable index of the evolution of pregnancy, unless low DAO levels have also been reported in uncomplicated pregnancies.[77]

The lack of serum DAO increase or the fall of the enzyme activity have been related to fetal wastage.[77] Also, decreasing serum DAO levels in amniotic fluids was observed in cases of severe rhesus isoimmunization, resulting in abortion.[78] The detection of high DAO activity in vaginal fluid was only associated with the loss of amniotic fluid due to ruptured fetal membranes.[79] After delivery, serum DAO promptly falls to normal levels.[80]

Heparin administration to normal subjects is followed by a rise of serum DAO activity.[81] The increase is dose dependent, and after the administration of the maximal allowed doses the rise has a bimodal behavior, reaching a first peak after few minutes and then slowly increasing to the maximal values in about 1 hr. The fall to basal levels occurs within a few hours.[81] The extent of the heparin-induced DAO increase is related to the basal serum DAO levels, and the bimodal increase may be due to the late inflow of DAO-rich lymph from the thoracic duct into the bloodstream.[81]

The intestinal origin of post-heparin serum DAO seems to be demonstrated by the higher DAO values in thoracic duct lymph,[82] and by post-heparin GI DAO activity. In fact, we demonstrated that a simultaneous post-heparin early increase in both plasma and GI DAO occurs, followed by a further plasma increase with GI decrease of the enzyme activity.[83] However, a contribution to post-heparin serum DAO increase by other organs cannot be excluded in man. Physiological DAO storages are very difficult to exhaust, since in 40 days full-dose heparin treatment in two patients suffering for phlebothrombosis, we observed an identical post-heparin DAO increase either before and after the treatment.[105]

A reduced post-heparin DAO rise has been observed in patients with acute and chronic liver diseases,[84] with chronic renal diseases,[85] during flogosis,[86] in hyperlipoproteinemia,[87] and after anaphylactic shock.[88]

High DAO activities have been reported in cancer tissues, namely in medullary carcinoma of the thyroid,[89] in small cell carcinoma of the lung[90] and in cancers of the digestive tract.[91]

In medullary carcinoma of the thyroid, which is believed to originate from calcitonin-producing C cells, the presence of DAO differentiates carcinoma cells from normal or hyperplastic C cells[92] (see also Chapter 17).

In large bowel cancers DAO activity was lower in neoplastic tissue than in surrounding normal mucosa.[66] In gastric carcinomas DAO activity was higher than in normal mucosa, but the tumor was surrounded by a non-neoplastic mucosa (with histological signs of gastritis or intestinal metaplasia) which had higher DAO activity than the cancer itself.[66]

Also, serum DAO was found to be increased in some patients with medullary carcinoma of the thyroid,[93,94] small cell carcinoma of the lung,[95] ovarian cancers,[96,98] and other malignancies,[97] especially involving the digestive tract.[91]

Increased DAO activity in serous effusions of subjects with cancer have also been reported.[91,99] The demonstration that high DAO activity was associated with the presence of neoplastic cells in the effusion was obtained by Ettinger and colleagues in ovaric cancers,[96] by Lin,[99] and in our laboratory in lung, pancreas and GI cancers.[91]

High serum DAO levels were also reported in patients with chronic renal failure,[100] in severely burned patients,[101] and in subjects with anaphilactic shock[16] and with cystic fibrosis.[102]

Employing a nonspecific assay method with histamine as substrate, a serum DAO increase was described in bronchial asthma and in other allergic conditions.[74,103,104] Using putrescine as substrate we were unable to demonstrate serum DAO modifications in a number of allergic or immunological disorders (including bronchial asthma, hay fever, serum sickness, urticaria, and insect puncture.[105])

In two cases of severe anaphylactic shock we found a several-fold increased serum DAO, which persisted for hours, while in six cases of anaphylactic-like shock (due to drug intolerance, without evidence of immunological disorders) no increase of serum DAO activity was found.[105]

A serum DAO elevation has also been reported in post-mortem specimens of a subject who died from anaphylactic shock.[82]

The correlation between DAO activity and allergic condition has been investigated with promising results by Herman,[67] who evaluated DAO directly in eosinophils. An increase of eosinophil DAO in patients with active asthma, urticaria, and parasitic infestion has been reported.[67] Finally, a urinary DAO increase was associated with active renal diseases.[71]

In conclusion, the clinical utilization of DAO assay may be confined to too few pathological conditions. The assessment of serum DAO as a cancer marker has been considered. However, DAO has a very short half-life in plasma, and even in cancers which produce elevated amounts of the enzyme, serum DAO levels may be normal. Nevertheless, in medullary carcinoma of the thyroid a rise of serum DAO after surgery may indicate the growth of metastases. Serous effusion DAO assays may help in recognizing the presence of neoplastic cells in cases of difficult interpretation of cytological specimens. In pregnancy, low serum DAO activity may suggest a placental disorder or, if associated with elevated chorionic gonadotropin levels, a placental idatiform degeneration. In the diagnosis of anaphilactic-like shock, serum DAO assays may provide further information on the immunological nature of the phoenomenon, also being helpful for medicolegal purposes. Furthermore, DAO assay in eosinophils and in urines may have useful future developments.

IV. CONCLUSIONS

Over 2 decades, investigations on copper-dependent AOs in humans have opened promising applications for AO assays in different clinical conditions, but further experience is required to consider HSAO and DAO assays as a useful routine analysis. However, these clinical investigations provided useful, but yet incomplete, information on HSAO and DAO biological role. As a final consideration we can observe that no spontaneous or drug-induced copper metabolism alteration directly affects the activity of the copper-dependent AOs, since even in Wilson's disease HSAO and DAO activities are not influenced by D-penicillamine treatment.[39]

ACKNOWLEDGMENT

The original studies reported here were supported in part by the Italian National Research Council, Special Project on the ''Chimica Fine'', Contract No. 83.00310.95.

REFERENCES

1. **Crabbe, M. J., Waight, R. D., Bardsley, W. G., Barker, R. W., Kelly, I. D., and Knowles, P. F.,** Human placental diamine oxidase, *Biochem. J.*, 155, 1, 1976.
2. **McEwen, C. M.,** Human plasma monoamine oxidase. Purification and identification, *J. Biol. Chem.*, 240, 2003, 1965.
3. **McEwen, C. M.,** The soluble monoamine oxidase of human plasma and sera, *Adv. Biochem. Psycopharm.*, 5, 151, 1972.
4. **Buffoni, F. and Blaschko, H.,** Benzylamine oxidase and histaminase: purification and crystallization of an enzyme from pig plasma, *Proc. R. Soc. B*, 161, 153, 1964.
5. **Yamada, H. and Yasunobu, K. T.,** Monoamine oxidease. II. Copper, one of the prosthetic groups of plasma monoamine oxidase, *J. Biol. Chem.*, 237, 3077, 1962.
6. **McEwen, C. M.,** Human plasma monoamine oxidase. II. Kinetic studies, *J. Biol. Chem.*, 240, 2011, 1965.
7. **Rauch, N. and Rauch, R. J.,** Isozymes of amine oxidase in human plasma and other tissues, *Experientia*, 29, 215, 1973.
8. **Lin, A. W. S. M., Davis, L. M., and Castell, D. O.,** Isoenzyme composition of human plasma monoamine oxidase in normal subjects and in fibrotic liver disease, *Proc. Soc. Exp. Biol. Med.*, 151, 40, 1976.
9. **Lin, A. W. S. M. and Castell, D. O.,** Multiple forms of monoamine oxidase in human plasma, *Biochem. Med.*, 13, 141, 1975.
10. **Inoue, F., Robinson, J. B., and Dost, K. A.,** Some observation on the attempted separation of isoenzymes of nonoamine oxidase, *J. Pharm. Pharmac.*, 28, 521, 1976.
11. **Houslay, M. D., Tipton, K. F., and Youdim, M. B. H.,** Multiple forms of monoamine oxidase: fact and artefact, *Life Sci.*, 19, 467, 1976.
12. **Bardsley, W. G., Crabbe, J. C., and Scott, I. V.,** The amine oxidases of human placenta and pregnancy plasma, *Biochem. J.*, 139, 169, 1974.
13. **Bieganski, T., Kusche, J., Feussner, K. D., Hesterberg, R., Richter, H., and Lorenz, W.,** Human intestinal diamine oxidase: substrate specificity and comparative inhibitor study, *Agents Actions*, 10, 108, 1980.
14. **Tufvesson, G.,** A comparison between diamine oxidases from human post-heparin blood serum, pregnancy blood serum and amniotic fluid, *Scand. J. Clin. Lab. Invest.*, 38, 473, 1978.
15. **Tufvesson, G.,** Purification and properties of human amniotic fluid diamine oxidase, *Scand. J. Clin. Lab. Invest.*, 38, 463, 1978.
16. **Buffoni, F.,** Histaminase and related amine oxidases, *Pharmacol. Rev.*, 18, 1163, 1966.
17. **Zeller, E. A.,** Identity of histaminase and diamine oxidase, *Fed. Proc.*, 24, 766, 1965.
18. **McEwen, C. M. and Cohen, J. D.,** An amine oxidase in normal human serum, *J. Lab. Clin. Med.*, 62, 766, 1963.
19. **Otsuka, S. and Kobayashi, Y.,** A radioisotopic assay for monoamine oxidase determinations in human plasma, *Biochem. Pharmacol.*, 13, 995, 1964.
20. **Tufvesson, G.,** Determination of monoamine oxidase activity in human blood serum with 14-C-labelled substrates, *Scand. J. Clin. Lab. Invest.*, 23, 71, 1969.
21. **Tufvesson, G.,** Determination of monoamine oxidase activity in human blood serum with 14-C-Butylamine as substrate, *Scand. J. Clin. Lab. Invest.*, 25, 231, 1970.
22. **Kobayashi, Y.,** The effect of three monoamine oxidase inhibitors on human plasma monoamine oxidase activity, *Biochem. Pharmacol.*, 15, 1287, 1966.
23. **Ono, T., Eto, K., Sakata, Y., and Takeda, M.,** A new colorimetric assay for monoamine oxidase in serum and its clinical application, *J. Lab Clin. Med.*, 85, 1022, 1975.
24. **Nakano, H., Yamamoto, Y., Ohnishi, S., Ito, K., and Imura, H.,** Serum monoamine oxidase assayed with a new synthetic benzylamine derivative as substrate. Its clinical significance with special reference to hepatic fibrosis, *Clin. Chim. Acta*, 88, 315, 1978.
25. **Tryding, N., Nilsson, S. E., Tufvesson, G., Berg, R., Carlstroem, S., Elmfors, B., and Nilsson, J. E.,** Physiological and pathological influences on serum monoamine oxidase level, *Scand. J. Clin. Lab. Invest.*, 23, 79, 1969.
26. **Robinson, D. S., Davis, J. M., Nies, A., Ravaris, C. L., and Sylvester, D.,** Relation of sex and aging to monoamine oxidase activity of human brain, plasma and platelets, *Arch. Gen. Psychiat.*, 24, 536, 1971.
27. **Robinson, D. S., Davis, J. M., Nies, A., Colburn, R. W., Davis, J. N., Bourne, H. R., Bunney, W. E., Shaw, D. M., and Coppen, A. J.,** Ageing, monoamines and monoamine-oxidase levels, *Lancet*, 5, 290, 1972.
28. **Robinson, D. S.,** Changes in monoamine oxidase and monoamines with human development and aging, *Fed. Proc.*, 34, 103, 1975.

29. **Murphy, D. L., Wright, C., Buchsbaum, M., Nichols, A., Costa, J. L., and Wyatt, R. J.,** Platelet and plasma amine oxidase activity in 680 normals: sex and age differences and stability over time, *Biochem. Med.*, 16, 254, 1976.
30. **Gilmore, N. J., Robinson, D. S., Nies, A., Sylvester, D., and Ravaris, C. L.,** Blood monoamine oxidase levels in pregnancy and during the menstrual cycle, *J. Psychos. Res.*, 15, 215, 1971.
31. **McEwen, C. M.,** Serum amine oxidases in pregnancy, *J. Lab. Clin. Med.*, 64, 540, 1964.
32. **Zeller, E. A.,** Oxidation of amines, in *The Enzymes*, Vol 2, Academic Press, New York, 1951, 536.
33. **Zeller, E. A.,** Diamine oxidases, in *The Enzymes*, Vol. 8, Academic Press, New York, 1963, 313.
34. **Klaiber, E. L., Kobayashi, Y., Broverman, D. M., and Hall, F.,** Plasma monoamine oxidase activity in regularly menstruating women and in amenorrheic women receiving cyclic treatment with estrogens and a progestin, *J. Clin. Endocrinol.*, 33, 630, 1971.
35. **Buffoni, F., Ignesti, G., Pirisino, R., and Cortesini, C.,** Platelet monoamine oxidase, plasma benzylamine oxidase and liver mixed function oxidase in cirrhotic patients, *Med. Biol.*, 55, 109, 1977.
36. **McEwen, C. M., and Castell, D. O.,** Abnormalities of serum monoamine oxidase in chronic liver disease, *J. Lab. Clin. Med.*, 70, 36, 1967.
37. **Rossi, L., Bombardieri, G., and De Rosa, G.,** La monoaminossidasi sierica dell'uomo in alcune condizioni morbose, *Gazz. Intern. Med. Chir.*, 73, 3343, 1968.
38. **Rossi, L., Bombardieri, G., and De Rosa, G.,** Rilievi sul comportamento della monoaminossidasi sierica in epatopazienti, *Il Fegato*, 15, 353, 1969.
39. **Bombardieri, G., Bevilacqua, E., Conti, L. R., Innocenti, P., and Perretti, M. A.,** Amine oxidases in Wilson's disease, *Life Sci.*, 29, 117, 1981.
40. **Ito, K., Nakagawa, J., Minakuchi, C., and Fukase, M.,** A clinical evaluation of serum monoamine oxidase, with special reference to hepatic fibrosis, *Digestion*, 4, 49, 1971.
41. **Kirchner, J. P., Nebel, O. T., and Castell, D. O.,** Serum monoamine oxidase, an index of hepatic fibrosis: review of clinical studies, *Mil. Med.*, 159, 952, 1974.
42. **Gressner, A. M.,** Validity of monoamine oxidase in serum for diagnosis of liver cirrhosis: estimation of predictive values, sensitivity and specificities, *J. Clin. Chem. Clin. Biochem.*, 20, 509, 1982.
43. **Borthakur, P. C., Nath, R., Chhuttani, P. N., and Datta, D. V.,** Serum monoamine oxidase levels in portal hypertension, *Ind. J. Med. Res.*, 61, 1165, 1973.
44. **McEwen, C. M. and Harrison, D. C.,** Abnormalities of serum monoamine oxidase in chronic congestive heart failure, *J. Lab. Clin. Med.*, 65, 546, 1965.
45. **Nilsson, S. E., Tryding, N., and Tufvesson, G.,** Serum monoamine oxidase in diabetes mellitus and some other internal diseases, *Acta Med. Scand.*, 184, 105, 1968.
46. **Bombardieri, G., De Rosa, G., Rossi, L., and Barbarino, A.,** La monoamino-ossidasi sierica nell'ipertiroidismo, *Rivista Cirt. Clin. Med.*, 69, 245, 1969.
47. **Oriente, P., Riccio, A., Scarpa, R., Postiglione, L., Spano', A., and Pucino, A.,** Evaluation of serum monoamine oxidase in patients with connective tissue diseases, *Clin. Chem.*, 26, 1514, 1980.
48. **Bombardieri, G., Rossi, L., Romano, A., and Zamperini, D.,** Studies on serum monoamine oxidase in carcinomas, lymphomas, and other malignant hemopathies, *Acta Med. Rom.*, 13, 186, 1975.
49. **Lewinsohn, R.,** Human serum amine oxidase. Enzyme activity in severely burnt patients and in patients with cancer, *Clin. Chim. Acta*, 81, 247, 1977.
50. **Bandziulene, S. K. and Murza, V.,** The activity of serum monoamine oxidase in rheumatic fever and rheumathoid arthritis, *Reumatologia (Warsawa)*, 11, 141, 1973.
51. **Shih, J. H. C. and Eiduson, S.,** Multiple forms of monoamine oxidase in developing tissues. The implications for mental disorders, *Adv. Behav. Biol.*, 1, 3, 1971.
52. **Zis, A. P., Gold, P. W., Paul, S. M., Goodwin, F. K., and Murphy, D. L.,** Elevation of human plasma and platelet amine oxidase activity in response to intravenous dopamine, *Life Sci.*, 28, 371, 1981.
53. **Briggs, M. and Briggs, M.,** Relationship between monoamine oxidase activity and sex hormone concentration in human blood plasma, *J. Reprod. Fertil.*, 29, 447, 1972.
54. **Robinson, D. S., Lovenberg, W., Keiser, H., and Sjoerdsma, A.,** Effects of drugs on human blood platelet and plasma amine oxidase activity in vitro and in vivo, *Biochem. Pharmacol.*, 17, 109, 1968.
55. **Rossi, L., Bombardieri, G., and Muratore, D.,** Osservazioni sulla monoamino ossidasi del siero dell'uomo in corso di trattamento con farmaci ad attivita' antiaminossidasica, *Gazz. Intern. Med. Chir.*, 73, 4726, 1968.
56. **Rosmus, J., Trnavsky, K., and Deyl, Z.,** Experimental lathyrism — a molecular disease, *Biochem. Pharmacol.*, 15, 1405, 1966.
57. **Groshong, R., Baldessarini, R. J., and Gibson, D. A.,** Activities of types A and B MAO and cathecol-*O*-methyltransferase in blood cells and skin fibroblasts of normal and chronic schizophrenic subjects, *Arch. Gen. Psychiatr.*, 35, 1198, 1978.
58. **Kershenobich, D., Uribe, M., Suarez, G. I., Mata, J. M., Perez-Tamayo, R., and Rojkind, M.,** Treatment of cirrhosis with colchicine, *Gastroenterology*, 77, 532, 1979.

59. **Maslinski, C.,** Histamine and its metabolism in mammals: catabolism of histamine and histamine liberation, *Agents Actions*, 5, 183, 1975.
60. **Crabbe, J.,** Summary of an international workshop on diamine oxidase held at the Department of Biogenic Amines, Polish Academy of Science, Lodz, *Agents Actions*, 11, 3, 1981.
61. **Kapeller-Adler, R.,** The estimation of histaminase by a micro-volumetric method involving a peroxydatic oxidation of indigo disulphonate, *Biochem. Biophys. Acta*, 22, 391, 1956.
62. **Aarsen, P. N. and Klemp, A.,** Rapid spectrophotometric micromethod for determination of histaminase activity, *Nature (London)*, 204, 1195, 1964.
63. **Bieganski, T., Blasinska, M. Z., and Kusche, J.,** Determination of histaminase (DAO) activity by O-dianisidine test: interference of ceruloplasmin, *Agents Actions*, 7, 85, 1977.
64. **Bardsley, W. G., Crabbe, M. J. C., Shindler, J. S., and Ashford, J. S.,** Oxidation of *p*-dimethylaminomethylbenzylamine by pig kidney diamine oxidase, *Biochem. J.*, 127, 875, 1972.
65. **Kusche, J., Lorenz, W., Stahlknecht, C. D., Friedrich, A., Schmidt, A., Boo, K., and Reichert, G.,** Diamine oxidase activity in gastric and duodenal mucosa of man and other mammals with special reference to the pyloric junction, *Agents Actions*, 8, 366, 1978.
66. **Kusche, J., Bieganski, T., Hesterberg, R., Stahlknecht, C. D., Feussner, K. D., Stahlenberg, I., and Lorenz, W.,** The influence of carcinoma growth on diamine oxidase in human gastrointestinal tract, *Agents Actions*, 10, 110, 1980.
67. **Herman, J. J.,** Eosinophil diamine oxidase activity in acute inflammation in humans, *Agents Actions*, 12, 46, 1982.
68. **Crabbe, M. J. C. and Kavanagh, J. P.,** The purification of four enzymes from human seminal plasma, *Biochem. Soc. Trans.*, 5, 735, 1977.
69. **Dahlback, O., Hansson, R., Tibbling, G., and Tryding, N.,** The effect of heparin on diamine oxidase and lipoprotein lipase in human lymph and blood plasma, *Scand. J. Clin. Invest.*, 21, 17, 1968.
70. **Tufvesson, G.,** Purification and properties of human amniotic fluid diamine oxidase, *Scand. J. Clin. Lab. Invest.*, 38, 463, 1969.
71. **Giarnieri, D., Costa, M. T., Giarnieri, V., and Mondovì, B.,** Diamine oxidase content in urine of patients with renal failure, *Agents Actions*, in press.
72. **Swanberg, H.,** Histaminase in pregnancy, with special reference to its origin and formation, *Acta Physiol. Scand.*, 23, (79), 1, 1950.
73. **Gennser, G., Soederberg, A., and Tryding, N.,** Family with a high activity of serum diamine oxidase, *Am. J. Obstet. Gynecol.*, 8, 1128, 1974.
74. **Kameswaran, L., Kanakambal, K., and Vijayasekaran, V.,** Studies on plasma histaminase levels in normal and allergic individuals, *Ind. J. Physiol. Pharmacol.*, 12, 159, 1968.
75. **Hansson, R., Tryding, N., and Toernqvist, A.,** Diamine oxidase (histaminase) in human pregnancy, *Acta Obstet. Gynecol. Scand.*, 48, 8, 1969.
76. **Beaven, M. A., Marshall, J. R., Baylin, S. B., and Sjordsma, A.,** Changes in plasma histaminase activity during normal early human pregnancy and pregnancy disorders, *Am. J. Obstet. Gynecol.*, 123, 605, 1975.
77. **Southren, A. L., Weingold, A. B., Kobayashi, Y., Sherman, D. H., Grimaldi, R., and Gold, E. M.,** Plasma diamine oxidase in pregnancy complicated by diabetes mellitus, *Am. J. Obstet. Gynecol.*, 101, 899, 1968.
78. **Ward, H., Whyley, G. A., and Millar, M. D.,** The value of amniotic fluid diamino oxidase estimations in the management of severe rhesus isoimmunization, *Acta Obstet. Gynecol. Scand.*, 55, 63, 1976.
79. **Elmforms, B., Tryding, N., and Tufvesson, G.,** The diagnosis of ruptured membranes by measurement of the diamino oxidase (DAO) activity in vaginal fluid, *J. Obstet. Gynecol. Br. Commun.*, 81, 361, 1974.
80. **Hanssons, R.,** Diamine oxidase isoenzymes in human blood plasma, *Scand. J. Clin. Lab. Invest.*, 25, 33, 1970.
81. **Hanssons, R., Holmberg, C. G., Tibbling, G., Tryding, N., Westling, H., and Wetterqvist, H.,** Heparin-induced diamine oxidase increase in human blood plasma, *Acta Med. Scand.*, 180, 533, 1966.
82. **Hansson, R.,** Heparin-induced increase of diamine oxidase/hisatminase in blood and lymph, *Scand. J. Clin. Lab. Invest.*, 31 (129), 4, 1973.
83. **Bombardieri, G., Rossi, L., Conti, L. R., Vecchioli, A., and Innocenti, P.,** Incremento dell'attivit' diaminossidasica posteparinica nella mucosa gastrica e digiunale, *Quad. Sclavo Diagn.*, 15, 1017, 1979.
84. **Gang, V.,** Postheparin-diaminooxidase (histaminase) bei akuten und chronischen Lebererkrankungen, *Zeitschr. f. Gastroenterol.*, 12, 407, 1974.
85. **Bevilacqua, E., Bombardieri, G., Conti, L. R., and Rossi, L.,** La diamino-ossidasi (DAO) posteparinica nelle epatopatie e nefropatie croniche, *Quad. Sclavo Diagn.*, 15, 1026, 1979.
86. **Gang, V., Gehring, G., and Schoenborn, J.,** Combined decrease of postheparin diamine oxidase (histaminase) and postheparin lipoprotein lipase in inflammatory diseases, *Klin. Wschr.*, 55, 169, 1977.

87. **Baylin, S. B., Beaven, M. A., Krauss, R. M., and Keiser, H. R.,** Response of plasma histaminase activity to small doses of heparin in normal subjects and patients with hyperlipoproteinemia, *J. Clin. Invest.*, 52, 1983, 1973.
88. **Gang, V., Gaubitz, W., and Gunzer, U.,** Postheparin diamine oxidase (histaminase) in anaphylaxis, *Klin. Wschr.*, 53, 285, 1975.
89. **Baylin, S. B., Beaven, M. A., Buja, L. M., and Keiser, H. R.,** Histaminase activity: a biochemical marker for medullary carcinoma of the thyroid, *Am. J. Med.*, 53, 723, 1972.
90. **Baylin, S. B., Weisburger, W. R., Eggleston, J. C., Mendelsohn, G., Beaven, M. A., Abeloff, M. D., and Ettinger, D. S.,** Variable content of histaminase, L-DOPA decarboxylase and calcitonin in small-cell carcinoma of the lung, *N. Engl. J. Med.*, 299, 105, 1978.
91. **Rossi, L., Bombardieri, G., Bevilacqua, E., and Innocenti, P.,** Significato clinico della diaminossidasi nei versamenti sierosi di natura neoplastica e non neoplastica, *Quad. Sclavo Diagn.*, 16, 1, 1980.
92. **Baylin, S. B., Mendelsohn, G., Weisburger, W. R., Gann, D. S., and Eggleston, J. C.,** Levels of histaminase and L-DOPA decarboxylase activity in the transition from C-cell hyperplasia to familial medullary thyroid carcinoma, *Cancer*, 44, 1315, 1979.
93. **Baylin, S. B., Beaven, M. A., Engelman, K., and Sjoerdsma, A.,** Elevated histaminase activity in medullary carcinoma of the thyroid gland, *N. Engl. J. Med.*, 283, 1239, 1970.
94. **Baylin, S. B., Beaven, M. A., Keiser, H. R., Tashjian, A. H., and Melvin, K. E. W.,** Serum histaminase and calcitonin levels in medullary carcinoma of the thyroid, *Lancet*, 1, 455, 1972.
95. **Baylin, S. B., Abeloff, M. D., Wieman, K. C., Tomford, J. W., and Ettinger, D. S.,** Elevated histaminase (DAO) activity in small cell carcinoma of the lung, *N. Engl. J. Med.*, 293, 1286, 1975.
96. **Ettinger, D. S., Rosenshein, N. B., Parmley, T. H., Mendelsohn, G., and Baylin, S. B.,** Tumor cell origin of histaminase activity in ascites fluid from patients with ovarian carcinoma, *Cancer*, 45, 2568, 1980.
97. **Borglin, N. E. and Willert, B.,** Increased histaminolytic power of plasma in endometrial adenocarcinoma, *Cancer*, 15, 271, 1962.
98. **Lin, C. W., Orcutt, M. L., and Stolbach, L. L.,** Elevation of histaminase and its concurrence with Regan isoenzyme in ovarian cancer, *Cancer Res.*, 35, 2762, 1975.
99. **Lin, C. W., Inglis, N. R., Rule, A. H., Turksoy, R. N., Chapman, C. M., Kirley, S. D., and Stolbach, L. L.,** Histaminase and other tumor markers in malignant effusion fluids, *Cancer Res.*, 39, 4894, 1979.
100. **Kopple, J. D., Tam, C. F., Wang, M., and Swendseid, M. E.,** Diamine oxidase in renal failure, *Kidney Intern.*, 13, suppl. 8, 20, 1978.
101. **Franco, S., Bombardieri, G., Rossi, L., and Lombardi, P.,** La diamino-ossidasi sierica e del liquido di flittena negli ustionati, *Min. Med.*, 66, 3990, 1975.
102. **Baylin, S. B., Rosestein, B. J., Marton, L. J.,and Lockwood, D. H.,** Age-related abnormalities of circulating polyamines and DAO activity in cystic fibrosis heterozygotes and homozygotes, *Pediatr. Res.*, 14, 921, 1980.
103. **Majumdar, S. K., Joshi, D. J., and Sachdev, K. S.,** Plasma histaminase in patients of bronchial asthma and the effect of prednisolone administration on it, *Ind. J. Med. Res.*, 57, 8, 1969.
104. **Kanakambal, K., Vijayasekaran, K., and Kameswaran,** Studies on plasma histaminase levels in some clinical conditions with possible allergic component, *Ind. J. Physiol. Pharmacol.*, 11, 8, 1967.
105. **Bombardieri, G. et al.** unpublished observations.

Chapter 19

QUALITATIVE ALTERATION (TRANSFORMATION) IN CATALYTIC ACTIVITY OF AMINE OXIDASES

V. Z. Gorkin

TABLE OF CONTENTS

I. INTRODUCTION

The term "transformation of enzyme" denotes the reversible qualitative alteration of the catalytic properties of enzymes elicted by treatment with various reagents.[1]

It is now well recognized that chemical modification of the active site of an enzyme or a change in the arrangement of the functional groups near this site, in a manner that does not inactivate the enzyme completely, may cause not only quantitative changes in activity (activation or inhibition), but also qualitative ones, including reversible alteration of substrate specificity. The capacity to undergo such reversible transformations is characteristic for many enzymes, including amine oxidases.[2]

II. TRANSFORMATION OF PURIFIED AO

The transformation of FAD-amine oxidases (AOs) has been demonstrated in experiments with highly purified preparations of these enzymes from various sources.[2,3] Reversible and qualitative alterations in the catalytic properties of these enzymes were initiated by partial oxidation of their SH groups, apparently to sulfenic acid residues (– SOH).[4] The transformed FAD-AOs acquired the ability of deaminating, via oxidative or hydrolytic pathways, diamines, ω-amino acids, nucleotides (e.g., AMP), and numerous other nitrogenous compounds which are not substrates of MAOs.[2,4] Structural analogs of these compounds blocked the activity of the transformed MAOs.[4,5] The transformations of FAD-AOs were not accompanied by alterations in the quaternary structure detectable by polyacrylamide gel electrophoresis[4,6] or by separation of a flavin component from an apoenzyme, although it has been possible to demonstrate kinetically that the transformation in catalytic properties of the bacterial MAO from *Sarcina lutea*[7] caused a particular alteration in the state of a flavin component of the MAO.[4] The transformations of MAOs were prevented in vitro by MAO inhibitors or SH reagents.[4]

Transformation of copper AO has also been observed.[8-10] Treatment of the highly purified enzyme from pig kidney with reducing agents caused reversible qualitative alterations in its properties. The ability of oxidizing histamine and aliphatic diamines decreased and, simultaneously, new properties appeared: the transformed enzyme catalyzed (in complete absence of flavins!) the oxidative deamination of monoamines (tyramine, tryptamine) and even of the secondary amine *N*-methyl-β-phenylethylamine.

It should be emphasized that the term "transformation" of AOs does not imply that MAOs are actually converted into DAOs (or vice versa). We are dealing, instead, with reversible modifications of the catalytic properties of AOs. As a result of a modification, MAOs, for example, acquire new properties which resemble some of those of native DAOs, but the MAOs do not acquire all the catalytic properties of DAOs. In contrast to transformed MAOs, there are no naturally occurring DAOs that catalyze hydrolysis of AMP, or the oxidation of lysine, spermine, etc.[1,2] The transformed MAOs thus differ qualitatively in their catalytic properties not only from the native MAOs, but also from all the naturally occurring DAOs or adenylate deaminases. This important property of the transformed MAOs, discovered in experiments with highly purified enzymes, enables one to monitor transformation of MAOs indirectly, without isolation and purification of the enzymes in each case. If a given treatment of a biological sample causes a decrease in MAO activity and, at the same time, there is an increase in or appearance of the abilities to deaminate AMP, substrates of DAOs, or ω-amino acids, and if these phenomena are prevented by MAOs inhibitors, it may be assumed that we are dealing with the transformation of MAOs.

III. TRANSFORMATION OF FAD-AOs IN MITOCHONDRIAL MEMBRANES

In mitochondrial membranes the transformation of MAO activity accompanied the stimulation of lipid peroxidation and was prevented either by inhibitors of free-radical chain oxidative processes butylhydroxytoluene and propyl gallate (which did not influence the MAO activity and did not chelate FE^{2+} ions used for stimulation of the chain reactions) or by specific MAO inhibitors pargyline and tranylcypromine.[4] Studies carried out with selective MAO inhibitors clorgyline and deprenyl[11] showed that only the MAO of the type A underwent the transformation in mitochondrial membranes in which lipid peroxidation was stimulated.[12]

It has been shown that relatively high concentrations of some amine[13] or hydrazine[14,15] MAO inhibitors (exceeding the concentrations required for complete blocking of the catalytically active sites of MAO) caused transformation of membrane-bound FAD-AOs apparently due to stimulation of free-radical lipid peroxidation processes in biomembranes.[2] The compounds *initiating qualitative alteration (instead of simple blocking!)* of the MAO catalytic activity exhibited the following common properties: (1) high affinity towards the active sites of the MAO of the type A; (2) potential ability to be oxidized on the catalytic surface of MAO with formation of molecules containing an electron-accepting azomethine bound $-C=N-$; (3) penetration to the places of MAO location *rapidly* (creating instantaneously a high concentration gradient instead of supply of discrete portions of the inhibitor), thus ensuring effective stimulation of free-radical oxidative chain reactions in the biomembrane lipids *before* the development of MAO inhibition in the course of preincubation of mitochondria with the inhibitor.[16]

IV. TRANSFORMATION OF MAO IN VIVO

In vivo transformation of MAO has been observed under numerous pathological conditions of quite different etiology but possessing one common feature: accumulation of lipid peroxides (stimulation of free-radical chain reactions) in tissues.[2] The phenomenon of FAD-AO transformation was noted not only in tissues of tumor-bearing animals, in experimental radiation injury, in hypervitaminosis D_2,[4] experimental tuberculosis, hyperbaric oxygen intoxication,[12] but also in many other intoxications,[2,13-15] experimental hypercholesterolemia and atherosclerosis,[17,18] hypoxia,[19] and stress caused by exposure to low temperature for 3 days.[20]

The transformation of FAD-AOs, resulting in marked increases in the deamination of AMP (and numerous nucleotides in molecules of which a residue of AMP is present) or of amino sugars (e.g., glucosamine, galactosamine), not only accompany the development of, but are probably pathogenetically important in these and related diseases.[2] Structural analogs of AMP (e.g., adenosine 3′-monophosphate) inhibited the deamination of nucleotides by the transformed MAO (but did not inhibit deamination of biogenic monoamines) and had a definite beneficial effect in the tumor-bearing animals and in experimental radiation injury.[2,4] Specific inhibitors of MAO, especially those which cause selective inhibition of the MAO of the type A, usually prevented the transformation of FAD-AO in vivo.[12] However, it has to be noted that, depending on the experimental conditions (concentration, velocity of supply, species and organ differences), a single given compound may either initiate the qualitative modification of catalytic activity of the MAO of the type A or, to the contrary, completely prevent the possibility of this modification.[2,13-15] Inhibitors of free-radical chain reactions (e.g., the derivatives of dihydropyridine), blocking lipid peroxidation, prevented the transformation of FAD-AOs in vivo.[2,17,18] Some reducing agents (e.g., sodium thiosulfate, isoniazid) caused a retransformation of modified MAO in vivo which was accompanied by beneficial effects in several pathological conditions.[17,18,21]

REFERENCES

1. **Gorkin, V. Z.,** Enzyme transformations, in *Horizons in Biochemistry and Biophysics,* Vol. 3, Quagliariello, E., Palmieri, F., and Singer, T. P., Eds., Addison-Wesley, Reading, 1977, chap. 1.
2. **Gorkin, V. Z.,** *Amine Oxidases in Clinical Research,* Pergamon, Oxford, 1983, 121.
3. **Moskvitina, T. A., Kamyshanskaya, N. S., Garishvili, T. G., and Gorkin, V. Z.,** On preparative separation of brain mitochondrial monoamine oxidases, *Prep. Biochem.,* 9, 171, 1979.
4. **Gorkin, V. Z.,** Monoamine oxidase inhibitors and the transformation of monoamine oxidases, in *Monoamine Oxidase and its Inhibition,* Wolstenholme, G. E. W. and Knight, J., Eds., Elsevier, Amsterdam, 1976, 61.
5. **Gorkin, V. Z.,** Monoamine oxidases: versatility of catalytic properties and possible biological functions, in *Advances in Pharmacology and Chemotherapy,* Vol. 11, Garattini, S., Goldin, A., Hawking, F., and Kopin, I., Eds., Academic Press, New York, 1973, chap. 1.
6. **Garishvili, T. G. and Gorkin, V. Z.,** Study of brain monoamine oxidases modified in lipid peroxidation processes. *Bull. Acad. Sci. Georgian SSR (Tbilisi),* 89, 185, 1978.
7. **Yamada, H., Uwajima, T., Kumagai, H., Watanabe, M., and Ogata, K.,** Bacterial monoamine oxidases. I. Purification and crystallization of tyramine oxidase of *Sarcina lutea, Agric. Biol. Chem. (Tokyo),* 31, 890, 1967.
8. **Stesina, L. N., Akopyan, Zh. I., and Gorkin, V. Z.,** Modification of catalytic properties of amine oxidases, *FEBS Lett.,* 16, 349, 1971.
9. **Gorkin, V. Z.,** Qualitative alterations in enzymatic properties of amine oxidases, in *Monoamine Oxidases — New Vistas. Advances in Biochemical Psychopharmacology,* Vol. 5, Costa, E. and Sandler, M., Eds., Raven Press, New York, 1972, 55.
10. **Zeller, E. A. and Subba Rao, K.,** On the phenomenon of the two linear Hammett relationships: its origin and its use as a tool in the analysis of diamine oxidase preparations, *Agents Actions* 3, 197, 1973.
11. **Knoll, J.,** Selective inhibition of B type monoamine oxidase in the brain: a drug strategy to improve the quality of life in senescence, in *Strategy in Drug Research,* Keverling Buisman, J. A., Ed., Elsevier, Amsterdam, 1982, 107.
12. **Gorkin, V. Z.,** Modifications of monoamine oxidases of type-A in some pathological conditions and the effects of drugs, in *Monoamine Oxidase: Structure, Function and Altered Functions,* Singer, T. P., Von Korff, R. W., and Murphy, D. L., Eds., Academic Press, New York, 1979, 347.
13. **Baumanis, E. A., Kalnina, I. E., Iriste, A. A., and Gorkin, V. Z.,** Modification of mitochondrial monoamine oxidase activity by some amines — inhibitors of the monoamine oxidases, *Vopr. Med. Khim.,* 27, 72, 1981.
14. **Kalnina, I. E., Baumanis, E. A., Kairane, Ch. B., and Gorkin, V. Z.,** Modification by some hydrazine derivatives of brain mitochondrial monoamine oxidase activity, *Vopr. Med. Khim.,* 27, 773, 1981.
15. **Baumanis, E. A., Kalnina, I. E., Kaminka, M. E., Mashkovsky, M. D., and Gorkin, V. Z.,** Modification of the activity of mitochondrial monoamine oxidases *in vitro* and *in vivo, Agents Actions,* 11, 685, 1981.
16. **Kairane, Ch. B., Baumanis, E. A., Burlakova, E.B., Gorkin, V. Z., Kalnina, I. E., and Molochkina, E. M.,** Changes in the anti-oxidative activity of lipids in mouse organs and of the monoamine oxidase activity by the action of the dibenzylhydrazide of D,L-malic acid, *Vopr. Med. Khim.,* 25, 776, 1979.
17. **Khuzhamberdiev, M., Mamadiev, M., and Gorkin, V. Z.,** On the mechanism of impairments in deamination of nitrogenous compounds in experimental hypercholesterolaemia and atherosclerosis, *Vopr. Med. Khim.,* 27, 829, 1981.
18. **Mamadiev, M., Khuzhamberdiev, M., and Gorkin, V. Z.,** On the approach to normalization of impairments in deamination of nitrogenous compounds in experimental atherosclerosis, *Vopr. Med. Khim.,* 29, 2, 83, 1983.
19. **Goroshinskaya, I. A., Bronovizkaya, Z. G., Krichevskaya, A. A., and Karabuhina, E. G.,** Effect of hypoxia on the activity and substrate specificty of rat brain monoamine oxidases, *Neurochemistry (Yerevan),* 1, 282, 1982.
20. **Goroshinskaya, I. A., Grabovskova, L. L., Bronovitskaya, Z. G., and Krichevskaya, A.A.,** Activities of the brain monoamine oxidase during cold adaptation and under combined effect of low temperature and hyperbaric oxygenation, *Physiol. J. USSR* 67, 1611, 1981.
21. **Gorkin, V. Z. and Yelistratova, N. A.,** Deamination of biogenic amines in experimental tuberculosis, *Vopr. Med. Khim.,* 25, 743, 1979.

Chapter 20

ADVANCES IN MONOAMINE OXIDASE ENZYMOLOGY

Kerry T. Yasunobu and Anthony Tan

TABLE OF CONTENTS

I. INTRODUCTION

Due to the research progress in the past 20 years, monoamine oxidase (MAO) (monoamine:O_2 oxidoreductase[deaminating], EC 1.4.3.4) has become a respectable member of the flavoenzyme family.[1-4] The enzyme was initially discovered by Hare[5] from liver and it catalyzes the oxidation of certain amines as shown in Equation 1:

$$RCH_2NH_2 + O_2 + H_2O = RCHO + NH_3 + H_2O_2 \quad (1)$$

Because of its localization in the outer membrane of the mitochondria,[6] purification and characterization of the enzyme was difficult until methods for isolation of membrane enzymes were developed.

In this review article, recent important developments concerning the various aspects of MAO are discussed. Included are: (1) structure-function investigation of bovine liver MAO, (2) the mechanism of amine oxidation based on kinetic studies, (3) the MAO-A and MAO-B relationship and related topics, (4) the effect of phospholipids, and (5) new methods of enzyme purification and the isolation of a new type of amine oxidase.

II. BASIC PROPERTIES OF BOVINE LIVER MAO-B

Very little is known about the basic molecular properties of any MAO. The most physicochemically characterized enzyme is bovine liver MAO-B. The amino acid composition of bovine liver MAO-B was reported by Minamiura et al.[7] and the data is summarized in Table 1.

The S-β-carboxymethylcysteinyl-derivative of MAO was analyzed. Thus, in the intact enzyme, there are four cysteine residues and two cystine residues from sulfhydryl titration studies.[8] The hydrophobic amino acid content, which included the amino acids proline, valine, methionine, isoleucine, leucine, tyrosine, phenylalanine, and tryptophan, was 193 out of 481 residues (40%) and is not as hydrophobic as some other membrane enzymes. The amino acid composition of the porcine liver enzyme is very similar.[9] However, the enzyme has a high content of leucine residues. The isoelectric point of the enzyme is about pH 4.7 to 5.0,[10] and MAO contains an excess of acidic over basic residues although the content of asparagine and glutamine residues are unknown. Histidine and cysteine residues are essential for activity and thus the number of these residues is important. Assuming the molecular weight of the monomer is 52,000, there are approximately 14 histidine residues and one of these residues is important or essential. There are approximately eight cysteine residues per 100,000 g of protein or four per 52,000 g of enzyme.[8] Assuming there are four residues, one is essential for activity and three are nonessential.

The NH_2-terminal residue has a blocked α-NH_2-group. This was shown from protein sequencer analysis of the *S*-β-carboxymethylcysteine derivative. When about 7 mg of the carboxymethylated protein was sequenced in the Beckman 890 Protein Sequencer for five cycles of Edman degradation, no Pth-amino acids were detected at any of these steps.[11] When carboxypeptidase A digestion was performed on the *S*-β-carboxymethylcysteine derivative, a number of amino acids were liberated but it appeared that leucine was the first residue to be liberated in significant amounts.[8] However, since hydrazinolysis did not liberate any C-terminal amino acids, probably because it was insoluble under the conditions of the experiment, the identity of the C-terminal amino acid must be considered tentative. It is well known that the use of carboxypeptidase kinetics alone may give misleading C-terminal sequence data.

Very little sequence data of MAO-B is available. Kearney et al.[12] have isolated the

Table 1
AMINO ACID COMPOSITION OF BOVINE LIVER MAO-B[7]

Amino acid	Residues/ 52,000 g MAO	Amino acid	Residues/ 52,000 g MAO	Amino acid	Residues/ 52,000 g MAO
Lysine	26	Proline	27	Leucine	51
Histidine	14	Glycine	43	Tyrosine	18
Arginine	23	Alanine	30	Phenylalanine	19
Aspartic acid	37	Cysteine	8	Tryptophan	4
Threonine	27	Valine	36	Total	481
Serine	31	Methionine	11		
Glutamic acid	49	Isoleucine	27		

pentapeptide Ser-Gly-Gly-Cys-8α-FAD-Tyr from a combined tryptic-chymotryptic digest of beef liver MAO-B.

III. CHEMICAL MODIFICATIONS OF MAO-B

A. Evidence for Essential Histidine Residue

Hiramatsu et al.[13] reported that there were 2 essential histidine residues per 100,000 g of enzyme but it now appears to be 1 histidine residue since the monomer molecular weight of the enzyme is about 53,000. Ethoxyformic acid anhydride, amino-1-H-tetrazole, and photooxidation using Rose Bengal inactivated MAO. Ethoxyformic acid anhydride reacted with three histidine residues per monomer and completely inactivated MAO-B. However, 0.2 *M* hydroxylamine reactivated the derivative to 100% of the enzyme treated with only 0.2 *M* hydroxylamine. Photooxidation using Rose Bengal in the presence and absence of substrate showed that there was one active site histidine residue. A kinetic study has suggested that the histidine residue is required for substrate cleavage.[14]

B. Evidence for Essential Cysteine Residue

Despite earlier conflicting reports, recent evidence by Zeidan et al.[15] indicate that bovine liver MAO-B contains one essential cysteine residue and three nonessential cysteines per monomer. The reactivity of these cysteine residues differed and depended on the sulfhydryl reagent used. When iodoacetic acid was used, the three nonessential cysteine residues were modified first and then the essential residue. When nitroxide-containing iodoacetic acid derivatives of varying size were used, it was possible to probe the environment of the essential cysteine residue by ESR spectral technique.[16] It was noted that the essential cysteine residue existed in two states: one in which it was buried in a shallow pocket and another in which the residue was exposed. The majority existed in a buried state. However, when the competetive inhibitor, *p*-hydroxyamphetamine, was added to the enzyme derivative, the cysteine residue became exposed. Thus, when substrate/competitive inhibitor is added to MAO-B, a conformational change of the essential cysteine residue occurs. Phenylalanine and tyrosine, which are analogs of phenethylamine and tyramine, but are neither substrates nor inhibitors of MAO, had no effect. A kinetic study[14] suggested the cysteine residue is required for substrate binding.

Singer has reported that phenylcyclopropylamine is a competetive inhibitor of MAO-B.[17] Paech et al.[18] have suggested that the enzyme is inhibited by a mechanism in which the essential cysteine residues react with the imine or corresponding ketone of phenylcyclopropylamine to form compound I or II shown in Figure 1 (see Chapter 21).

IV. FLAVIN PROSTHETIC GROUP

Bovine liver MAO-B was first shown to contain covalently linked FAD by Nara et al.[19]

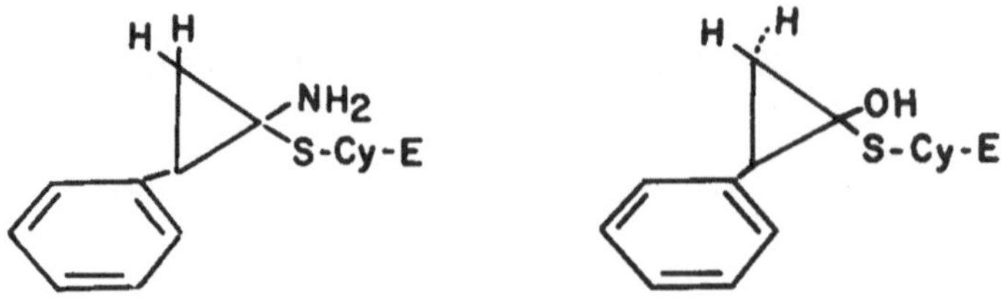

(I) (II)

FIGURE 1. Proposed structures of the adduct formed between phenylcyclopropylamine and the essential sulfhydryl group of the monoamine oxidase.[17] (I) stands for proposed imine adduct and (II) the corresponding ketone adduct.

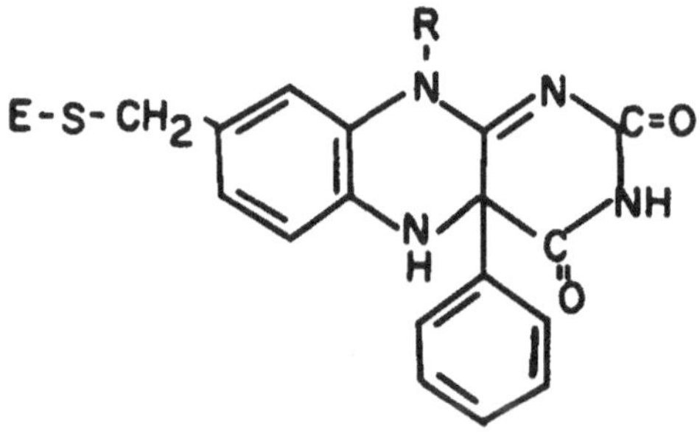

FIGURE 2. Proposed structure of adduct formed when phenylhydrazine was reacted with monoamine oxidase.[17]

In a very elegant study, Walker et al. have established the structure of the flavin as 8-a-cysteinyl-FAD.[20]

Singer et al.[17] have studied the mechanism of inhibition by a variety of irreversible inhibitors of MAO-B. They formed adducts involving the 5- and 4α-positions of the isoalloxazine ring of the FAD. Of importance here is the inhibition of the enzyme by phenylhydrazine. Singer[17] proposed the following mechanism of inhibition (Equations 2 to 5) where F-FAD has the structure shown in Figure 2.

$$E-FAD + \O NHNH_2 = E^{FAD}_{\O NHNH_2} \tag{2}$$

$$E^{-FAD}_{\O NHNH_2} = E^{-FADH_2}_{\O N=NH} \tag{3}$$

$$E^{-FADH_2}_{\O N=NH} + O_2 = E^{-FAD}_{-\O N=NH} + H_2O_2 \tag{4}$$

$$E^{-FAD}_{-\O N=NH} = E-FADH\O + N_2 \tag{5}$$

In order to probe the environment of the FAD portion of the active center of MAO-B, Tan et al.[21] synthesized the following two nitroxide-containing hydrazine derivatives (Figure 3) whose structures are shown along with structures of the adducts (Figure 4).

The ESR spectra of these derivatives indicated that the FAD was exposed since the nitroxide moiety was only partially immobilized.

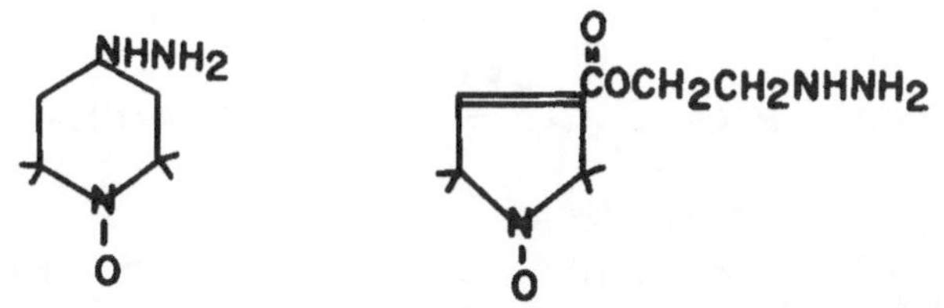

FIGURE 3. Structures of nitroxide spin labeled hydrazine probes. (I) is 1-oxy-2,2,6,6-tetramethylpiperidino-4-hydrazine and (II) is 1-oxyl-2,2,5,5-tetramethylpyrroline-3-carbonyl ethyl hydrazine.

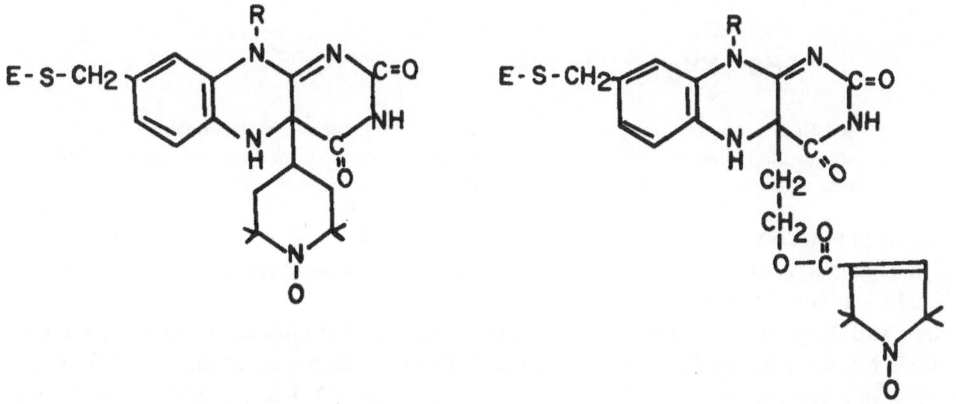

FIGURE 4. Proposed structures of adducts formed between monoamine oxidase and compounds I and II shown in Figure 3.

V. MECHANISTIC INVESTIGATIONS OF MAO-B

There have been a limited number of kinetic studies conducted with highly purified bovine liver MAO-B. These include the studies of Oi et al.[14,22] and the more recent study of Husain et al.[23] utilizing both steady state and presteady state kinetic studies. The latter investigators proposed the substrate-dependent reaction pathway of amine oxidase reaction pathway shown in Figure 5. Contrary to the mechanism proposed by Oi et al.,[22] benzylamine was reported to be oxidized by the lower ternary complex mechanism. Phenethylamine was oxidized by the upper binary complex mechanism. Substrates which are oxidized by both pathways can thus exist. At the time the mechanism was proposed by Oi et al.,[22] it had not been shown that the imine intermediate is produced during the reaction. (Pateck et al.[24] and Oi et al.[25] proposed that the ammonia and aldehyde were liberated sequentially.)

Returning to the matter of the mechanism of benzylamine oxidation, Oi and Yasunobu[25] reported that radioactive ethylamine was liberated anaerobically when *N*-methylbenzylamine was used as substrate. If traces of oxygen were present, imine production could have occurred or if *N*-methylbenzylamine was metabolized differently from benzylamine, the different conclusions concerning the mechanism can be reconciled.[22,23]

VI. NEW OR MODIFIED ENZYME PURIFICATION SCHEMES

Three reports have appeared which are of interest. Stadt et al.[26] have recently isolated rat liver MAO with both enzymes A and B which could not be distinguished by polyacrylamide gel electrophoresis (PAGE). A new detergent, octylglucoside, was used to extract the iso-

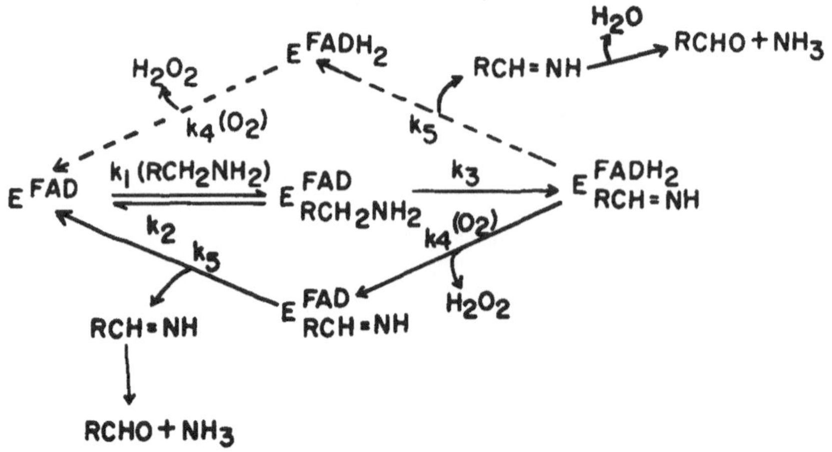

FIGURE 5. Mechanism of substrate oxidation by MAO proposed by Husain et al.[28] The lower pathway (solid line) pertains to route of benzylamine oxidation and the upper pathway (dashed lines) is the route of phenylethylamine oxidation.

zymes from mitochondrial membrane fragments which were collected by ultracentrifugation. Dialysis resulted in vesicle formation, which were collected by ultracentrifugation. However, the yield was low (about 8%).

Naoi and Yagi have purified bovine heart MAO using modifications of earlier reported methods but with the addition of a substrate affinity column step at the end.[27] Very good yields of enzyme are reported. Again, enzyme with both MAO-A and MAO-B activity was isolated to a state where a single band is observed by SDS-PAGE, but the observed molecular weight was only 28,000 or about half of the value usually found for other MAO enzymes. Elution with buffer containing substrates eluted smaller amounts of enzyme but led to some separation of MAO-A and MAO-B.

A new type of MAO which is unlike MAO-A or MAO-B has recently been isolated from pig liver by Yagi and Naoi.[28] The procedure used was a modification of the procedure used to purify the bovine heart MAO reported above and included several additional steps such as octyl-sepharose chromatography and crystallization. The enzyme differed from the classical MAO in that the FAD was not covalently linked as 8-a cysteinyl-FAD and the enzyme was not inhibited by clorgyline or deprenyl. However, the enzyme was able to oxidize both MAO-A and MAO-B type of substrates, such as benzylamine and serotonin.

VII. MAO-A/MAO-B DILEMMA: ONE OR TWO ENZYMES?

Classification of MAO into two forms, A and B, first suggested by Johnston,[29] is based on the selective inhibition of the activity of these forms by clorgyline and deprenyl. Since then there has been a debate as to whether this is due to the presence of one enzyme with two active sites or two separate enzymes. The current theory favors the two-enzyme hypothesis since Denney et al.[30] have shown that MAO-B specific monoclonal antibody does not react with human liver MAO-A. Buckman et al.[31] have also recently summarized their reasons for the two separate enzyme hypothesis. Table 2 summarizes the distribution of the two forms of the enzyme in different tissues in different organs. It is not meant to be an exhaustive list including all reported cases but just a partial list of the type of variations observed.

VIII. EFFECT OF PHOSPHOLIPID

This is currently a controversial topic.[28,31,37,38] The varied effects reported may be due to

Table 2
THE DISTRIBUTION OF THE TWO TYPES OF MONOAMINE OXIDASE IN VARIOUS SPECIES

Animal	Organ	% Type A	% Type B	Animal	Organ	% Type A	% Type B
Rat	Liver[32-33]	40	60	Pig	Brain[32-34]	0	99
	Spleen[32]	95	4		Liver[33-34]	0	99
	Brain[32-33]	55	45	Cat	Liver[33-34]	8	94
	Lung[32]	50	50		Kidney[33-34]	16	87
	Kidney[32-33]	70	30		Brain[33-34]	31	83
	Intestine[32-33]	70	30		Intestine[33-34]	54	38
Mouse	Liver[33-34]	2	96	Dog	Liver[33-34]	0	94
	Kidney[33-34]	66	34		Kidney[33-34]	8	92
	Brain[33-34]	44	54		Brain[33-34]	31	83
	Intestine[33-34]	72	31		Intestine[33-34]	69	34
Guinea pig	Liver[33-34]	39	62	Bovine	Liver[32]	3	97
	Kidney[33-34]	58	47		Brain[35]	15	85
	Brain[33-34]	20	79	Human	Placenta[36]	90	
	Intestine[33-34]	84	21				

species differences, mitochondria or outer membrane fragments used, or to experimental conditions adopted. Various methods have been used for delipidating the mitochondria or enzyme: e.g., the use of phospholipases, organic solvent extraction, sodium perchlorate, and sodium dodecylsulfate (SDS) treatment. The general conclusion made was the delipidation of mitochondria or enzyme resulted in loss of MAO-A or MAO-B activity. Except for one report for a general nonspecific activation of MAO in delipidated mitochondria, various specific phospholipids were reported to reactivate one form of MAO, e.g., phosphatidyl inositol activated delipidated MAO-A activity.[31,37] Specific lipids were reported to inhibit; e.g., phosphatidylethanolamine inhibited MAO-A activity and phosphatidylserine inhibited MAO-B.[31,37]

Only a limited number of investigations have been reported with the enzyme. Naoi and Yagi reported phospholipase A_2 treatment of the beef heart MAO results in loss of MAO-A activity while treatment of the enzyme with phospholipase D results in loss of MAO-B activity.[38,39] The activity was reconstituted by the formation of liposomes made from egg yolk phosphatidylcholine. Similar activation of beef heart MAO delipidated with sodium perchlorate or SDS were reported to be reactivated by phosphatidylcholine-derived liposomes.[27]

Further definite studies are needed to show what phospholipids are activators and inhibitors as well as whether or not the level of MAO-A/MAO-B is regulated in vivo by phospholipids.

IX. FUTURE OUTLOOK

It is desirable to know the complete structures of MAO-A and MAO-B. As a step in determining the complete structure by single crystal X-ray study, the amino acid sequence is useful. Our lab attempted to solve the sequence of the bovine liver MAO-B. However, the peptides produced by various methods could not be separated due to the hydrophobic nature of the enzyme. However, with the development of high performance liquid chormatography (HPLC), these peptides can now be resolved. Quicker methods for sequence determination exist and involves nucleic acid sequence determination of the c-DNA or of the m-RNA. Sagara and Ito[40] have developed the procedure for isolating the rat liver MAO m-RNA and it is only a matter of time until an investigator will determine the sequence of the m-RNA or the c-DNA. Crystallization of MAO is possible as Yagi and Naoi have reported for the pig liver MAO.[28]

ACKNOWLEDGMENT

The authors wish to thank the National Institutes of Health Grant No. MH 21539 for their support.

REFERENCES

1. **Kamijo, K., Usdin, E., and Nagatsu, T., Eds.,** *Monoamine Oxidase: Basic and Clinical Frontiers,* Exceprta Medica, Amsterdam 1982, 1.
2. **Singer, T. P., Von Korff, R. W., and Murphy, D. L., Eds.,** *Monoamine Oxidase: Structure, Function and Altered Functions,* Academic Press, New York, 1979, 1.
3. **Wolstenholme, G. E. W. and Knight, J., Eds.,** *Monoamine oxidase and its inhibition, Ciba Foundation Symp.,* 39, 1, 1976.
4. **Costa, E. and Sandler, M., Eds.,** *Monoamine oxidase — new vistas, Adv. Biochem. Psychopharmacol.,* 5, 1, 1972.
5. **Hare, M. L. C.,** Tyramine oxidase, *Biochem. J.,* 22, 968, 1928.
6. **Greenawalt, J. W.,** Localization of monoamine oxidase in rat liver mitochondria, *Arch. Biochem. Biophys.,* 146, 410, 1971.
7. **Minamiura, N. and Yasunobu, K. T.,** Bovine liver monoamine oxidase, *Arch. Biochem. Biophys.,* 189, 481, 1978.
8. **Gomes, B., Kloepfer, H. G., Oi, S., and Yasunobu, K. T.,** The reaction of sulfhydryl reagents with hepatic monoamine oxidase, *Biochim. Biophys. Acta,* 438, 347, 1976.
9. **Oreland, L.,** Purification and properties of pig liver monoamine oxidase, *Arch. Biochem. Biophys.,* 146, 410, 1971.
10. **Igaue, I. and Yasunobu, K. T.,** Unpublished data.
11. **Minamiura, N. and Yasunobu, K. T.,** Unpublished data.
12. **Kearney, E. B., Salach, J. I., Walker, J. I., Seng, R. L., Zeszotck, E., and Singer, T. P.** Covalent-bound flavin of hepatic monoamine oxidase, *Eur. J. Biochem.,* 24, 321, 1971.
13. **Hiramatsu, A., Tsurushiin, S., and Yasunobu, K. T.,** Evidence for essential cysteine residues in bovine liver mitochondrial monoamine oxidase, *Eur. J. Biochem.,* 57, 587, 1975.
14. **Oi, S., Yasunobu, K. T., and Westley, J.,** The effect of pH on the kinetic parameters and mechanism of beef liver monoamine oxidase, *Arch. Biochem. Biophys.,* 145, 557, 1971.
15. **Zeidan, H., Watanabe, K., Piette, L. H., and Yasunobu, K. T.,** ESR studies of spin labeled bovine liver monoamine oxidase B, in *Frontiers in Protein Chemistry,* Liu, T. -Y., Mamiya, G., and Yasunobu, K. T., Eds., Elsevier/North-Holland, New York, 1980, 131.
16. **Yasunobu, K. T., Watanabe, K., and Zeidan, H.,** Monoamine oxidase: some new findings, in *Monoamine Oxidase: Structure, Function and Altered Functions,* Singer, T. P., Van Korff, R. W., and Murphy, D. L., Eds., Academic Press, New York, 1979, 251.
17. **Singer, T. P.,** Active-site-directed, irreversible inhibitors of monoamine oxidase, in *Monoamine Oxidase: Structure, Function and Altered Functions,* Singer, T. P., Von Korff, R. W., and Murphy, D. L., Eds., Academic Press, New York, 1979, 7.
18. **Paech, C., Salach, J. I., and Singer, T. P.,** Suicide inactivation of monoamine oxidase by trans-phenylcyclopropylamine, in *Monoamine Oxidase: Structure, Function and Altered Function,* Singer, T. P., Von Korff, R. W., and Murphy, D. L., Eds., Academic Press, New York, 1979, 39.
19. **Nara, S., Igaue, I., Gomes, B., and Yasunobu, K. T.,** The prosthetic groups of animal amine oxidases, *Biochem. Biophys. Res. Commun.,* 23, 324, 1966.
20. **Walker, W. H., Kearney, E. B., Seng, R., and Singer, T. P.,** The covalently bound flavin of hepatic monoamine oxidase. II. Identification and properties of cysteinyl flavin, *Eur. J. Biochem.,* 24, 328, 1971.
21. **Tan, A., Glantz, M., Piette, L. H., and Yasunobu, K. T.,** ESR analysis of the FAD in bovine liver monoamine oxidase, *Biochem. Biophys. Res. Commun.,* 117, 509, 1983.
22. **Oi, S., Shimada, K., Inamasu, M., and Yasunobu, K. T.,** Mechanistic studies of beef liver mitochondrial amino oxidase, *Arch. Biochem. Biophys.,* 139, 28, 1970.
23. **Husain, M., Edmundson, D. E., and Singer, T. P.,** Kinetic studies in the catalytic mechanism of liver monoamine oxidase, *Biochemistry,* 21, 595, 1982.
24. **Patek, D. R., Chuang, H. Y. K., and Hellerman, L.,** Aldimine intermediates in oxidation of amines catalyzed by mitochondrial monoamine oxidase, *Fed. Proc., Fed. Am. Soc. Exp. Biol.,* 31, 420, 1972.

25. **Oi, S. and Yasunobu, K. T.,** Mechanistic aspects of the oxidation of amines by monoamine oxidase, *Biochem. Biophys. Res. Commun.*, 53, 631, 1973.
26. **Stadt, M. A., Banks, P. A., and Kobes, R. D.,** Purification of rat liver monoamine oxidase by octyl glucoside extraction and reconstitution, *Arch. Biochem. Biophys.*, 214, 223, 1982.
27. **Naoi, M. and Yagi, K.,** Effect of phospholipids on beef liver mitochondrial monoamine oxidase, *Arch. Biochem. Biophys.*, 20, 18, 1980.
28. **Yagi, K. and Naoi, M.,** Crystallization of a monoamine oxidase purified from pig liver mitochondria, *Biochem. Int.*, 4, 457, 1982.
29. **Johnston, J. P.,** Some observations upon a new inhibitor of monoamine oxidase in brain tissue, *Biochem. Pharmacol.*, 17, 285, 1968.
30. **Denny, R. M., Fritz, R. R., Pate, M. T., and Abell, C. W.,** Human liver MAO-A and MAO-B separated by immunoaffinity chromatography with MAO-B specific monoclonal antibody, *Science*, 215, 1400, 1982.
31. **Buckman, T. D., Eiduson, S., Sutphim, M. A., and Chang, R.,** Selective effects on catalysis by the multiple forms of monoamine oxidase produced by interaction of acidic phospholipids with mitochondrial membranes, *J. Biol. Chem.*, 258, 8670, 1983.
32. **Tipton, K. F., Houslay, M. D., and Mantle, T. J.,** The nature and locations of the multiple forms of monoamine oxidase, *Ciba Found. Symp.*, 39, 5, 1976.
33. **Squires, R. F.,** Multiple forms of monoamine oxidase in intact mitochondria as characterized by selective inhibitors and thermal stability: a comparison of eight mammalian species, *Adv. Biochem. Psychopharmacol.*, 5, 355, 1972.
34. **Hall, D. W. R., Logan, B. W., and Parsons, G. H.,** Further studies on the inhibition of monoamine oxidase by M D B 9320 (clorgyline) - I substrate specificity in various mammalian species, *Biochem. Pharmacol.*, 18, 1447, 1969.
35. **Hidaka, H., Hartamn, B., and Udenfriend, S.,** Comparison of mitochondrial monoamine oxidase from bovine brain and liver using antibody to purified liver monoamine oxidase, *Arch. Biochem. Biophys.*, 147, 805, 1971.
36. **Oreland, L., Fowler, C. J., and Wilberg, Å.,** Titration of monoamine oxidase-A and -B and its clinical application, in *Monoamine Oxidase Inhibitors, The State of the Art*, Youdim, M. B. H. and Paykel, E. S., Eds., 1981, 77.
37. **Huang, R. H. and Faulkner, R.,** The role of phospholipid in the multiple functional forms of brain monoamine oxidase, *J. Biol. Chem.*, 256, 9211, 1981.
38. **Navarro-Welch, C. and McCauley, R. B.,** An evaluation of phospholipids as regulators of monoamine oxidase A and monoamine oxidase B activities, *J. Biol. Chem.*, 257, 13645, 1982.
39. **Naoi, M. and Yagi, K.,** Modification of the reactivity of beef heart mitochondrial monoamine oxidase by phospholipases, *Biochem. Int.*, 1, 371, 1980.
40. **Sagara, Y. and Ito, A.,** In vitro synthesis of rat liver outer mitochondrial membrane, *Biochem. Biophys. Res. Commun.*, 109, 1102, 1982.

Chapter 21

INHIBITORS OF FAD-CONTAINING MONOAMINE OXIDASES

Thomas P. Singer

TABLE OF CONTENTS

I. INTRODUCTION

Mitochondrial monoamine oxidase MAO is a tightly bound component of the outer membrane, which occurs in two forms: A and B. The two differ in substrate specificity, although there is some overlap and some substrates are processed by both forms. A more clean-cut distinction is in terms of sensitivity to certain acetylenic inhibitors. MAO-A is irreversibly inhibited by very low concentrations of clorgyline, but is unaffected by low concentrations of deprenyl and pargyline. The reverse is true of MAO-B. Certain tissues contain only MAO-A (e.g., human placenta), others only MAO-B (e.g., bovine liver), but many tissues contain both MAO-A and MAO-B in varying proportions. Although there have been many speculations in the literature that the two forms of MAO represent different active sites on the same protein or different lipid complexes of one enzyme, none of the evidence offered for those notions is persuasive, while evidence for distinct proteins, differing in the structure of the substrate site, is more impressive (cf. recent reviews [1,2]) (see Chapter 20).

Both MAO-A and MAO-B contain the same covalent, amino acid adduct of FAD, 8α-*S*-cysteinyl FAD (Figure 1) and the peptide sequence at their flavin sites is also identical. There have been two reports in the literature of the occurrence of mitochondrial MAO in which the flavin is noncovalently bound: one in pig brain and the other in pig liver.[3,4] However, as fully discussed in several reviews, these reports are very likely to be erroneous. In the case of the enzyme recently isolated in Yagi's laboratory, the evidence strongly suggests that the activity measured was not due to MAO[1,2] (see Chapter 20).

II. REVERSIBLE INHIBITORS

The most widely used reversible MAO inhibitors are amphetamines and tricyclic antidepressants (Figure 2). They tend to be selective for MAO-A. Thus, (+) amphetamine inhibits the oxidation of serotonin, an A-type substrate, in rat brain preparation with K_i = 8 μM, while the oxidation of phenylethylamine, a predominantly B-type substrate in this source, is inhibited with a K_i = 475 μM. The (−) enantiomer is reported to be less specific, as are many substituted (+) amphetamines. There are notable exceptions, however. Thus, *p*-methoxyamphetamine is 20 times more potent an inhibitor of serotonin oxidation by MAO from mouse brain (K_i = 0.22 μM) than is (+) amphetamine.

One of the most potent reversible inhibitors of MAOs is 3-[4-(3-cyanophenylmethoxy)phenyl]-5-(methoxymethyl)-2-oxazolidinone.[7] Like other reversible inhibitors of MAO, it is selective for MAO-A. The inhibition of serotonin oxidation by rat brain homogenates is 1.5 orders of magnitude greater than that of phenylethylamine oxidation. Similarly, in vivo the oxidation of MAO-A is much more profoundly affected than the processing of MAO-B.

A particularly fascinating aspect of the reaction of reversible inhibitors with MAO-B concerns the different kinetics observed with benzylamine and phenylethylamine as substrates. Roth[8] reported several years ago that tricyclic psychoactive drugs, such as chlorpromazine, imipramine, as well as (+) amphetamine, show purely competitive inhibition in Lineweaver-Burke plots when benzylamine is the substrate, but show mixed inhibition when phenylethylamine is oxidized. This phenomenon is illustrated for a preparation of MAO-B from human brain in Figure 3.

The clue to this puzzling behavior came from the observation that as the O_2 concentration in the assay was measured, the kinetics of the inhibition of phenylethylamine oxidation gradually shifted from mixed to competitive type.[8]

Although Roth[8] correctly surmised that taken together, their observations may suggest that different forms of the enzyme were inhibited when benzylamine and phenylethylamine were used, respectively, this seemed to be in seeming contradiction with his conclusion,

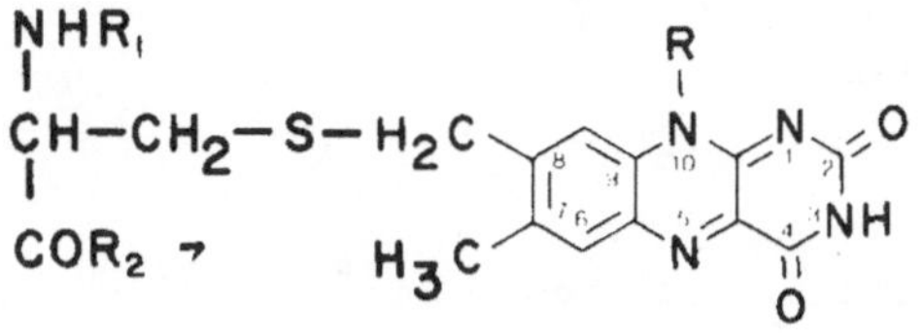

FIGURE 1. Structure of 8α-*S*-cysteinyl FAD.

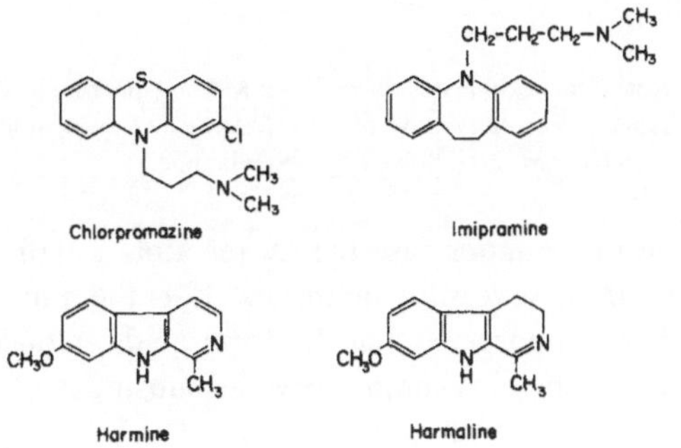

FIGURE 2. Some tricyclic reversible inhibitors of MAO.

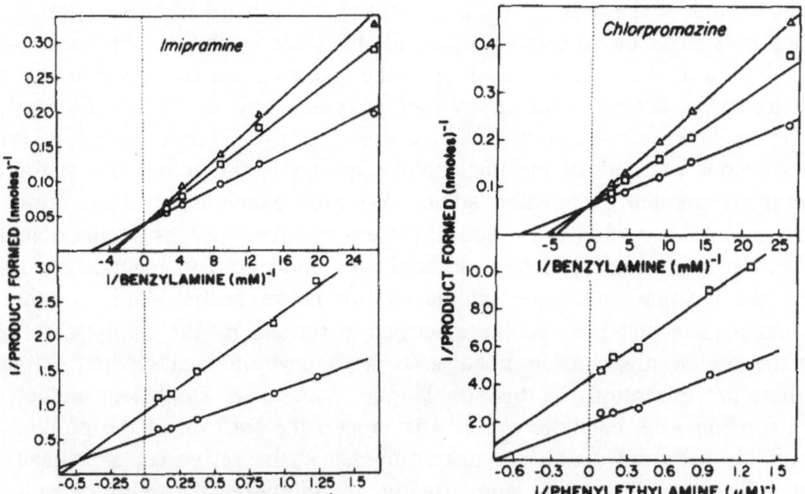

FIGURE 3. Lineweaver-Burke plots for the inhibition of benzylamine and phenylethylamine oxidation by imipramine and chlorpromazine. Concentrations; with the benzylamine as substrate: squares, 4×10^{-4} *M:* triangles, 7.5×10^{-4} *M;* with phenylethylamine as substrate, squares, 5×10^{-5} *M.*[8]

based on parallel line kinetics in Lineweaver-Burke plots, that with both of these substrates a ping-pong mechanism operates.

The resolution of this paradox came from a detailed study of the kinetic mechanism of MAO-B some years later by Husain et al.[9] Based on presteady state kinetic analysis and kinetic isotope effects, they showed that with phenylethylamine a ping-pong mechanism is operative (Figure 4, upper loop). During benzylamine oxidation an ordered mechanism functions (Figure 4, lower loop). This means that in the former case O_2 reacts with the *free*

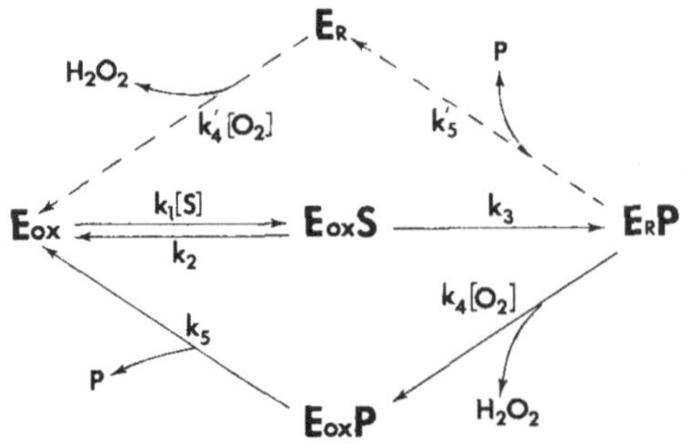

FIGURE 4. Reaction mechanisms of beef liver MAO. Upper loop, binary complex mechanism, as followed by phenylethylamine. Lower loop, tertiary complex mechanism, as followed by benzylamine.

reduced enzyme (E_R), while in the latter case the O_2 reaction is with the reduced enzyme-product complex. Therefore, if the reversible inhibitors affect the reoxidation of the enzyme, as Roth has suspected, it is of importance that the form of the reduced enzyme is not the same when benzylamine and phenylethylamine serve as substrates.

III. IRREVERSIBLE INHIBITORS

Since nearly all irreversible inhibitors of MAO are suicide substrates, it seems appropriate to refresh the mind of the reader of the general characteristics of this important class of enzyme inhibitors. Suicide substrates (also called suicide inhibitors, mechanism-based irreversible inhibitors), are structural analogs of the substrate and are processed by the target enzyme. Prior to the action of the appropriate enzyme, they are chemically inert, but the action of the enzyme creates a highly reactive structure, usually an electrophile, which will be attacked by one or more of the nucleophilic groups at the active site of the enzyme, resulting in the formation of a covalent adduct. With few exceptions (such as thiobemiacetal type of enzyme-inhibitor adducts), the resulting enzyme derivative is nondissociable. If the nucleophile involved in the reaction is essential for activity or if its reaction creates a steric hindrance to the access of substrates, the inactivation is irreversible. An important corollary of this mechanism is that the reactive electrophile formed by the catalytic action of the enzyme on the suicide substrate does not disassociate and recombine elsewhere on the protein, although there are exceptions to this simple rule. Moreover, a covalent adduct does not necessarily form in each catalytic cycle. The greater the fraction of the products formed from the suicide substrate which remains combined at the active center instead of being disassociated from the protein, the more effective the inhibitor. This partition ratio, in other words, equals the number of catalytic cycles the enzyme undergoes while acting on the suicide substrate before being inactivated. With the most potent ones the ratio is one.

It should be apparent from the above that loss of activity during suicide inhibition is a biphasic process. The first phase is simple, competitive inhibition, which is reversible and is prevented by the presence of normal substrate. The ensuing second phase is irreversible and, once it starts, cannot be prevented by substrate. The irreversible inactivation should thus be a first order reaction. Commonly, but not always, the inactivated enzyme contains 1 molecule of the product formed from the suicide substrate per active site. Using appropriately labeled suicide substrates, it is often possible to proteolyze the inactivated enzyme, isolate the radioactive peptide(s), and determine their structure — in other words, identifying the nucleophile involved in the suicide inactivation. This has been accomplished with all major types of suicide substrates in the case of MAO.[10,11] In the case of arylhydiazines more

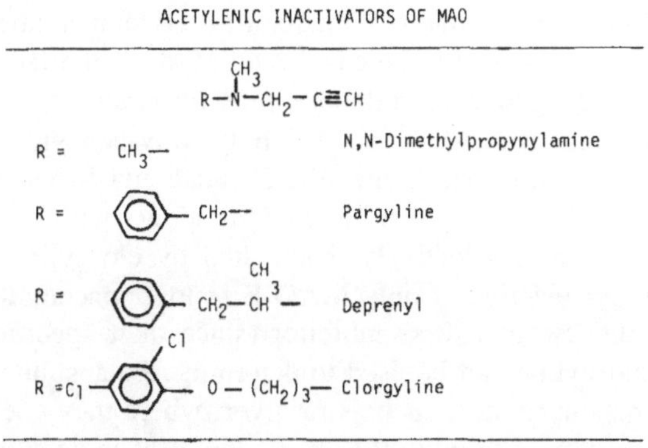

FIGURE 5. Structures of some acetylenic suicide inhibitors of MAO.

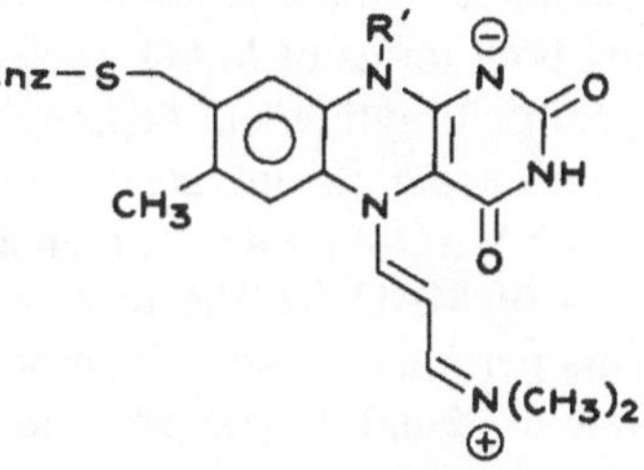

FIGURE 6. Structure of flavocyanine adduct.

than one nucleophile can attack the aryldiazine formed by the action of MAO; thus, the inhibitor is partitioned between two types of nucleophiles, both at the active site. Since covalent reaction with either one leads to immediate and complete loss of activity, a given molecule of inactivated MAO can have no more than 1 molecule of the inhibitor bound.

A. Acetylenic Suicide Inhibitors

Among irreversible inhibitors of MAO, those containing an acetylenic linkage in the β position to the amine have been most widely investigated from the biochemical standpoint and have received the widest attention as potential or actual therapeutic agents. Figure 5 illustrates the structures of the best known compounds in this group. The facts that the inactivation of MAO by acetylenic amines is biphasic, i.e., an instantaneous reversible inhibition followed by a time-dependent irreversible one[12] and that the irreversible phase hinges on prior oxidation of the amine[13] clearly indicated that these compounds are suicide substrates. Hellerman and Erwin,[14] in fact, had demonstrated that the inactivation involves combination with the flavin component of the enzyme several years before the existence of mechanism-based inhibitors was recognized.

Oxidation of acetylenic amines by MAO creates highly reactive products, which react avidly with nearby nucleophiles at the active site. The fact that the nucleophilic group involved in the inactivation of MAO-B by dimenthylpropynlamine is N(5) of the covalently bound flavin was established in a collaborative research by Abeles' and the author's laboratories,[15] which resulted in the isolation and characterization of a flavocyanine adduct (Figure 6). Subsequently, the irreversible inactivation of MAO-A and MAO-B by a variety of acelylenic amines, including pargyline, clorgyline, and deprenyl, was shown to involve the formation of analogous products.[11]

Although the product of the enzymatic oxidation of acetylenic amines might be presumed to be an allene, this is not necessarily the case when MAO is the target enzyme. This fact came to light from an investigation in collaboration with Krantz's laboratory,[10,16] showing that while the modification of MAO-B by *N*-2-butynyl-*N*-benzylmethylamine yields the expected flavocyanine, the tautomeric allene, *N*-2,3-butadienyl-*N*-benzylmethylamine yields a different product, most likely the cyclic N(5)-C(4a) adduct with the flavin.

The form of MAO which is selectively inactivated by clorgyline has been defined as MAO-A,[17] while deprenyl inactivates only MAO-B at low concentrations.[18] With suitable precautions, avoiding the use of excess inhibitor (since their specificity is not absolute), ^{14}C-deprenyl and ^{14}C-clorgyline can be used to determine the absolute concentration of the two forms of the enzymes in tissues, such as rat liver, where they coexist in the same cell. Since the binding is 1:1 with both inhibitors, the amount of radioactive inhibitor bound to the preparation is a convenient measure of the concentration of the form of MAO for which it is selective. Note, however, that at moderately higher concentrations clorgyline is bound to MAO-B and deprenyl to MAO-A, in amounts in excess of the enzyme concentration, and this "unspecific binding" bears no simple relation to loss of activity.[10]

Tissue preparations containing both forms of MAO, such as rat liver mitochondria, show a biphasic inhibition curve by both clorgyline and deprenyl when A-B substrates are used (i.e., amines oxidized by both forms at significant rates). At low concentrations of clorgyline the inhibition is due to blockage of MAO-A, while at higher concentrations of the inhibitor it can be ascribed to inactivation of MAO-B. The reverse is true with MAO-B-selective inhibitors, such as deprenyl. This permits an approximation of the relative activities of the two enzymes toward the substrate used and, by extrapolation, of the relative concentrations of the two forms.

The reactivity of MAO-A and MAO-B, respectively, with suicide inhibitors and substrates may be a basis for predicting the structural basis of differences between the two forms. In view of growing evidence from isolation of MAO-A and MAO-B in pure form,[19-21] and from immunochemical studies,[22,23] it seems clear that the two represent different discrete proteins. With the molecular weights and subunit composition essentially identical, and in view of the demonstration[24] that both MAO-A and MAO-B contain the same covalently bound flavin, attached to the same amino acids at the active site, it seems very probable that differences are at the substrate binding site.[1] This would, of course, explain why the two forms differ in specificity toward substrates and suicide substrates.

Figure 7 emphasizes the structural differences between substrates (and suicide substrates) of MAO-A and MAO-B. These authors contend that compounds in which the aromatic nucleus is separated by 1 or 2 carbons from the oxidizable N, react preferentially with MAO B. Typical examples are benzylamine, phenylethylamine, deprenyl, pargyline, U1424, AGN1133, and AGN1135. When the distance is equivalent to three or more carbons, the substrate or inhibitor bind preferentially to MAO-A. This is interpreted to suggest that the active sites of the two forms differ by the distance between the binding sites for the N and of the aromatic ring. Although a number of irreversible inhibitors of MAO follow this pattern,[25] this cannot be the only difference in the activity of the two enzymes, since some substrates (e.g., noradrenaline), do not seem to follow the pattern.

B. Cyclopropylamines

Starting with the extensive studies of Zeller and Sarkar in 1962,[26] which demonstrated that cyclopropylamines, as exemplified by *trans*-2-phenylcyclopropylamine (tranylcypromine), are powerful and selective inhibitors of MAO, several laboratories have studied the kinetics of the inactivation[12,14] and the in vivo pharmacological effects[27] of this class of inhibitors. McEwen et al.[12] were the first to point out that the loss of activity is biphasic, a fast, reversible, and competitive phase, followed by a time-dependent irreversible one.

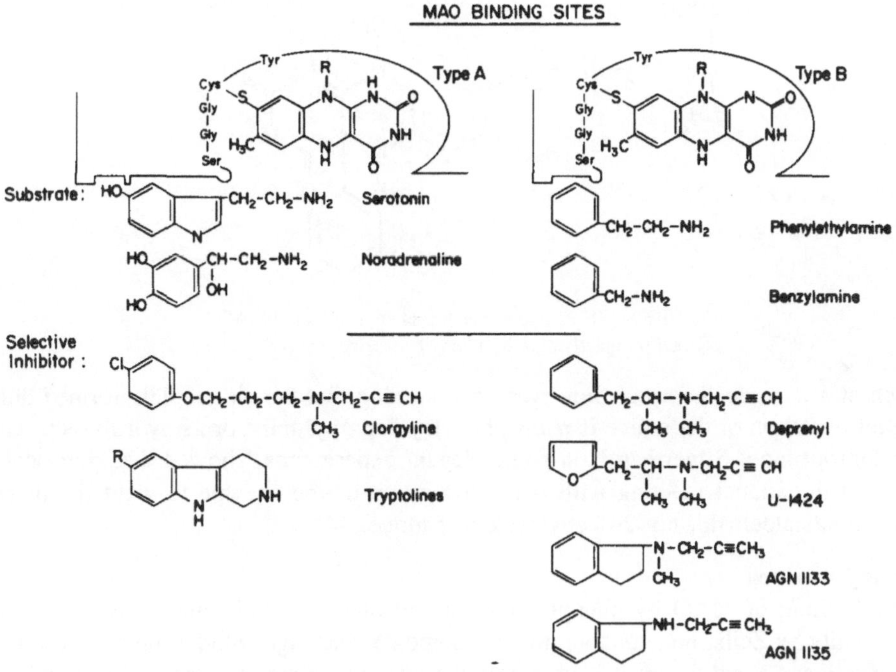

FIGURE 7. Schematic representation of the active sites of MAO A and B, the binding sites of their substrates, and of selective inhibitors. (From Youdim, M. B. H. and Fineberg, J., *Neurotransmitters and their Receptors*, Littauer, V. Z. et al., Eds., John Wiley & Sons, London, 1980, 73. With permission.)

This presaged the later demonstration in the author's laboratory[28] that tranylcypromine and related cyclopropylamines are suicide inhibitors of MAO-B. The effect on MAO-A is similar.

In the late 1970s two laboratories, that of Richard Silverman and the author's, undertook to determine the mechanism of the inactivation. Both laboratories concluded that the initial step in the irreversible phase is oxidation of the compound by MAO, contradicting an earlier report,[14] which claimed that the oxidized form of the enzyme was not required for inactivation. Since the flavin is the electron acceptor in the oxidation of tranylcypromine, the enzyme is bleached by this event. MAO, however, is auto-oxidizable, so that in the normal catalytic cycle the reduced (bleached) form is rapidly reoxidized. The fact that MAO *remains* bleached after inactivation by tranylcypromine lead Silverman and Hoffman[29] to the conclusion that the nucleophile involved in the formation of the covalent adduct with the inhibitor is the N(5) group of the flavin ring system. This was logical because covalent adducts with N(5) are formally reduced flavins and, as such, are decolorized.

This conclusion could not be verified, however, with the pure enzyme.[28,30] For one thing, the difference in spectrum observed on inactivation did not resemble that observed on reduction with normal substrates or with dithionite. Further, on denaturation of MAO inactivated with [^{14}C]-labeled tranylcypromine, radioactivity was released under conditions where N(5) adducts would be stable, so that the adduct could not involve this nitrogen atom. Third, using mild denaturants, it could be shown that the yellow color of the flavin returned much faster than was radioactivity released from the protein. Hence, the flavin ring system could not be the target of the suicide inhibition. Instead, the $-SH$ group at the active site was proposed to be the nucleophile involved in the inactivation.[28,30] This appeared to be consistent with the dissociability of the thiohemiacetal, which was thought to be formed between the thiol group and the oxidation product of the inhibitor, once stabilization by the tertiary structure of the protein was lost on unfolding. This interpretation now seems to have been accepted by Silverman.[31]

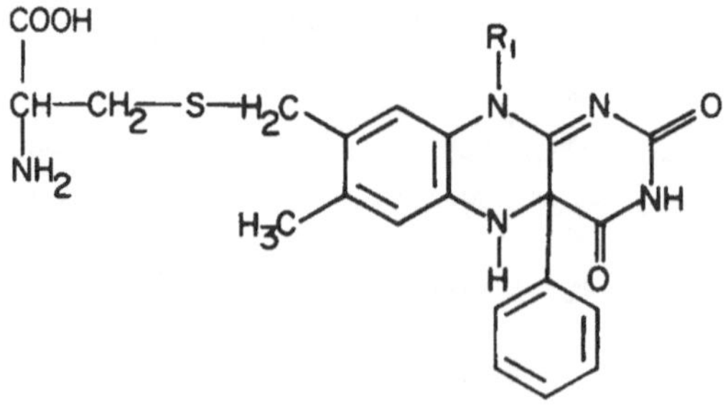

FIGURE 8. Structure of the 4(a) phenyl adduct of MAO formed on inactivation by phenylhydrazine.

Paech et al.[30] were incorrect, however, in assuming that the electrophile formed during the initial oxidation of the amine is *trans*-phenylcyclopropylimine or its hydrolysis product, the cyclopropanone. Silverman,[31] in some elegant experiments, showed that ring opening occurs and the adduct reacting with the −SH group of the enzyme to yield the inactive form is cinnamaldehyde, not 2-phenylcyclopropanone.

C. Arylhydrazines

The inhibition of MAO by phenylhydrazine and other arylhydrazines was first studied systematically by Zeller and collaborators[26,32] some 25 years ago. Studies on the mechanism of the inactivation had to await isolation of the enzyme in highly purified form and were undertaken by Hellerman and colleagues,[14,33] who showed that the initial event is dehydrogenation of the inhibitor to the highly reactive aryldiazine. They proposed that the diazine then combines with an unspecified site on the flavin.

With the availability of ring-labeled phenylhydrazine and of methods for isolating a peptide containing the covalently bound flavin at the active site, a more precise localization became possible. The investigation was greatly aided by prior studies in the author's laboratory on the suicide inactivation of bacterial trimethylamine dehydrogenase by phenylhydrazine.[34] This enzyme resembles mammalian MAO in many respects, although its substrate specificity is quite different. Thus, it is inhibited by a variety of typical MAO inhibitors and, like MAO, contains covalently labeled FAD in thioether linkage to a cysteine residue, although while in MAO the site of attachment is the 8α-methyl group, in trimethylamine dehydrogenase it is C(6) of the aromatic nucleus.

The inactivation of trimethylamine dehydrogenase by phenylhydrazine has been shown to involve initial dehydrogenation to phenyldiazene, and culminates in the formation of a phenyl adduct at C(4a) of the isoalloxazine ring system.[34] The inactivation goes hand-in-hand with bleaching of the yellow color of the oxidized flavin, since combination at C(4a), as at N(5), is formally a reduction.

MAO is also decolorized during inactivation by phenylhydrazine, but in this case loss of activity and color do not run a parallel course. Moreover, the difference spectrum is not that of a reduced flavin. On denaturation of MAO which had been inactivated with [^{14}C]-phenylhydrazine, followed by proteolysis and recovery of the flavin peptide fraction, only about 50% of the radioactivity of the inactive enzyme was recovered; the rest was bound to nonflavin peptide residue(s). Further only about 50% of the flavin was present in bleached form in such aliquots: the rest appeared to be present as unmodified, oxidized flavin.

Analysis of this complex reaction between MAO-B and phenylhydrazine suggests that in this case the phenyldiazine reacts in part with the flavin at C(4a) (Figure 8), in part with another nucleophile, probably the −SH group at the substrate site.[35] The partition ratio is about equal; when the hydrazine is separated by one or more carbons from the phenyl group, the partition ratio seems to be different.

Qualitatively, the same mechanism seems to be involved in the inactivation of MAO-A from human placenta by phenylhydrazine: the inactivation is again accompanied by the incorporation of 1 mol of phenyl group, but in the case of MAO-A the primary target seems to be the −SH group.[36]

In conclusion, it is clear that the irreversible inactivation of MAO-A and MAO-B is usually a suicide inhibition but not all suicide substrates react with the same nucleophile in MAO. N(5) of the flavin is the target of all acetylenic inhibitors studied, C(4a)-N(5) of allenic amines, the −SH group of arylcyclopropylamines, while arylhydrazenes seem to yield adducts with both C(4a) of the flavin and with another group in the protein, most likely the −SH at the substrate site.

D. Combined Use of Reversible and Irreversible Inhibitors

An ingenious instance of the combined use of harmalene, an A-selective reversible inhibitor and of pargyline, which inactivates both MAO-A and MAO-B, at least in rat heart and brain, though the latter enzyme is more sensitive to pargyline, has been reported by Fuller et al.[37] They found that when pargyline was injected by itself into rats, the oxidation of both MAO-A and MAO-B was inhibited, but when harmalene was administered just before pargyline, it prevented the inactivation in vivo of MAO-A but not MAO-B.

ACKNOWLEDGMENT

The original studies reported here were supported in part by the National Institutes of Health (HL-16251), the National Science Foundation (PCM 81-19609), and the Veterans Administration.

ADDENDUM

Since the submission of this manuscript, important and far-reaching findings have been reported concerning the role of MAO in the neurotoxicity of MPTP, the development of experimental Parkinsonism, and the inhibition of MAO by MPTP and its biological oxidation products. Langston et al.[38] noted that injections of the illicit drug "new heroin" produced Parkinsonism in humans and traced it to the presence of 1-methyl-4-phenyl-1,2,3,6-tetrahydropyridine (MPTP) as an impurity. MPTP was soon shown in several laboratories to elicit Parkinsonian symptoms in primates. Chiba et al.[39] reported in early 1984 that the neurotoxicity of MPTP was probably due to its oxidation by MAO-B in brain. Inhibitors of MAO-B were then found to prevent the toxicity of MPTP in animals. Using pure MAO-A and MAO-B, we have shown that both enzymes oxidized MPTP and that the process is accompanied by the time-dependent irreversible inactivation of both enzymes.[40,41] Moreover, $MPDP^+$ and MPP^+, the 2- and 4-electron oxidation products of MPTP, are potent competitive inhibitors of MAO-A and MAO-B, particularly the former.[41]

REFERENCES

1. **Singer, T. P.,** Demise and resurrection of misconceptions about monoamine oxidase, in *Neuropharmacology '85*, Kelemen, K., Magyar, K., and Vizi, E. S., Eds., Akademiai Kiado, Budapest, 1985, 29.
2. **Singer, T. P.,** Mitochondrial monoamine oxidase, in *Advances in Biochemical Pharmacology*, Vol. 1, Vessey, D. A. and Zakim, D., Eds., John Wiley & Sons, New York, in press.
3. **Tipton, K. F.,** The purification of pig brain mitochondrial monoamine oxidase, *Eur. J. Biochem.*, 4, 103, 1968.
4. **Yagi, K. and Naoi, M.,** Crystallization of a monoamine oxidase purified from pig liver mitochondria, *Biochem. Int.*, 4, 457, 1982.

5. **Fuller, R. W. and Hemrick-Luecke, S. K.,** Influence of ring and side-chain substituents on the selectivity of amphetamine as a monoamine-oxidase inhibitor, *Res. Commun. Substances Abuse,* 3, 159, 1982.
6. **Green, A. L. and El Hait, M. A. S.,** *p*- Methoxyamphetamine, a potent reversible inhibitor of type-A monoamine oxidase in vitro and in vivo, *J. Pharmacol.*, 32, 262, 1979.
7. **Benedetti, M. S.,** A new specific reversible type A monoamine oxidase inhibitor: MD 780515, in *Monoamine Oxidase: Structure, Function, and Altered Functions,* Singer, T. P., Von Korff, R. W., and Murphy, D. L., Eds., Academic Press, New York, 1979, 335.
8. **Roth, J. A.,** Effect of drugs on inhibition of oxidized and reduced forms of MAO, in *Monoamine Oxidase: Structure, Function, and Altered Functions,* Singer, T. P., Von Korff, R. W., and Murphy, D. L., Eds., Academic Press, New York, 1979, 153.
9. **Husain, M., Edmundson, D. E., and Singer, T. P.,** Kinetic studies on the Catalytic Mechanism of Liver Monoamine Oxidase, *Biochemistry,* 21, 595, 1982.
10. **Singer, T. P.,** Active site-directed irreversible inhibitors of monoamine oxidase, in *Monoamine Oxidase: Structure, Function, and Altered Functions,* Singer, T. P., Von Korff, R. W., and Murphy, D. L., Eds., Academic Press, New York, 1979. 7.
11. **Singer, T. P., Edmundson, D. E., and Salach, J. I.,** Suicide inhibitors of the central nervous system, in *Essays in Neurochemistry and Neuropharmacology,* Youdim, M. B. H., Lovenberg, W., Sharman, D. F., and Lagnado, J. F., Eds., John Wiley & Sons, London, 1981, 131.
12. **McEwen, C. M., Jr., Sasaki, G., and Jones, D. C.,** Human liver mitochondrial monoamine oxidase. III. Kinetic studies concerning time-dependent inhibitions, *Biochemistry,* 8, 3963, 1969.
13. **Rando, R. R.,** Bromoallylamine induced irreversible inhibition of monoamine oxidase, *J. Am. Chem. Soc.*, 95, 4438, 1973.
14. **Hellerman, L. and Erwin, V. G.,** Mitochondrial monoamine oxidase, *J. Biol. Chem.*, 243, 5237, 1968.
15. **Maycock, A. L., Abeles, R. H., Salach, J. I., and Singer, T. P.,** The structure of the covalent adduct formed by the interaction of 3-dimethylamino-1-propyne and the flavine of mitochondrial amine oxidase, *Biochemistry,* 15, 114, 1976.
16. **Krantz, A., Kobel, B., Salach, J. I., Singer, T. P., Claesson, A., and Sahlberg, D.,** Inactivation of monoamine oxidase by allenic amines, in *Flavins and Flavoproteins,* Yagi, K. and Yamano, T., Eds., Japan Scientific Societies Press, Tokyo, 1980, 67.
17. **Johnston, J. P.,** Some observations upon a new inhibitor of monoamine oxidase in brain tissue, *Biochem. Pharmacol.*, 17, 1285, 1968.
18. **Knoll, J., Ecsery, Z., Kelemen, K., Nievel, J. G., and Knoll, B.,** Phenylisopropylmethylpropinylamine (E-250) a new spectrum psychic energizer, *Arch. Int. Pharmacodyn. Ther.*, 155, 154, 1965.
19. **Salach, J. I.,** Monoamine oxidase from beef liver mitochondria: simplified isolation procedure, properties, and determination of its cysteinyl flavin content, *Arch. Biochem. Biophys.*, 192, 128, 1979.
20. **Minamiura, N. and Yasunobu, K. T.,** Bovine liver monoamine oxidase, *Arch. Biochem. Biophys.*, 189, 481, 1978.
21. **Zeller, E. A.,** Classification and nomenclature of monoamine oxidases and other amine oxidases, in *Monoamine Oxidase: Structure, Function, and Altered Functions,* Singer, T. P., Von Korff, R. W., and Murphy, D. C., Eds., Academic Press, New York, 1979, 531.
22. **Denney, R. M., Fritz, R. R., Patel, N. T., and Abell, C. W.,** Human liver MAO-A and MAO-B separated by immunoaffinity chromatography with MAO-B-specific monoclonal antibody, *Science,* 215, 1400, 1980.
23. **Pintar, J. E., Levitt, P., Salach, J. I., Weyler, W., Rosenberg, M. B., and Breakefield, X. O.,** Specificity of antisera prepared against pure bovine MAO-B, *Brain Res.*, 276, 127, 1983.
24. **Nagy, J. and Salach, J. I.,** Identity of the active site flavin-peptide fragments from the human "A"-form and bovine "B"-form of monoamine oxidase, *Arch. Biochem. Biophys.*, 208, 388, 1981.
25. **Youdim, M. B. H. and Finberg, J.,** Sites of action of monoamine oxidase inhibitors, in *Neurotransmitters and their Receptors,* Littauer, V. Z., Dudai, Y., Silman, I., Teichberg, V. I., and Vogel, Z., Eds., John Wiley & Sons, London, 1980, 73.
26. **Zeller, E. A. and Sarkar, S.,** Amine oxidases, *J. Biol. Chem.*, 237, 2333, 1962.
27. **Pare, C. M. B.,** Introduction to clinical aspects of monoamine oxidase inhibitors in the treatment of depression, in *Monoamine Oxidase and Its Inhibition,* Wolstenholme, G. E. W. and Knight, J., Eds., Elsevier, Amsterdam, 1976, 271.
28. **Paech, C., Salach, J. I., and Singer, T. P.,** Suicide inactivation of monoamine oxidase by *trans*-phenylcyclopropylamine, in *Monoamine Oxidase: Structure, Function, and Altered Functions,* Singer, T. P., Von Korff, R. W., and Murphy, D. C., Eds., Academic Press, New York, 1979, 39.
29. **Silverman, R. B. and Hoffman, J. T.,** Mechanism of inactivation of monoamine oxidase by *N*-cyclopropyl-*N*-arylalkyl amines, in *Monoamine Oxidase: Structure, Function, and Altered Functions,* Singer, T. P., Von Korff, R. W., and Murphy, D. L., Eds., Academic Press, New York, 1979, 71.

30. **Peach, C., Salach, J. I., and Singer, T. P.,** Suicide inactivation of monoamine oxidase by *trans*-phenylcyclopropylamine, *J. Biol. Chem.*, 255, 2700, 1980.
31. **Silverman, R. B.,** Mechanism of inactivation of monoamine oxidase by *trans*- 2-phenylcyclopropylamine and the structure of the enzyme-inactivator adduct, *J. Biol. Chem.*, 258, 14766, 1983.
32. **Barsky, J., Pacha, W. L., Sarkar, S., and Zeller, E. A.,** Amine oxidases, *J. Biol. Chem.*, 234, 389, 1959.
33. **Patek, D. R. and Hellerman, L.,** Mitochondrial monoamine oxidases, *J. Biol. Chem.*, 249, 2373, 1974.
34. **Nagy, J., Kenney, W. C., and Singer, T. P.,** The reaction of phenylhydrazine with trimethylamine dehydrogenase and with free flavins, *J. Biol. Chem.*, 254, 2684, 1979.
35. **Kenney, W. C., Nagy, J., Salach, J. I., and Singer, T. P.,** Structure of the phenylhydrazine adduct of monoamine oxidase, in *Monoamine Oxidase: Structure, Function, and Altered Functions*, Singer, T. P., Von Korff, R. W., and Murphy, D. L., Eds., Academic Press, New York, 1979, 25.
36. **Singer, T. P.,** Suicide inhibitors: an overview, in *Molecular Basis of Drug Action*, Singer, T. P. and Ondarza, R., Eds., Elsevier/North-Holland, Amsterdam, 1981, 97.
37. **Fuller, R. W. and Hemrich, S. K.,** Enhanced selectivity of pargyline as an inhibitor of type B monoamine oxidase in harmaline-treated rats, *Life Sci.*, 22, 1083, 1978.
38. **Langston, J. W., Ballard, P., Tetrud, J. W., and Irwin, I.,** Chronic Parkinsonism in humans due to a product of meperidine-analog synthesis, *Science*, 219, 979, 1983.
39. **Chiba, K., Trevor, A., and Castagnoli, N., Jr.,** Metabolism of the neurotoxic amine, MPTP, by brain monoamine oxidase, *Biochem. Biophys. Res. Commun.*, 120, 574, 1984.
40. **Salach, J. I., Singer, T. P., Castagnoli, N., Jr., and Trevor, A.,** Oxidation of the neurotoxic amine 1-methyl-4-phenyl-1,2,3,6-tetrahydropyridine (MPTP) by monoamine oxidases A and B and suicide inactivation of the enzymes by MPTP, *Biochem. Biophys. Res. Commun.*, 125, 831, 1984.
41. **Singer, T. P., Salach, J. I., and Crabtree, D.,** Reversible inhibition and mechanism-based irreversible inactivation of monoamine oxidases by 1-methyl-4-phenyl-tetrahydropyrine (MPTP), *Biochem. Biophys. Res. Commun.*, 127, 707, 1985.

Chapter 22

PHYSIOLOGICAL AND CLINICAL ASPECTS OF MONOAMINE OXIDASE

D. M. Kuhn, D. L. Murphy, and Moussa B. H. Youdim

TABLE OF CONTENTS

I. INTRODUCTION

The amine neurotransmitters epinephrine (E), norepinephrine (NE), and dopamine (DA), and serotonin (5-HT) and other amines with false neurotransmitter or sympathomimetic properties, e.g., octopamine, tyramine, phenylethylamine (PEA), and tryptamine, are considered classic substrates for the ubiquitous mitochondrial monoamine oxidase (MAO) (amine:oxygen oxidoreductase, deaminating, flavin-containing; E.C. 1.4.3.4.). Many of the above amines have been implicated in the pathogenesis of neuropsychiatric diseases. Inhibition of MAO in vivo with irreversible MAO inactivators results in increased brain levels of the above amines with a concomitant decrease in their deaminated metabolites. A number of MAO inhibitors have been shown to have true antidepressant activity and some have found their way into clinical use.[1-4] Animal behavioral studies have shown that treatment with the amine precursors L-tryptophan, L-5-hydroxtryptophan, or L-dihydrophenylalanine following MAO inhibition can result in overt behavioral syndromes that can be attributed to the accumulation and release of amine neurotransmitters.[2-4] Therefore, a regulatory role has been assigned to MAO in the brain. This chapter is by no means intended to be a comprehensive review of the physiological and clinical aspects of MAO. We will, however, concentrate our discussion on the role of intraneuronal MAO in regulating the "functional pool" of neurotransmitters, with special emphasis on the role of multiple forms of MAO in serotonin (5-HT) deamination.

The readers' attention is directed to other extensive recent reviews which deal with biochemical,[3] pharmacological,[1,4] clinical,[2] and physiological aspects of this important enzyme.

II. MULTIPLE FORMS OF MAO

The existence of multiple forms of MAO, described early in the 1960s,[5] was reinforced by the discovery of clorgyline, an irreversible inhibitor of this enzyme.[6] Johnston[6] was unable to distinguish two enzyme forms in mitochondrial preparations from a number of sources, including the brain. With 5-HT as a substrate, much lower clorgyline concentrations were required to inactive MAO than with benzylamine as the substrate. With tryamine as substrate, a double sigmoid inhibition curve was obtained, where half the enzyme activity was inhibited at concentrations which fully inhibited 5-HT and the other half at concentrations which inhibited benzylamine deamination. Johnston attributed this phenomenon to the presence of two enzyme forms in the mitochondria, with MAO-A having activity for 5-HT and being sensitive to inhibition by clorgyline; MAO-B was relatively resistant to inhibition by clorgyline and showed higher affinity to benzylamine than to 5-HT. Tyramine and DA were considered substrates for both forms. Subsequently, another propargyl "suicide" inhibitor, deprenyl, was described which selectively inhibited MAO-B[7] (see Chapters 20 and 21). Since these original findings, the affinity of mitochondrial MAO for various monoamine substrates (Table 1) and selective inhibition by various drugs (Table 2) have been described.

III. EVIDENCE FOR IN VIVO PRESENCE OF MAO-A AND MAO-B

Ex vivo experiments in which animals are pretreated with the selective MAO-A (clorgyline) or MAO-B (L-deprenyl) inhibitors and the tissue is subsequently removed, homogenized, and assayed for MAO activity with various monoamine substrates substantiate the presences of two enzyme forms[8] as outlined by Johnston.[6] Of great interest has been the question of whether the enzyme forms (A and B), when inactivated selectively in vivo, would lead to the selective accumulation of their respective neurotransmitter and non-neurotransmitter amine substrates in the brain. Thus, selective inhibition of MAO-A has been shown to result in increased brain levels of 5-HT and norepinephrine as well as dopamine,[9] while selective

Table 1
SELECTIVITY OF MITOCHONDRIAL MAO FOR SOME COMMON SUBSTRATES

Substrates for MAO-A	Substrates for MAO-B	Substrates for both MAO forms
Epinephrine	β-Phenylethylamine (low concentration)	Tyramine
Norepinephrine	Benzylamine	Dopamine
5-Hydroxytryptamine	*N-tele*-Methylhistamine	Kynuramine
Octopamine	Tryptamine	
3-*O*-Methylnorepinephrine	5-Methoxytryptamine	
3-*O*-Methylepinephrine		

Note: This information is intended as a guide and may not hold rigidly in every case. Selectivity for one MAO form or other may depend on experimental conditions, particularly substrate concentration and enzyme source.

From Youdim, M. B. H. and Finberg, J. P. M., *Monoamine Oxidase Inhibitor Antidepressants in Psychopharmacology*, Grahame-Smith, D. G., Hippius, H. and Winoker, G., Excerpta Medica, Amsterdam, 1982. 38. With permission.

Table 2
SOME SELECTIVE INHIBITORS OF MITOCHONDRIAL MAO-A AND MAO-B

Type A inhibitor	*Type B inhibitors*
Irreversible inhibitors	
Clorgyline	(−)-Deprenyl
M & B 9303	Pargyline
MO 1671	AGN 1135
LY 51641	AGN 1278
	U 1424
Reversible inhibitors	
Amphetamine	Tricyclic antidepressants
β-Methyltryptamine	MD 780236
Harmaline	
MD 780515	
FLA 336	
RO 11-1163	
β-Carbolines	

Note: The above is a much-abridged list of some of the most potent and/or most selective compounds of many that have been described in the literature.

From Youdim, M. B. H. and Finberg, J. P. M., *Monoamine Oxidase Inhibitor Antidepressants in Psychopharmacology*, Grahame-Smith, D. G., Hippius, H. and Winoker, G., Excerpta Medica, Amsterdam, 1982. 38. With permission.

inhibition of brain MAO-B leads to elevation in brain phenylethylamine and dopamine.[9] Corresponding decreases in acidic or alcoholic deaminated products have also been very well documented, whereas *O*-methylated metabolites, e.g., normetanephrine, are increased. While clorgyline-induced MAO inhibition has been shown to decrease brain levels of 5-hydroxyindoleacetic acid, 3-methoxy-4-hydroxyphenylglycolaldehyde, 3-4-dihydroxyphenylacetic acid, and homovanillic acid, inhibition of MAO-B with deprenyl reduces 3,4-dihydroxyphenylacetic acid and phenylacetic acid. However, it should be noted that the increases in brain monoamines observed after the selective inhibition of one or the other form of MAO are not as high as those obtained using a nonselective MAO inhibitor such as tranylcypromine.[10-12] These data have been interpreted as suggesting that while MAO-A or MAO-B show a certain degree of specificity for various monoamines, this specificity is not absolute.[14] Therefore, at high substrate concentrations, either enzyme form can deaminate amines which are not normally its substrate.[10-13] In vitro experiments have substantiated this data.[15] These studies are, therefore, consistent with the view that 5-HT and NE are type A substrates in rat brain in vivo, while DA is a substrate for both enzyme forms. In the human brain the situation is rather similar but the proportion of MAO-A to MAO-B is significantly different. It has been suggested that the major form of the enzyme is MAO-B[16] and DA is a major substrate for this enzyme form.

IV. NEURONAL LOCALIZATION OF MAO

The fact that 5-HT and NE are type A substrates in vivo does not prove that this form of MAO is localized in the amine-containing neurons and that MAO-B is absent from them. Even if both enzymes were present in the neurons, as has been suggested,[9,17] 5-HT and NE would be preferred substrates of MAO-A. It is unlikely that in the rat brain the major form of neuronal MAO is type B since, if this were so, selective inhibition with deprenyl should result in an elevation of all neurotransmitter amine levels along with a decrease in their deaminated metabolites.[10]

With regards to the physiological role of MAO, the central question is whether the enzyme form showing greater affinity for the neurotransmitters 5-HT and NE (MAO-A) is selectively localized within the aminergic neuron. Most of the evidence available is of an indirect nature. In the peripheral nervous system, studies have shown that sympathetic denervation in the vas deferens resulted in a preferential reduction of MAO-A activity.[18] The major form of MAO in a number of cell types of neural origin grown in culture have been shown to be type A.[3] It should be noted, however, that in other cells of non-neural origin, MAO-A has also been found. To date the only tissue of neural origin having entirely MAO-B is the isolated cultured adrenal chromaffin cell,[19] a property shared with the human platelet.[20] It is interesting to note that the latter two cell types have the highest concentrations of NE and 5-HT in the body. In contrast, pheochromocytoma-derived PC-12 cells have primarily MAO-A. These cells have been used as a model for adrenergic nerve endings since they show a profound similarity in their catecholamine metabolism to catecholaminergic neurons.

Until recently, there was no adequate histochemical method for visualizing MAO-A and MAO-B in intact tissue, especially the brain. Therefore, it was extremely difficult to establish the neuronal or extraneuronal distribution of the two enzyme forms and whether they are associated with the outer mitochondrial membrane from tissues that contains both forms of the enzyme. Using a peroxidation histochemical method,[21] MAO activities were precisely localized within the cytoplasm of certain neuronal cells in the locus coeruleus and median raphe of guinea pig brain. Despite the known substrate specificities of the two enzyme forms, no difference was found in the pattern of distribution of MAO-containing neurons. The histochemical data indicated that the majority of neuronal cells in the locus coeruleus and median raphe contained both enzyme forms, with MAO-A being predominant. The glial (astrocytes) cells contained predominantly MAO-B. Immunocytochemical studies,[22] where antiserum to purified MAO-B was used to establish the distribution of MAO in adult rat brain, indicated that MAO-B was present in astrocytes and serotonin-containing neurons of the dorsal raphe nucleus. Double immunofluorescence experiments using antisera to 5-HT and MAO-B enabled Levitt et al.[22] to positively identify these two molecules within the same cells of the nucleus raphe dorsalis. The localization of MAO-B in 5-HT-containing neurons is not unexpected, since earlier pharmacological studies had provided evidence for the metabolism of 5-HT by MAO-B when MAO-A was totally inactivated with the selective inhibitor clorgyline. Immunohistochemical studies confirm what was demonstrated by pharmacological and animal behavioral studies. The most surprising aspect of this work, however, is that no other neurones in the rat brain, including the substantia nigra, caudate nucleus, and locus coeruleus, were stained for MAO-B. There is ample evidence to suggest that in rat brain, NE and DA can be deaminated by both MAO forms with markedly different rates,[10-12,23] Although Levitt et al.[22] could not find MAO-B in NE-containing neurones, this does not mean it is absent from the latter neurones.

V. THE FUNCTIONAL POOL OF SEROTONIN

As the major catabolic enzyme for serotonin in brain, MAO-A is in a pivotal position to influence not only the absolute levels of this transmitter within neurons but also the amounts of 5-HT available for release into the synapse. Obviously, many factors ultimately determine the amount of neurotransmitter released from the neuron and, while some of these factors are related to the transmitter itself (e.g., synthesis, uptake), many are not. We shall focus on those factors which are directly related to 5-HT and will discuss the roles that MAO can play in determining both the amount of 5-HT which is available for release and that which is actually released.

Insofar as MAO inhibitors (MAOI) increase the presynaptic levels of 5-HT, it is more than possible that greater amounts of 5-HT could be released as a result. However, this possibility can only be considered within the context of where in the presynaptic neuron the MAOI-induced increases in 5-HT take place and from which intracellular location release occurs. It is generally accepted that 5-HT exists within neurons in separate pools or compartments.[24,25] One pool is quite large, containing 80 to 85% of total brain 5-HT, and is referred to as a "reserve" or "storage" pool. Release is not thought to occur from the reserve pool except under extreme or "emergency" conditions.[26] A second 5-HT pool is small (15 to 20% of total brain 5-HT) and is thought to be comprised of newly synthesized 5-HT. Since newly synthesized 5-HT is preferentially released,[24,27] the small pool of 5-HT is logically assumed to be the primary transmitter pool from which release occurs. These compartments of 5-HT have not been clearly associated with defined anatomical (subcellular) structures, although the most compelling evidence in support of large vs. small pools comes from studies with reserpine. Reserpine reduces brain 5-HT by 80 to 90%, and based on the well-known ability of reserpine to disrupt the vesicular storage of 5-HT,[28] it is easy to ascribe the large 5-HT pool to a vesicular compartment. Furthermore, that amount of 5-HT remaining in the brain after reserpinization represents the newly synthesized pool of 5-HT since acute treatment with reserpine does not alter tryptophan hydroxylation,[29,30] the initial and rate-limiting step in the biosynthesis of 5-HT. This small pool of newly synthesized 5-HT is a cytoplasmic or extravesicular pool, at least initially. Synthesis of 5-HT takes place in two enzymatic steps. The 5-HT precursor L-tryptophan is first converted to L-5-hydroxytryptophan (5-HTP) by tryptophan hydroxylase, the initial and rate-limiting enzyme in 5-HT biosynthesis,[31] and then 5-HTP is subsequently decarboxylated to 5-HT by the action of aromatic L-amino acid decarboxylase. Since both the hydroxylase and the decarboxylase enzymes are cytoplasmic, the synthesis of new 5-HT occurs in the extravesicular space. Once decarboxylation of 5-HTP takes place, the fate of newly synthesized 5-HT follows one of three major paths and the particular path followed is largely determined by MAO.

First, the 5-HT could be enzymatically inactivated by MAO; second, it could be incorporated into storage vesicles from out of the cytoplasm; third, the transmitter could be released. The mechanisms controlling the intracellular movement of 5-HT (into vesicles or into the synapse) are not understood but it is clear that for 5-HT to reach its storage site or for it to be released, it must be protected from MAO. Since MAO is a mitochondrial enzyme, 5-HT is effectively protected from deamination once it is transported into storage vesicles after synthesis. However, any cytoplasmic 5-HT is subject to attack by MAO. How then is the cytoplasmic or the small pool of newly synthesized 5-HT protected from MAO? Certain metabolic compartments[32] may simply restrict access of 5-HT to MAO or the intrinsic activity of the enzyme could change. A variety of 5-HT binding proteins have also been identified,[33] and these may protect 5-HT from MAO. In fact, proteins of this type were proposed by Halaris and Freedman[34] to account for a reserpine-resistant, "juxtavesicular" fraction of 5-HT observed in their studies on the effects of LSD on 5-HT metabolism and compartmentation. More recently, Small and Wurtman[35] demonstrated that an actin-like protein isolated from synaptosomes binds 5-HT in a saturable and selective manner so that some 5-HT binding proteins are being positively identified. Thus, although MAO avidly deaminates 5-HT, mechanisms are apparently operative within neurons to protect cytoplasmic 5-HT from MAO.

It is clear from the foregoing discussion that the impact of MAO on the releasable pool of 5-HT is entirely dependent on which pool of 5-HT is the one from which release occurs. It is often proposed that 5-HT release in brain is exocytotic,[36] i.e., vesicles containing 5-HT fuse with the synaptosomal membrane and subsequently empty their contents into the synapse. If 5-HT release is indeed exocytotic (and there is actually little *direct* evidence to indicate that it is), MAO would have little impact on the releasable pool per se once 5-HT is within the vesicles. On the other hand, it is also assumed that newly synthesized 5-HT is preferentially released and the small, extravesicular pool of 5-HT contains the newly synthesized transmitter. Thus, if 5-HT release is not exocytotic but occurs from the cytoplasm into the synapse, MAO would play a very important role in determining the amount of 5-HT available for release.

VI. ASSESSMENT OF SEROTONIN RELEASE IN VIVO FROM THE FUNCTIONAL POOL

It is obvious that the release of 5-HT in the brain is quite complicated and probably involves some combination of exocytocis (from vesicle into synapse), direct release from the cytoplasm with or without a carrier, and release from vesicular sites into the cytoplasm then into the synapse. In order to gain more insight into in vivo 5-HT release, we have recently used a behavioral syndrome which has been extensively characterized as a reliable and specific indicator of 5-HT release,[37,38] and in an attempt to determine which pool of 5-HT is the functional or releasable pool, we studied this 5-HT behavioral syndrome in reserpinized rats. The reserpine model seemed ideal since vesicle function is drastically compromised and 5-HT levels are markedly reduced but, at the same time, the rate of tryptophan hydroxylation (or 5-HT synthesis) is unaltered.[29,30] Thus, 5-HT continues to be synthesized *de novo* (the small pool), but the newly synthesized transmitter is not incorporated into vesicular storage sites. Furthermore, post-synaptic receptors for 5-HT, as measured by ^{3}H-5-HT binding, are not altered significantly by reserpine.[39]

The 5-HT behavioral syndrome consists of a variety of symptoms which are easily observed and quantified. These include forepaw treading, hind-limb abduction, head weaving, salivation, piloerection, and tail lashing. This syndrome can be produced in normal rats by drugs which are direct post-synaptic 5-HT receptor agonists (e.g., 5-methoxy-*N,N*,dimethyltryptamine), by drugs which release 5-HT from the presynaptic neuron (e.g., parachloroamphetamine, fenfluramine), and by the combined treatment with an irreversible

Table 3
IN VIVO RELEASE OF 5-HT IN RESERPINIZED RATS

Treatment	Syndrome
PCA (5 mg/kg)	35/35
PCPA + PCA (250 mg/kg + 5 mg/kg)	0/19
Metergoline + PCA (5 mg/kg + 5 mg/kg)	0/8

Note: Rats were injected with the indicated doses and observed for appearance of the 5-HT behavioral syndrome. The expression of the symptoms was scored as a positive response and data are presented as a number showing the syndrome per number injected.[81]

MAO inhibitor plus the 5-HT precursor L-tryptophan. The importance of MAO in determining 5-HT release is quite well illustrated by this latter treatment. L-Tryptophan by itself does not cause the syndrome (or 5-HT release) since the newly formed 5-HT is readily deaminated by MAO. Only after MAO inhibition will L-tryptophan cause the 5-HT-behavioral syndrome. With this brief introduction to the 5-HT behavioral syndrome, we will present recent results from our laboratory where this syndrome was studied in detail in reserpinized rats.[81,82]

Our initial experiments were directed at determining if the 5-HT syndrome could be produced in reserpinized rats by the 5-HT-releasing drug, parachloroamphetamine (PCA). These results are presented in Table 3. Treatment of reserpinized rats (5 mg/kg 24 hr prior to PCA) with 5.0 mg/kg of PCA produced a syndrome which was, if anything, more severe than that produced by the same dose of PCA in normal animals. Determinations of 5-HT and 5-HIAA by HPLC with fluorescence detection[40] indicated that reserpine lowered 5-HT levels by 90 to 95% in the hypothalamus, striatum, and the thoracic spinal cord whereas the dorsal raphe nucleus was depleted of 75% of its normal 5-HT content by reserpine. Levels of 5-HIAA showed the characteristic reserpine-induced increase. Thus, despite drastic reductions in 5-HT concentrations by reserpine, PCA still produced the entire behavioral syndrome. If reserpinized rats are injected with parachlorophenylalanine (PCPA), an irreversible inhibitor of tryptophan hydroxylase, to inhibit any additional synthesis of 5-HT, PCA no longer produced the syndrome. Neurochemical analysis revealed that PCPA further reduced already lowered 5-HT levels, and in some brain areas neither 5-HT nor 5-HIAA was detectable. This observation indicates that the 5-HT remaining after reserpine treatment, and that which is released by PCA, is newly synthesized. Furthermore, these results rule out a direct receptor effect by PCA in producing the syndrome. Finally, if PCA is indeed releasing 5-HT from this reserpine-resistant pool, then a 5-HT receptor antagonist should prevent the PCA effect. Pretreatment of reserpinized rats with metergolin (5 mg/kg) did in fact prevent the PCA-induced behavioral syndrome. Taken together, these results indicate that the pool of 5-HT which survives reserpine treatment, as small as it is (approximately 5 to 10% of total brain 5-HT), is more than sufficient to allow production of the complete 5-HT behavioral syndrome by PCA. These results also indicate that the pool of 5-HT which is released by PCA is not contained in vesicles. Although we can never rule out the possibility that some small population of vesicles is resistant to reserpine and that it is this vesicular pool or a newly synthesized vesicle population from which PCA is releasing 5-HT, this possibility seems unlikely since PCA continues to produce the characteristic syndrome after repeated reserpine treatment.[82]

Additional drugs and drug combinations were given to reserpinized rats in order to more fully characterize the role of MAO in determining the in vivo release of 5-HT.[81] These

Table 4
ROLE OF MAO IN THE IN VIVO RELEASE OF 5-HT

Treatment	Syndrome
L-Tryptophan (100 mg/kg)	0/8
Tranylcypromine (10 mg/kg)	8/8*
L-Tryptophan + tranylcypromine	8/8
Pargyline (10 mg/kg)	4/4*
Clorgyline (5 mg/kg)	0/4
Deprenyl (5 mg/kg)	0/4
Clorgyline + deprenyl	4/4*
L-Tryptophan + clorgyline + deprenyl	8/8

Note: Rats were scored for expression of the 5-HT behavioral syndrome as described above. The asterisk indicates that the animals displayed some symptoms and were scored as positive, yet in the indicated groups the syndrome was much weaker than in other positive groups and the symptoms only appeared sporadically.[81]

treatments are summarized in Table 4. Injections of L-tryptophan were no more effective in producing the 5-HT syndrome in reserpinized rats than in normal rats. However, high does of the MAO inhibitors tranylcypromine (TCP; 10 mg/kg) and pargyline (75 mg/kg) produce some intermittent signs of the syndrome by themselves. The combined treatment of reserpinized rats with L-tryptophan + TCP produced a very potent behavioral syndrome. Separate injections of the specific MAO-A and MAO-B inhibitors, clorgyline and deprenyl, respectively, did not produce the syndrome, but the combination of these two drugs produced effects much like that observed after TCP or pargyline. Although the MAO inhibitors (TCP, pargyline, or clorgyline + deprenyl) clearly reverse many of the overt symptoms of reserpine, other signs of the 5-HT syndrome, including forepaw treading, head weaving, and salivation, were observed intermittently as well. Administration of L-tryptophan to reserpinized rats previously treated with clorgyline plus deprenyl also resulted in a profound behavioral syndrome. Thus, it appears that many treatments which produce the 5-HT syndrome in normal rats also produce the syndrome in reserpinized rats. As such, similar interpretations apply to the mechanism by which MAO inhibitors plus L-tryptophan produce the behavioral syndrome: the increase in 5-HT produced by L-tryptophan alone does not result in the syndrome since MAO deaminates the newly formed transmitter prior to its release. The syndrome does appear, however, after MAO is inhibited since the 5-HT now builds up in the presynaptic neuron then ''spills over''[41] into the synapse as storage capacity is exceeded. Once again these results point to the very important role that MAO plays in modulating 5-HT release.[42] It appears that precursor-induced increases in 5-HT do not cause the release of 5-HT if MAO is intact. In fact, both forms of MAO must be inhibited before the syndrome appears (see above). Results of preliminary neurochemical studies in our laboratories with clorgyline and deprenyl in reserpinized rats agree quite well with previous work in normal animals.[10] Even though MAO-A is specific for 5-HT, complete inhibition of the A form with clorgyline does not increase 5-HT levels as much as a nonselective inhibitor of MAO (e.g., TCP or pargyline), nor as much as clorgyline plus deprenyl. These results indicate that even after MAO-A has been inactivated, MAO-B can deaminate 5-HT as its intraneural concentration increases. The recent localization of MAO-B in 5-HT-containing cells in the raphe nuclei by immunocytochemical means[43] provides strong support to the biochemical results which indicate that 5-HT is also a substrate for MAO-B.[10]

Although the influence of MAO on maintaining 5-HT at certain levels is easy to observe after injections of L-tryptophan (i.e., when 5-HT synthesis is "pushed"), its role in modulating 5-HT levels and release in the absence of exogenous L-tryptophan is not as easy to appreciate but, upon inspection, is just as profound. For example, the drastic reductions in 5-HT and large increases in 5-HIAA produced by reserpine result from disruption of vesicular storage — 5-HT escapes from vesicles in large amounts and is apparently deaminated without being released. In fact, if even very small amounts of released 5-HT can produce the behavioral syndrome (as the data with PCA in reserpinized rats indicates), one must conclude that the loss of 5-HT after reserpine is primarily to MAO, not release, since rats do not display the syndrome after reserpine. This is further substantiated by our observation that injections of MAO inhibitors shortly before or after reserpine lead to the apperance of the complete 5-HT behavioral syndrome.[82] Treatment of rats with selective 5-HT uptake blockers before or after reserpine, however, does not produce the behavioral syndrome. If the 5-HT was being released into the synapse by reserpine as previously suggested,[44] inhibition of the re-uptake of even a small amount of the 5-HT should produce the syndrome, but it does not. On the other hand, the combined treatment with MAOI and an uptake blocker in normal animals produces the 5-HT behavioral syndrome.[45]

Taken together, the pharmacological results with the 5-HT behavioral syndrome in reserpinized rats suggest several conclusions regarding 5-HT release and its regulation by MAO. First, PCA apparently elicits 5-HT release from the small cytoplasmic pool into the synapse. Second, although reserpine abolishes the vesicular pool of 5-HT (90% of total brain 5-HT, the 5-HT does not reach the synapse as long as MAO is active. As a further indication of how important MAO is in determining the size of the cytoplasm pool in reserpinized rats, injections of MAOIs increase 5-HT concentrations ten fold over reserpine levels. Thus, even in the absence of proper vesicle function, presynaptic neurons have quite a large storage capacity. That this cytoplasmic pool is releasable is further substantiated by our observation that the PCA dose response curve for inducing the 5-HT behavioral syndrome in reserpinized rats is shifted to the left by MAO inhibitors. Third, while various 5-HT binding proteins might protect cytoplasmic 5-HT from MAO, it remains to be demonstrated that PCA liberates 5-HT from its binding proteins or if these proteins function as carriers for 5-HT in the release process.

In summary, we have attempted to gain a better understanding of which 5-HT pool is the functional compartment so that the influence of MAO on release of 5-HT can be properly studied. Many in vitro experiments which examine 5-HT release often involve the induced release of ^{3}H-5HT from prelabeled synaptosomes and, almost without exception, MAO inhibitors are added to these preparations to prevent the deamination of the ^{3}H-5HT once it is taken up into the nerve ending. Under these conditions, the role of MAO cannot be appreciated. In addition to the behavioral data discussed above, a few other papers have appeared in the literature which show the important role that MAO plays in modulating the release of 5-HT. Perhaps the most impressive of these was presented by Elks et al.[27] These investigators demonstrated that the release of 5-HT from electrically stimulated slices of brain was not dependent on the amount of tryptophan in the tissue or incubation media as long as MAO was intact. Rather high concentrations of tryptophan do not increase the release of 5-HT. However, in the presence of pargyline, increases in 5-HT synthesis with increases in tryptophan led to a much larger release of 5-HT.[27] Thus, when MAO activity is normal, increasing the synthesis of 5-HT with precursor loading does not augment release. Finally, another study supports our claim that the release of 5-HT can be induced from a nonvesicular pool. Theoharides et al.[46] demonstrated that under conditions where the exocytotic release of histamine from the granules of mast cells is inhibited, 5-HT can still be released by a Ca^{2+}-dependent, energy-requiring process. Thus, at least in one secretory organ, it appears that 5-HT release can occur in a nonexocytotic manner.[81,82]

VII. CLINICAL ASPECTS OF THE REGULATORY FUNCTIONS OF MAO

When MAO activity is normal, the neuronal enzyme functions in two major ways, as described above: (1) by deaminating intracellular neurotransmitter amines, it regulates the concentrations of those amines available for release from vesicular and nonvesicular pools, and thus secondarily has feedback regulatory effects on the synthesis of new neurotransmitter amines;[47] (2) normally functioning MAO also rapidly degrades nontransmitter amines such as phenylethylamine, octopamine, and tyramine. While these functions are most clearly and dramatically evident in animal studies such as those discussed above, they are also reflected in clinical treatment studies of MAO inhibitors as antidepressants.

Post-morten studies have shown that clinically used doses of nonselective MAO inhibitors (e.g., phenelzine, tranycypromine, and isocarboxizid) as well as selective MAO-A inhibitors (e.g., clorgyline) and MAO-B inhibitors (e.g., deprenyl) result in marked inhibition of human brain MAO activity and elevations in brain concentrations of serotonin, norepinephrine, and dopamine. These changes are also evidenced in reductions of the major amine metabolites such as 5-hydroxyindoleacetic acid, vanylmandelic acid, and homovanilic acid in the cerebrospinal flud and urine of patients receiving these drugs.,[48] That these metabolic changes have functional significance is most clearly evidenced in some of the side effects and toxic effects which may occur in patients treated with MAO inhibitors upon exposure to indirect-acting sympathomimetic drugs or to tyramine in some food stuffs. Under these circumstances, acute hypertensive reactions which may be so severe as to be life-threatening may develop. Similarly, reserpine administration or treatment with amine-uptake inhibiting tricyclic antidepressants can precipitate severe hyperthermic crises with changes in motor function and other neurologic abnormalities.[1,4,49]

The therapeutic antidepressant effects of the MAO inhibitors have similarly been attributed to the increased brain amine concentrations available for release.[50] In the case of serotonin, for example, there is evidence that 5-hydroxytryptophan potentiates the antidepressant effects of MAO inhibitors, while inhibition of serotonin synthesis by PCPA, in contrast, has been shown to produce an acute return of depressive symptoms in patients who had improved in response to tranylcypromine.[51] Similarly, additive therapeutic effects in Parkinson's disease have been observed when l-dopa and carbidopa treatment are combined with deprenyl.[52]

A process of reinterpretation of the likely mode of therapeutic action of both the MAO-inhibiting antidepressants as well as the tricyclic and heterocyclic antidepressants has occurred in the last few years. In large part, this has resulted from wider recognition of the fact that the beneficial clinical effects produced by these drugs develop much more slowly — over a period of weeks — than do the most evident acute metabolic changes. Current emphasis has been placed on a series of secondary, longer-term adaptive changes which occur in monoamine receptor function whose time course and greater universality (at least for the tricyclics) seem to approximate more closely the beneficial actions of antidepressant drug treatment.

Greatest emphasis has been placed upon an association between antidepressant drug efficacy and reductions in brain β adrenergic receptor density, which is accompanied by a reduction in cyclic adenosine monophosphate production in response to adrenergic stimuli.[53] Associated changes in α-2 adrenoceptors and in serotonin receptors have also been described.[54] Of special pertinence to the evidence discussed above is evidence that an intact brain serotonergic system is a necessary prerequisite for β adrenergic receptor down regulation with antidepressant drug treatment.[55]

VIII. STUDIES WITH SELECTIVE MAO INHIBITORS AS AN APPROACH TO EVALUATING THE MODE OF THERAPEUTIC ACTION OF THE MAO-INHIBITING ANTIDEPRESSANTS

The earlier clinical literature on the efficacy and the biological consequences of the MAO inhibitors has been frequently reviewed (e.g., References 1, 56, and 57). Some more recent developments on the mode of action of these drugs has emerged from studies using substrate-selective MAO inhibitors, and this chapter will focus in some detail on these investigations.

Conventional MAO-inhibiting antidepressants currently available as prescription drugs in the U.S. (phenelzine, isocarboxazid, and tranylcypromine) have negligible substrate selectivity. While substrate-selective agents have thus far only been available for investigational use, the recent introduction of four new selective reversible MAO inhibitors (amiflamine, cimoxatone, moclobemide, and CGP-11035A) into clinical investigational study phases, along with additional clinical studies with the "older" selective MAO inhibitors, clorgyline, deprenyl, and pargyline, have provided a new window into the therapeutic actions of the MAO inhibitors.[1]

As discussed above, the basis for the substrate-selective actions of the newer MAO inhibitors is the existence of two subtypes of MAO, MAO-A and MAO-B. While questions of additional subtypes of the enzyme or even a whole family of closely related MAOs with a spectrum of affinities for different monoamines have been raised, the evidence currently available is reasonably compatible with two main enzyme forms which vary in immunogenicity, in electrophoretically identified fragments, in molecular weight, and in their substrate selectivity. The most clinically relevant distinction is their different sensitivity to certain inhibitors. The two enzyme forms are also differentially localized in certain cells and tissues.

In regard to current thinking about the mode of action of antidepressants, presumably the most important characteristic of the different selective MAO inhibitors is their effects on specific monoamine neurotransmitter or neuromodulator systems. As indicated in Table 1, MAO-A is relatively selective in functioning to inactivate serotonin, norepinephrine, and epinephrine, while MAO-B selectively deaminates phenylethylamine, dopamine (in primates but not in rodents), phenylethanolamine, *tele*-methylhistamine, and *ortho*-tyramine. Additional substrates such as *para*-tyramine, octopamine, and others show lesser selectivity for MAO-A or MAO-B. While inhibitors have been grouped as MAO-A or MAO-B selective in Table 1, it should be noted that the degree of selectivity (i.e., the ratio of MAO-A to MAO-B inhibition) varies among the different MAO inhibitors. Also, there is some data for within-class differences in that the MAO-A inhibitor amiflamine has been supported to be preferentially accumulated within serotonin nerve terminals and consequently to have greater effects on the serotonergic than the noradrenergic neurotransmitter system in the brain.[58] Clorgyline, on the other hand, reduces the major norepinephrine metabolite (3-methoxy-4-hydroxyphenylglycol [MHPG]) in brain to a greater extent than it does the major serotonin metabolite (5-hydroxyindoleacetic acid), and hence its effects have been suggested to be most closely correlated with noradrenergic changes in brian.[59]

In considering how these drugs may act, an initial caveat is needed concerning the selective inhibitors being regarded as necessarily like the conventional MAO inhibitors, and different only in having a narrower spectrum of action. We have presented some evidence elsewhere and reviewed the results from other studies suggesting that different classes of MAO inhibitors may have different clinical profiles in keeping with their different chemical and pharmacologic properties in vitro and in vivo.[57] This should not be surprising, as there is some evidence that two of the most well-studied conventional MAO inhibitors, tranylcypromine and phenelzine, may not be equivalent drugs,[56,60] although this question has not yet been adequately investigated by direct comparisons using modern assessment and rating instruments in the same population of patients.

A. Clinical Antidepressant Responses to MAO-A Selective Inhibitors

Unlike the conventional inhibitor phenelzine, for which there is contradictory evidence regarding its efficacy in severely depressed patients, but fairly strong evidence for its usefullness in "atypical", predominantly out-patient populations,[61,62] several studies have suggested that clorgyline may be effective across a fairly broad spectrum of patient populations. In the first random assignment, a double blind study conducted with clorgyline, Herd reported that significantly more clorgyline-treated patients (14 of 16) than amitriptyline-treated individuals (7 of 15) were rated as "improved or recovered".[63] Mean drug doses were not reported but ranged between 20 to 30 mg/day for clorgyline and 100 to 150 mg/day for amitriptyline. A second study compared clorgyline (30 mg/day) with imipramine (100 mg/day) in 92 depressed outpatients using a double blind, random-assignment design.[64] Both drugs were found to be equally effective and to yield a similar incidence of side effects.

In the only other double blind study with clorgyline in the literature, which included a cross-over comparison with pargyline, Lipper et al. reported that clorgyline treatment yielded significant reductions in Hamilton, Beck, and other depression rating scale inventories, while pargyline did not.[65] In all of the comparisons of the various rating scales in the cross-over design where significant differences in treatment response were found, the results uniformly favored clorgyline. In our extended study, which totaled 23 depressed patients treated with clorgyline, a significant association was demonstrated between initial severity of depression (as assessed by Hamilton scale ratings and several self-rating scales) and therapeutic benefit.[66] This finding supported the observation made in our initial report that the most striking antidepressant responses to clorgyline, as indicated by a reduction of greater than 50% on the Hamilton rating scale, occurred in a subgroup of patients who had pretreatment Hamilton depression scale scores over 30 and who exhibited predominatly endogenous-type depressive symptoms, including weight loss, diurnal variation in mood, and psychomotor retardation.[65]

Potter and co-workers described five bipolar women with unusually treatment-resistant, severe, incapacitating symptoms of recurrent depressions. All had been given standard treatments including lithium, tricyclics, psychotherapy, thyroxine, propranolol, carbamazepine, and neuroleptics, and had been studied and treated for more than 6 months as inpatients before clorgyline administration in low dosage (2.5 to 10 mg/day) was begun.[67] Each of the five patients experienced a longer duration of euthymic periods and a lesser severity of depressive symptoms upon the addition of clorgyline to their optimal conventional therapy (usually lithium). Some moderate lengthening of cycle duration had been observed with phenelzine and tranylcypromine, but depressive episodes were qualitatively more rather than less severe and included psychotic features. Clorgyline in low doses was suggested to have anticycling as well as antidepressant effects in this small but extremely refractory patient group. In higher doses, however, clorgyline treatment may be associated with the precipitation of manic episodes in bipolar patients, a common propensity of all MAO-inhibiting drugs.[57] Because there have been other suggestions that bipolar patients may be less responsive to tricyclics and more responsive to MAO inhibitors, especially tranylcypromine,[60] further study is needed to evaluate whether clorgyline possesses definite advantages in treatment-resistant bipolar depressed patients.

One study with clorgyline and pargyline noted that several patients who completely recovered during clorgyline administration had mild to moderate hypertensive patients became normotensive during clorgyline administration and tended to respond especially favorably to the drug.[69] Correlations between both pretreatment vs. depression rating reductions were fairly robust. Similar positive correlations were not observed with pargyline, although the small number of responders to pargyline and the smaller number of patients included in the pargyline trials lessen the meaningfulness of this information. Comparable studies of other MAO inhibitors are needed to evaluate whether this is a general finding with all MAO inhibitors or one specific to MAO-A inhibitors such as clorgyline. Our current interpretation the effect of clorgyline is that the drug leads to diminished central sympathetic outflow, associated with decreased plasma norepinephrine concentrations when measured both at rest and standing. As the blood pressure changes tend to develop only slowly during the course of chronic drug administration, and, in fact, parallel the course of clinical improvement and the delay required for adrenergic receptor numbers to change, we have hypothesized that these events may be interrelated.[69]

B. Clinical Antidepressant Responses to MAO-B Selective Inhibitors

Early uncontrolled studies with the MAO-B inhibitor, deprenyl, reported beneficial effects in anergic, negativistic patients, several of whom had not responded to imipramine, electroconvulsive therapy, and other treatments.[70] Deprenyl dosages in these early studies were in the range now known to be non-MAO-B selective. The first double blind, random assignment study of deprenyl in mild to moderately depressed outpatients using final doses of 20 mg/day showed greater effects of deprenyl compared to placebo over the 4-week treatment period.[71] A second, more recent controlled study of a small group of predominantly unipolar inpatients given 15 mg of deprenyl per day for 40 days revealed significantly greater Hamilton depression scale reductions than in the placebo-treated patients.[72] Prior treatment or possible treatment-resistant status was not discussed in these reports. One other open study with deprenyl (15 to 20 mg/day) in patient groups with endogenous depression suggested treatment efficacy,[73] but a more recent study by the same investigators indicated that primarily bipolar patients with endogenous-type symptoms and not endogenous unipolar patients responded.[74] In fact, the highest response rate in this open study was in nonendogenous patients, who were described as nonresponsive to tricyclic antidepressants and to other experimental drugs under evaluation in the same clinic, which argues against a placebo effect in these patients. Mann et al.[74] suggested that their nonendogenous depressed groups might represent a subpopulation of patients with phenylethylamine or dopamine deficiency states who would benefit from MAO-B inhibition, in contrast to more typical unipolar patients with possible norepinephrine or serotonin abnormalities who might preferentially benefit from MAO-A inhibitors, in keeping with the hypothesis of a biochemical heterogeneity of affective disorders.[74]

Our results with deprenyl in a small number of mixed depressive subjects, most with severe unipolar depression and histories of nonresponse to standard treatments, have thus far revealed no evidence for therapeutic efficacy in MAO-B selective doses of 10 mg/day. Small numbers of patients receiving 30 mg/day showed negligible changes, but several patients receiving larger doses (60 mg/day) have shown partial antidepressant responses. This large deprenyl dose reduced MHPG excretion to nearly as great an extent as clorgyline or tranylcypromine, and hence would appear to be acting also on MAO-A. These data, although preliminary, would seem to be compatible with our results from the clorgyline studies which indicate that some severe, treatment-resistant depressed patients may benefit from drugs which inhibit MAO-A. MAO-B inhibitors, in contrast, need to be studied further to evaluate whether possible beneficial effects might be found in nonendogenous patients.[56]

Low doses of deprenyl appear to have lesser tyramine-potentiation effects than found with the usual doses of conventional MAO-inhibitors or clorgyline,[71,75] and would have some advantage in this regard if antidepressant efficacy can be demonstrated in a patient subgroup. Like clorgyline, pargyline, and conventional MAO-inhibitors, no direct anticholinergic effects occur, and many troublesome side effects are avoided. Deprenyl may also possess fewer side effects of other types, including less orthostatic hypotension, and so may be of use in older patient populations, again if efficacy can be demonstrated. Deprenyl has been used safely in potentiating some effects of L-dopa therapy in Parkinson's disease,[52] and Knoll has suggested that its use may theoretically be of greatest value in older patients not only in regard to depression and Parkinsonism but also other age-related behavioral changes.[76] His hypothesis is predicated upon reports of MAO-B increasing with age and in certain types of age-related neuropathology — changes which may impair the modulation of dopamine and trace amine systems in the brain.

IX. ADAPTATIONAL CONSEQUENCES OF CHRONIC TREATMENT WITH MAO INHIBITORS

As discussed above, the most prominent acute consequence of MAO inhibition is a rapid increase in the intracellular concentrations of monoamines. Brain serotonin concentrations are raised to a greater extent than are those of two catecholamines, norepinephrine and dopamine, following acute, high-dose MAO inhibitor administration. After amine concentrations have increased, secondary adaptive consequences of MAO inhibition begin to occur. These include a reduction in amine synthesis via an apparent feedback mechanism which has been most clearly demonstrated within the noradrenergic system.[47] Within a period of several hours after drug administration a slowing of neuronal firing rates develops in both serotonin-containing neurons in the median raphe and in norephinephrine containing neurons in the locus ceruleus.[77,78] Both effects are thought to represent a direct consequence of the increased amines available to presynaptic feedback autoreceptors.

After several weeks of treatment with MAO inhibitors, reductions in β-adrenoreceptor numbers and in β-receptor functional activity — as measured by norepinephrine stimulated cyclic-AMP formation — as well as in α_2-adrenergic binding sites and serotonin binding sites, but not those for dopamine, have been observed in a number of studies.[79] These changes, for the most part, resemble those which follow tricyclic antidepressant treatment in animals, and which have been hypothesized to be importantly involved in the therapeutic actions of antidepressant drugs. Receptor binding site changes follow treatment with both nonselective MAO inhibitors (e.g., nialamide, tranylcypromine, and phenelzine) and the selective MAO-A inhibitor clorgyline at low doses, but generally occur only at high, probably nonselective doses of the partially selective MAO-B inhibitor, pargyline,[42,79] Conflicting reports exist regarding possible reductions in β-adrenoceptors and cyclic AMP responses following low doses of the MAO-B selective inhibitor deprenyl (see References 42, 55, and 80).

The neurochemical alterations induced by MAO-A inhibition are accompanied by a reduction in the sensitivity of cortical neurons to iontophoretically applied norepinephrine and serotonin, presumably reflecting the changes in amine receptor numbers and function reviewed above. The physiologic consequences of these adaptive neurochemical changes have not yet been as widely studied. In studies from our group, they did appear to be reflected in behavioral changes in animals; chronic, but not acute, clorgyline treatment led to an attenuation of the changes in locomotor activity and brain self-stimulation behavior produced by the α_2-adrenergic agonist clonidine, and by the serotonergic agonist metachlorophenylpiperazine.[54] The effects of clonidine, under these circumstances, are believed to be mediated through an α_2-adrenoreceptor presynaptic inhibitor system. These latter results in rodents are in agreement with findings from a recent clinical study indicating that the hypotensive responses of clonidine, which are also believed to involve α_2-adrenoreceptors, are significantly reduced after treatment of psychiatric patients with clorgyline for 21 days.[79]

It is extremely difficult to attribute biological or behavioral changes during MAO inhibition to a particular change in a specific monoamine neurotransmitter system without very extensive studies of the series of changes outlined above in several monoamine systems. In addition, treatment with MAO inhibitors in combination with selective antagonists for each neurotransmitter system is usually necessary to examine whether blockade of a specific metabolic change also antagonizes the physiologic or behavioral change under investigation. Because of the complex studies required, it is not yet clear how the fairly well-defined neuronal events which follow MAO inhibition can be related to the integrated physiologic and behavioral responses of the whole organism to MAO inhibition as observed clinically, but some strategies such as those used in defining the behavioral, cardiovascular, and neuroendocrine actions of selective MAO inhibitors like clorgyline may prove of value in further delineation of the consequences of MAO inhibition by other drugs, and in gaining further understanding of the mode of therapeutic activity in psychiatric disorders.

REFERENCES

1. **Youdim, M. B. H. and Finberg, J. P. M.,** *Monoamine Oxidase Inhibitor Antidepressants in Psychopharmacology I,* Grahame-Smith, D. G., Hippius, H., and Winoker, G., Eds., Excerpta Medica, Amsterdam, 1982, 38.
2. **Murphy, D. L., Garrick, N. A., and Cohen, R. M.,** Monoamine oxidase. Biochemical and physiological aspects relevant to human psychopharmacology, in *Antidepressants,* Burrow, N., Norman, N., and Davies, D., Eds., Elsevier, Amsterdam, 1982, 209.
3. **Finberg, J. P. M. and Youdim, M. B. H.,** Monoamine oxidases, in *Handbook of Neurochemistry,* Lajath, A., Ed., Plenum Press, New York, 1983, 293.
4. **Youdim, M. B. H. and Finberg, J. P. M.,** Monoamine oxidase inhibitor antidepressants, in *Psychopharmacology,* Grahame-Smith, D. G., Ed., Excerpta Medica, Amsterdam, 1985, in press.
5. **Youdim, M. B. H. and Collins, G. G. S.,** Properties and physiological significance of multiple forms of monoamine oxidase, in *Isozymes,* Markert, C. O., Ed., Academic Press, New York, 1975, 619.
6. **Johnston, J. P.,** Some observations upon a new inhibitor of monoamine oxidase in brain tissue, *Biochem. Pharmacol.,* 17, 1285, 1968.
7. **Knoll, J. and Magyor, K.,** Some puzzling pharmacological effects of monoamine oxidase inhibitors, *Adv. Biochem. Psychopharmacol.,* 5, 393, 1972.
8. **Tipton, K. T., Mantle, T. J., and Houslay, M. D.,** Some problems concerning the apparent multiplicity of monoamine oxidase, *Psychopharmacol. Bull.,* 13, 46, 1977.
9. **Neff, N. H. and Fuentes, J. A.,** The use of selective monoamine oxidase inhibitor drugs for evaluating pharmacological and physiological mechanisms, in *Monoamine Oxidase and Its Inhibition,* Ciba Foundation Symposium No. 39, Elsevier, Amsterdam, 1976, 163.
10. **Green, A. R. and Youdim, M. B. H.,** Effect of monoamine oxidase inhibition by clorgyline deprenil on tranylcypromine on 5-HT concentration in rat brain and hyperactivity following subsequent tryptophan administration, *Br. J. Pharmacol.,* 55, 415, 1975.
11. **Green, A. R., Mitchell, B., Tordoff, A., and Youdim, M. B. H.,** Evidence of dopamine deamination by both type A and type B monoamine oxidase in rat brain *in vivo* and for degree of inhibition of enzyme necessary for increased functional activity of dopamine and 5-hydroxytryptamine, *Br. J. Pharmacol.,* 60, 343, 1977.
12. **Youdim, M. B. H.,** In vivo noradrenaline is a substrate for monoamine oxidase A and B, *Br. J. Pharmacol.,* 79, 477, 1983.
13. **Ashkenazi, R., Finberg, J. P. M., and Youdim, M. B. H.,** Behavioral hyperactivity in rats treated with selective monoamine oxidase inhibitors and a selective 5-hydroxytryptamine uptake blocker, *Br. J. Pharmacol.,* 79, 765, 1983.
14. **Garrick, N. and Murphy, D. L.,** Monoamine oxidase type A: differences in selectivity towards L-norepinephrine compared to serotonin, *Biochem. Pharmacol.,* 31, 4061, 1982.
15. **Fowler, C. J. and Tipton, K. F.,** Deamination of 5-hydroxytryptamine by both forms of monoamine oxidase in the rat brain, *J. Neurochem.,* 38, 733, 1982.
16. **O'Carroll, A. M., Fowler, C. J., Phillips, J. P., Tobbia, I., and Tipton, K. F.,** The deamination of dopamine by human brain monoamine oxidase. Specificity of the two forms in seven brain regions, *Arch. Pharmacol.,* 322, 198, 1983.
17. **Jarrott, B. and Iversen, L. L.,** Noradrenaline metabolizing enzymes in normal and sympathetically denervated vas deferens, *J. Neurochem.,* 18, 1, 1971.
18. **Goridis, C. and Neff, N. H.,** Monoamine oxidase in sympathetic nerve: a transmitter specific enzyme type, *Br. J. Pharmacol.,* 43, 814, 1971.
19. **Youdim, M. B. H., Barierjee, D. K., and Pollard, H. B.,** Isolated chromaffin cells from adrenal medulla contain primarily monoamine oxidase B, *Science,* 224, 619, 1984.
20. **Murphy, D. L. and Donnelly, C. D.,** Monoamine oxidase in nani enzyme characteristics in platelets, plasma, and other human tissues, *Adv. Biochem. Psychopharmacol.,* 12, 71, 1974.
21. **Kishimoto, S., Kimura, H., and Maeda, T.,** Histochemical demonstration of monoamine oxidases by a new peroxidation method, *Cell. Mol. Biol.,* 29, 61, 1983.
22. **Levitt, P., Pinter, J. E., and Breakefield, X. O.,** Immunocytochemical demonstration of monoamine oxidase B in brain astrocytes and serotonergic neurons, *Proc. Natl. Acad. Sci. U.S.A.,* 79, 6385, 1982.

23. **Fowler, C. J. and Strolin Benedetti, M.,** The metabolism of dopamine by both forms of monoamine oxidase in the rat brain and its inhibition by Cimoxatone, *J. Neurochem.*, 40, 1534, 1983.
24. **Shields, P. J. and Eccleston, D.,** Evidence for the synthesis and storage of 5-hydroxytryptamine in two separate pools in the brain, *J. Neurochem.*, 20, 881, 1973.
25. **Lane, J. D. and Aprison, M. H.,** The flux of radioactive label through components of the serotonergic system following the injection of [^{3}H]tryptophan: product-precursor anomalies providing evidence that serotonin exists in multiple pools, *J. Neurochem.*, 30, 671, 1978.
26. **Tracqui, P., Morot-Gaudry, Y., Staub, J. F., Brezillon, P., Perault-Staub, A. M., Bourgoin, S., and Hamon, M.,** Model of brain serotonin metabolism. II. Physiological interpretation, *Am. J. Physiol.*, 244, R206, 1983.
27. **Elks, M. L., Youngblood, W. W., and Kizer, J. S.,** Serotonin synthesis and release in brain slices: independence of tryptophan, *Brain Res.*, 172, 471, 1979.
28. **Shore, P. A. and Giachetti, A.,** Reserpine: basic and clinical pharmacology, in *Handbook of Psychopharmacology*, Iversen, L. L., Iversen, S. D., and Snyder, S. H., Eds., Plenum Press, New York, 1978, 197.
29. **Morot-Gaudry, Y., Bourgoin, S., and Hamon, M.,** Kinetic characteristics of newly synthesized ^{3}H-5-HT in the brain of control and reserpinized mice. Evidence for the heterogenous distribution of 5-HT in serotonergic neurons, *Naunyn-Schmied. Arch. Pharmacol.*, 316, 311, 1981.
30. **Long, J. B., Youngblood, W. Y., and Kizer, J. S.,** Regional differences in the response of serotonergic neurons in the rat CNS to drugs, *Eur. J. Pharmacol.*, 88, 89, 1983.
31. **Jequier, E., Lovenberg, W., and Sjoerdsma, A.,** Tryptophan hydroxylase inhibition: the mechanism by which *p*-chlorophenylalanine depletes rat brain serotonin, *Mol. Pharmacol.*, 3, 274, 1967.
32. **van Gelder, N. M.,** Principles of compartmentation, in *Handbook of Neurochemistry*, Lajtha, A., Ed., Plenum Press, New York, 1982, 183.
33. **Tamir, H.,** Serotonin binding proteins, *J. Histochem. Cytochem.*, 30, 837, 1982.
34. **Halaris, A. E. and Freedman, D. X.,** Vesicular and juxtavesicular serotonin: effects of lysergic acid diethylamide and reserpine, *J. Pharmacol. Exp. Ther.*, 203, 575, 1977.
35. **Small, D. H. and Wurtman, R. J.,** Serotonin binds specifically and saturably to an actin-like protein isolated from rat brain synaptosomes, *Proc. Natl. Acad. Sci. U.S.A.*, 81, 959, 1984.
36. **Sanders-Bush, E.,** Regulation of serotonin storage and release, in *Serotonin in Biological Psychiatry*, Ho, B. T., Schoolar, J. C., and Usdin, E., Eds., Raven Press, New York, 1982, 17.
37. **Green, A. R. and Grahame-Smith, D. G.,** The effect of drugs on the precesses regulating the functional activity of brain 5-hydroxytryptamine, *Nature (London)*, 260, 487, 1976.
38. **Jacobs, B. L.,** An animal behavioral model for studying central serotonergic synapses, *Life Sci.*, 19, 777, 1976.
39. **Bennett, J. P. and Snyder, S. H.,** Serotonin and lysergic acid diethylamide binding in rat brain membranes: relationship to post-sypnaptic serotonin receptors, *Mol. Pharmacol.*, 12, 373, 1976.
40. **Wolf, W. A. and Kuhn, D. M.,** Simultaneous determination of 5-hydroxytryptamine, its amino acid precursors and acid metabolite in discrete brain regions by high performance liquid chromatography with fluorescence detection, *J. Chromatogr.*, 275, 1, 1983.
41. **Grahame-Smith, D. G.,** Studies in vivo on the relationship between brain tryptophan, brain 5-HT synthesis and hyperactivity in rats treated with a monoamine oxidase inhibitor and L-tryptophan, *J. Neurochem.*, 18, 1053, 1971.
42. **Youdim, M. A. H., Fineberg, J. P. M., Kuhn, D. M., and Wolf, W. A.,** The role of monoamine oxidase A in the metabolism and function of noradrenaline and serotonin, in *IUPHAR IX International Congress of Pharmacology Proceedings*, Vol. 2, Paton, W., Mitchell, J., and Turner, P., Eds., Macmillan Press, London, 1984, 203.
43. **Levitt, P., Pintar, J. E., and Breakefield, X. O.,** Immuocytochemical demonstration of monoamine oxidase B in brain astrocytes and serotonergic neurons, *Proc. Natl. Acad. Sci. U.S.A.*, 79, 6385, 1985.
44. **Reinhard, J. F. and Wurtman, R. J.,** Relation between brain 5-HIAA levels and the release of serotonin into brain synapses, *Life Sci.*, 21, 19741, 1977.
45. **Ashkenazi, R., Finberg, J. P. M., and Youdim, M. B. H.,** Behavioral hyperactivity in rats treated with selective monoamine oxidase inhibitors and LM5008, a selective 5-hydroxytryptamine uptake blocker, *Br. J. Pharmacol.*, 79, 765, 1983.
46. **Theoharides, T. C., Bondy, P. K., Tsakalos, N. D., and Askenase, P. W.,** Differential release of serotonin and histamine from mast cells, *Nature (London)*, 297, 229, 1982.
47. **Neff, N. H. and Costa, E.,** The influence of monoamine oxidase inhibition on catecholamine synthesis, *Life Sci.*, 5, 951, 1966.

48. **Murphy, D. C., Pickar, D., Jimerson, D., Cohen, R. M., Garrick, N. A., Karoum, F., and Wyatt, R. J.,** Biochemical indices of the effects of selective MAO inhibitors (clorgyline, pargyline and deprenyl) in man, in *Clinical Pharmacology in Psychiatry,* Usdin, E., Dahl, S., Gram, L. F., and Lingjaerde, O., Eds., Macmillan Press, London, 1981, 307.
49. **White, K. and Simpson, G.,** Combined MAOI-tricyclic antidepressant treatment: a reevaluation, *J. Clin. Psychopharmacol.,* 11, 264, 1981.
50. **Squires, R. F.,** Monoamine oxidase inhibitors: animal pharmacology, in *Handbook of Psychopharmacology, Affective Disorders: Drug Actions in Animals and Man,* Vol. 14, Iversen, L. L., Iversen, S. D., and Snyder, S. H., Eds., Plenum Press, New York, 1978.
51. **Murphy, D. L. and Kalin, N. H.,** Biological and behavioral consequences of alterations in monoamine oxidase activity, *Schizophr. Bull.,* 6, 355, 1980.
52. **Lees, A., Kohout, L. J., Shaw, K. M., Stern, G. M., Elsworth, J. D., Sandler, M., and Youdim, M. B. H.,** Deprenyl in Parkinson's disease, *Lancet,* 1, 791, 1977.
53. **Sulser, F., Vetulani, J., and Mobley, P.,** Mode of action of antidepressant drugs, *Biochem. Pharmacol.,* 27, 257, 1978.
54. **Murphy, D. L., Garrick, N. A., Aulakh, C. S., and Cohen, R. M.,** . New contributions from basic science to understanding the effects of monoamine oxidase inhibiting antidepressants, *J. Clin. Psychiatry,* in press.
55. **Mishra, R., Gillespie, D. D., Youdim, M. B. H., and Sulser, F.,** Effect of selective monoamine oxidase inhibition by clorgyline and deprenyl on the norepinephrine receptor-coupled adenylate cyclase system in rat cortex, *Psychopharmacology,* 81, 220, 1983.
56. **Quitkin, F., Rifkin, A., and Klein, D. F.,** Monoamine oxidase inhibitors. A review of antidepressant effectiveness, *Arch. Gen. Psychiatry.,* 36, 749, 1979.
57. **Murphy, D. L., Garrick, N. A., and Cohen, R. M.,** Monoamine oxidase inhibitors and monoamine oxidase: biochemical and physiological aspects relevant to human psychopharmacology, in *Drugs in Psychiatry,* Vol. 1, *Antidepressants,* Post, R. M. and Ballenger, J. C., Eds., Elsevier/North-Holland, Amsterdam, 1983, 209.
58. **Ask, A-L., Fagervall, I., and Ross, S. B.,** Selective inhibition of monoamine oxidase in monoaminergic neurons in the rat brain, *Naunyn-Schmiedeberg's Arch. Pharmacol.,* 324, 79, 1983.
59. **Beckmann, H. and Riederer, P., Eds.,** *Monoamine Oxidase and Its Selective Inhibitors: New Concepts in Therapy and Research,* Karger, Basel, 1983.
60. **Himmelhoch, J. M., Fuchs, C. Z., and Symons, B. J.,** A double-blind study of tranyl-cypromine treatment of major anergic depression, *J. Nerv. Ment. Dis.,* 170, 628, 1982.
61. **Paykel, E. S., Rowan, P. R., Parker, R. R., and Bhat, A. V.,** Response to phenelzine and amitriptyline in subtypes of outpatient depression, *Arch. Gen. Psychiatr.,* 39, 1041, 1982.
62. **Nies, A.,** Clinical applications of MAOI's, in *Drugs in Psychiatry,* Vol. 1, Burrows, G. D., Norman, T. R., and Davies, B., Eds., Elsevier, Amsterdam, 1983, 229.
63. **Herd, J. A.,** A new antidepressant — M and B 9302. A pilot study and double-blind controlled trial, *Clin. Trials,* 6, 219, 1969.
64. **Wheatly, D.,** Comparative trial of a new monoamine oxidase inhibitor in depression, *Br. J. Psychiatr.,* 117, 1970.
65. **Lipper, S., Murphy, D. L., Slater, S., and Buchsbaum, M. S.,** Comparative behavioral effects of clorgyline and pargyline in man: a preliminary evaluation, *Psychopharmacology,* 62, 123, 1979.
66. **Murphy, D. L., Lipper, S., Pickar, D., Jimerson, D., Cohen, R. M., Garrick, N. A., Alterman, I. S., and Campbell, I. C.,** Selective inhibition of monoamine oxidase type A: clinical antidepressant effects and metabolic changes in man, in *Monoamine Oxidase Inhibitors. The State of the Art,* Youdim, M. B. H. and Paykel, E. S., Eds., John Wiley & Sons, New York, 1981, 189.
67. **Potter, W. Z., Murphy, D. L., Wehr, T. A., Linnoila, M., and Goodwin, F. K.,** Clorgyline: a new treatment for refractory rapid cycling patients, *Arch. Gen Psychiatr.,* 39, 505, 1982.
68. **Pickar, D., Murphy, D. L., Cohen, R. M., Campbell, I. C., and Lipper, S.,** Selective and nonselective monoamine oxidase inhibitors: behavioral disturbances during their administration to depressed patients, *Arch. Gen. Psychiatr.,* 39, 535, 1982.
69. **Roy, B. F., Murphy, D. L., Lipper, S., Siever, L., Alterman, I. S., Jimerson, D., Lake, C. R., and Cohen, R. M.,** Cardiovascular effects of the selective monoamine oxidase inhibiting antidepressant, clorgyline: correlations with clinical responses and changes in catecholamine metabolism, *J. Clin. Psychopharmacol.,* in press.
70. **Varga, E. and Tringer, L.,** Clinical trial of a new type promptly acting psychoenergic agent (phenyl-isopropyl-methylpropinyl-HCl, "E-250"), *Acta Med. Acad. Sci. Hung.,* 23, 289, 1967.
71. **Mendis, N., Pare, C. M., Sandler, M., Glover, V., and Stern, G. M.,** Is the failure of (−)deprenyl, a selective monoamine oxidase B inhibitor, to alleviate depression related to freedom from the cheese effect, *Psychopharmacology,* 73, 87, 1981.

72. **Mendlewicz, J. and Youdim, M. B. H.,** L-Deprenil, a selective monoamine oxidase type B inhibitor, in the treatment of depression: a double-blind evaluation, *Br. J. Psychiatr.*, 142, 508, 1983.
73. **Mann, J. J. and Gershon, S.,** L-Deprenyl, a selective monoamine oxidase type-B inhibitor in endogenous depression, *Life Sci.*, 26, 877, 1980.
74. **Mann, J. J., Frances, A., Kaplan, R. D., Kocsis, J., Peselow, E. D., and Gerhson, S.,** The relative efficacy of l-deprenyl, a selective monoamine oxidase type B inhibitor, in endogenous and nonendogenous depression, *J. Clin. Psychopharmacol.*, 2, 54, 1982.
75. **Pickar, D., Cohen, R. M., Jimerson, D. C., and Murphy, D. L.,** Tyramine infusions and selective monoamine oxidase inhibitor treatment. I. Changes in pressor sensitivity, *Psychopharmacology*, 74, 4, 1981.
76. **Knoll, J.,** Selective inhibition of B type monoamine oxidase in the brain: a drug strategy to improve the quality of life in senescence, in *Strategy in Drug Research*, Keverling-Buisman, J. A., Ed., Elsevier, Amsterdam, 1982.
77. **Aghajanian, G. K.,** Influence of drugs on the firing of serotonin-containing neurons in brain, *Fed. Proc. Am. Soc. Exp. Biol.*, 31, 91, 1972.
78. **Campbell, I. C., Murphy, D. L., Gallager, D. W. et al.,** Neurotransmitter-related adaptation in the central nervous system following chronic monoamine oxidase inhibition, in *Monoamine Oxidase: Structure, Function, and Altered Functions*, Academic Press, New York, 1979.
79. **Siever, L. J., Ulhde, T. W., and Murphy, D. L.,** Possible subsentization of alpha$_2$-adrenergic receptors by chronic monoamine oxidase inhibitor treatment in psychiatric patients, *Psychiatr. Res.*, 6, 171, 1982.
80. **Zsilla, G., Barbaccia, M. L., Gandolfi, O., Knoll, J. and Costa, E.,** (−)-Deprenyl, a selective "B" inhibitor; increased (^{3}H)-imprimine binding and decreases β-adrenergic receptor function, *Eur. J. Pharmacol.*, 89, 111, 1983.
81. **Kuhn, D. M., Wolf, W. A., and Youdim, M. B. H.,** 5-Hydroxytryptamine release *in vivo* from a cytoplasmic pool: studies on the 5-HT behavioral syndrome in reserpinized rats, *Br. J. Pharmacol.*, 84, 121, 1985.
82. **Wolf, W. A., Youdim, M. B. H., and Kuhn, D. M.,** Does brain 5-HIAA indicate serotonin release or monoamine oxidase activity, *Eur. J. Pharmacol.*, 109, 381, 1985.

Chapter 23

HUMAN PLATELET MONOAMINE OXIDASE

Christopher J. Fowler and J. Sääf

TABLE OF CONTENTS

I. INTRODUCTION

One of the major difficulties associated with the measurement of biochemical parameters that may be associated with certain psychiatric and pathological disease states in man is that these measurements have to be by indirect means. Most studies involve the use of either post-mortem human brain material or alternatively involve analysis of body fluids taken from the patient. The first approach has the advantage that the measurement is direct, but any conclusions made from such studies are always tempered by a large number of parameters (e.g., cause of death, death-autopsy delay, drug treatment, etc.) that must be taken into account. In addition, biochemical changes that can be detected in this sort of study may reflect changes found when the disease has reached an advanced stage, and may not reflect the biochemical state of the brain at the onset of the disease. However, this approach has yielded much useful information as to brain biochemistry changes in disease. As an example, a large number of studies have demonstrated cholinergic deficits in senile dementia of Alzheimer type, a disease for which there is at present no satisfactory animal model.[1]

An alternative approach, which avoids some of the pitfalls associated with autopsy material, has been the analysis of biochemical parameters in body fluids (urine, blood, and cerebrospinal fluid). This approach has the advantage that the biochemical parameters can be studied during different stages of the disease process, but suffers from the disadvantage that it is at best an indirect measure of central function. However, such studies have proven useful in the elucidation of biochemical changes associated with disease processes. For example, it has been demonstrated that patients attempting suicide by active means (e.g., hanging, shooting, etc.) have lower levels of the 5-hydroxytryptamine metabolite 5-hydroxyindoleacetic acid than patients using passive methods (e.g., drug overdose).[2]

The human platelet has been the subject of much study with respect to psychiatric illnesses. This is partly due to the ease of collection of blood platelets, and partly due to the observation that human blood platelets and central serotoninergic systems have many processes (e.g., active 5-hydroxytryptamine uptake) in common.[3,4] It has even been speculated that the platelet and neuronal tissue may have a common embryological origin.[5] Such a speculation would predict that genetically determined deficiencies in the central serotoninergic system might be reflected in the biochemical properties of the human platelet. Although such a suggestion is very controversial, and there is as yet no unequivocal evidence in support of such a contention (for review with respect to platelet monoamine oxidase, see Reference 6), it would afford an explanation for the altered platelet 5-HT uptake properties in, for example, migraine[7] and Down's Syndrome.[8]

Another biochemical parameter studied in the human platelet is the enzyme monoamine oxidase (MAO, EC 1.4.3.4). In the present article, the biochemical properties of this enzyme are briefly reviewed, together with a discussion of the studies into the activity of platelet MAO in psychiatric illnesses.

II. BIOCHEMISTRY OF HUMAN PLATELET MAO

A. Comparison with Plasma Amine Oxidase

Amine oxidizing activity is found in both human platelet and human plasma.[9] However, these activities appear to be rather different in nature. The platelet enzyme metabolizes a variety of substrates, such as tyramine, 2-phenylethylamine, dopamine, and benzylamine, and is inhibited in an irreversible manner by the acetylenic MAO inhibitors clorgyline, deprenyl, pargyline, and J-508 (also known as AGN 1133, Su- 11,739). The platelet MAO activity is not, however, inhibited by the carbonyl reagent semicarbazide. The plasma amine oxidase activity (SAO), on the other hand, has a much more limited substrate specificity, and is inhibited by semicarbazide.[9] This difference has allowed for the determination of

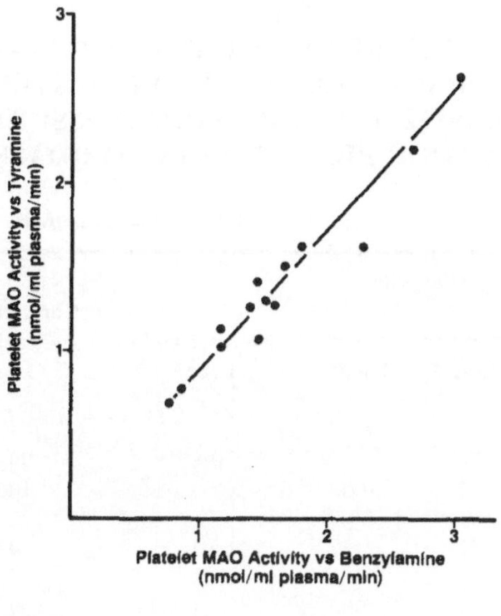

FIGURE 1. Correlation between the MAO activity in the platelet-rich plasma in 14 female nonsmokers, assayed with either tyramine (ordinate) or benzylamine in the presence of semicarbazide (in order to inhibit the SAO activity) (abscissa). The regression line of best fit (correlation coefficient $r = 0.95$) is shown. Each point represents the means of duplicate determinations of enzyme activity at a substrate concentration of 50 μM, assayed at pH 7.8.[12]

platelet MAO activity in samples of platelet-rich plasma: with substrates like tyramine, dopamine, tryptamine, and 2-phenylethylamine, the platelet MAO alone is measured. Benzylamine is deaminated by both the MAO and the SAO, but only the MAO activity is measured if semicarbazide is added to the assay tube. With such an approach, it has been demonstrated that there is a very good intersubstrate correlation for platelet MAO activity[10-12] (see Figure 1).

It has been suggested that the SAO resembles the semicarbazide-sensitive, clorgyline-resistant amine oxidase (SSCRAO) found in a variety of rat tissues.[9] However, these two enzymes are qualitatively different: SSCRAO is not inhibited at all by clorgyline, even at concentrations as high as 1 m*M*,[13] whereas the SAO from ox plasma is inhibited in a reversible manner by this compound.[14] On the other hand, SSCRAO is inhibited by FLA 1042 (2-amino-1-(4-bromo-2-methyl-phenoxy)propane hydrochloride) in a mixed fashion $K_i^{slope} = 12\ \mu M$, $K_i^{int} = 6\ \mu M$) whereas human SAO is only inhibited to a small extent (20%) by 1 m*M* FLA 1042[15] (see Table 1).

B. Substrate Specificity of Human Platelet MAO

MAO Activity has been demonstrated to exist as two catalytically distinguishable forms, which are generally termed MAO-A and MAO-B. The MAO-A form of the enzyme is inhibited by very low (around nanomolar) concentrations of clorgyline, whereas the MAO-B form of the enzyme is not inhibited until micromolar concentrations of this inhibitor are used.[16] The reverse is true when the enzyme forms are inhibited by 1-deprenyl[17] (see Chapters 20 and 21). By use of these inhibitors, Donnelly and Murphy[18] have demonstrated that only MAO-B is present in human platelets. The substrate specificity of human platelet MAO-B resembles closely that of human brain MAO-B, but differs considerably from that of human brain MAO-A (Figure 2). Tyramine, 2-phenylethylamine, and benzylamine are readily metabolized by human brain and platelet MAO-B whereas 5-hydroxytryptamine is very poorly

Table 1
INHIBITORY CHARACTERISTICS OF RAT BRAIN MAO-A AND MAO-B, HUMAN PLATELET MAO-B, RAT HEART CLORGYLINE-RESISTANT, SEMICARBAZIDE-SENSITIVE AMINE OXIDASE (SSCRAO) AND HUMAN PLATELET-POOR PLASMA AMINE OXIDASE (SAO)

	% Inhibition of activity by:					
	Clorgyline		Semicarbazide	Clorgyline + semicarbazide	FLA 1042	
	30 n*M*	1 m*M*	1 m*M*	(1 m*M*	10 μ*M*	100 μ*M*
Rat brain						
100 μ*M* 5-HT (MAO-A)	96	98	0.8	99.7	76	96
100 μ*M* Bz (MAO-B)	8	100	7	100	4	28
Rat heart						
1 μ*M* Bz (SSCRAO)	2	2	95	99.6	64	90
Human platelet						
100 μ*M* Bz (MAO-B)	3	100	11	99.9	12	46
Human plasma						
1 μ*M* Bz (SAO)	19	39	100	100	0	2

Note: Samples were preincubated for 30 min at 37° with the inhibitors before assay for activity. Values given are means (n = 3) of the percent inhibition of enzyme activity with respect to samples preincubated for 30 min with distilled water. Key: 5-HT, 5-hydroxytryptamine; Bz, benzylamine.[15]

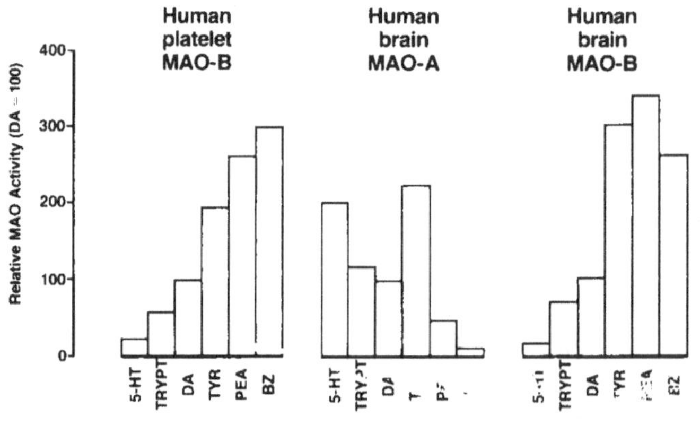

FIGURE 2. Substrate specificity of human platelet and human brain MAO. The values in all cases are expressed as the relative activity with respect to that of dopamine (set to 100). Platelet MAO values represent V_{max} values (with the exception of 5-HT, which was taken from Reference 18 and related to the activities towards the other substrates given in that study). The relative components of MAO-A and MAO-B for human brain were defined as the activities remaining (towards 50 μM substrate) after preincubation for 180 min with either 20 n*M* l-deprenyl or 6 n*M* clorgyline, respectively (for a discussion of the validity of this method, see Reference 19). Absolute activities towards dopamine, in nmol/mg protein/min were: human platelet, 1.06; human brain MAO-A, 0.33; human brain MAO-B, 0.45.[12,18,20,21]

metabolized by MAO-B from these tissues. It has been demonstrated that the ratio of activities of MAO-A:MAO-B are higher intraneuronally than extraneuronally.[22] Thus, although human platelets have a great many features in common with human serotoninergic neurones,[3-5] the MAO in the platelets appears to be qualitatively different from that found in nerve endings.

C. Subcellular Localization of MAO in Platelets

In most tissues, MAO appears to be localized to the outer membrane of the mitochondria.[23] This localization is rather different from that of SAO (soluble) and of rat SSCRAO, which appears to be localized to the plasma membrane.[24] A mitochondrial localization has been demonstrated for rabbit platelet MAO,[25] and, although no detailed subcellular distribution studies of human platelet MAO have been reported, a mitochondrial localization is not inconsistent with the finding that there is a threefold higher specific activity of human platelet MAO when a mitochondrial fraction is compared with a crude homogenate.[11]

D. Molecular Properties of Human Platelet MAO

Collins and Sandler[26] have reported that the molecular weight of purified human platelet MAO-B is 235,000. More recent studies utilizing SDS-gel electrophoresis of ^{3}H-pargyline-labeled MAO have demonstrated a subunit size of about 65,000 for the human platelet enzyme,[27,28] so that the result of Collins and Sandler[26] may represent the polymeric size of human platelet MAO-B. The enzyme almost certainly contains a flavine prosthetic group, since it binds ^{3}H-pargyline[27,28] and is inhibited by clorgyline, 1-deprenyl, and J-508 (also called AGN 1133 and Su-11,739).[11,18] These compounds act by first forming a reversible complex with the active site of the enzyme followed by covalent reaction with the flavine prosthetic group.[29] In line with studies in a variety of tissues of the rat,[30] human platelet MAO is dependent upon iron for its activity,[31] although it is not as yet clear how this dependence is brought about.

The detailed reaction pathway of human platelet MAO-B has not been studied in detail. In other tissues, however, MAO has been shown to follow a ping-pong or double-displacement reaction of the type shown below:

$$\text{E–FAD} + \text{RCH}_2\text{NH}_2 \rightleftharpoons \text{E–FAD–RCH}_2\text{NH}_2 \rightleftharpoons \text{E–FADH}_2\text{–RCH=NH}$$

$$\rightleftharpoons \text{E–FADH}_2 + \text{RCH=NH}$$

$$\text{E–FADH}_2 + \text{O}_2 \rightleftharpoons \text{E–FADH}_2\text{–O}_2 \qquad \text{E–FAD–H}_2\text{O}_2 \rightleftharpoons \text{E–FAD} + \text{H}_2\text{O}_2$$

$$\text{RCH=NH} + \text{H}_2\text{O} \rightleftharpoons \text{RCHO} + \text{NH}_3$$

Adapted from Reference 32

Recent studies have suggested that this mechanism is followed by most monoamine substrates, although benzylamine appears to follow a more complicated pathway where ternary complexes are involved.[33,34] As a result of the above reaction mechanism an increase in the concentration of the second substrate, oxygen, will produce an uncompetitive increase in the MAO activity towards the amine substrate.[32] Such an uncompetitive increase is indeed found for human platelet MAO.[35,36] Lineweaver-Burk plots of human platelet benzylamine and tryptamine deamination assayed under atmospheres of air and oxygen are shown in Figure 3. From the V_{max} values in air and oxygen, it is possible to estimate the limiting K_m values towards the amine and oxygen substrates, respectively (Table 2). Although these values must be considered as crude estimates, this approach has yielded values for rat liver MAO which are in good agreement with those obtained when more complete kinetic experiments (with 6 amine substrate and 6 oxygen concentrations) are undertaken.[38] The K_o (limiting K_m for oxygen) values for both benzylamine and tryptamine are relatively high

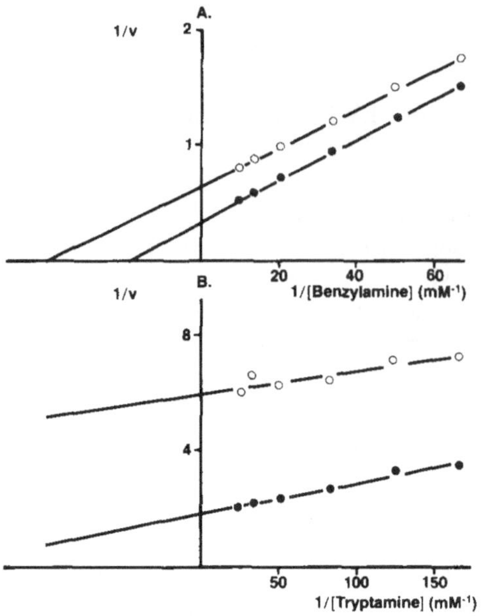

FIGURE 3. Double reciprocal plots of the activity of human platelet MAO towards (A) benzylamine and (B) tryptamine, assayed under atmospheres of air (○) and oxygen (●). Abscissae, 1/(substrate concentrations in m*M*; ordinates, 1/(initial velocity in nmol/mg protein/min). Values are means of determinations in six platelet preparations.[11]

Table 2
KINETIC PROPERTIES OF HUMAN PLATELET MAO

	Substrate used to assay for activity:	
	Benzylamine (n = 9)	Tryptamine (n = 7)
Samples assayed in oxygen		
K_m (μM)	62 ± 30	6.9 ± 2.4
V_{max} (nmol/mg protein/min)	2.9 ± 1.3	0.52 ± 0.13
Samples assayed in air		
K_m (μM)	26 ± 4	2.3 ± 1.5
V_{max} (nmol/mg protein/min)	1.4 ± 0.4	0.17 ± 0.05
Ratio $V_{max}^{oxygen}:V_{max}^{air}$	2.01	3.03
Ratio $K_m^{oxygen}:K_m^{air}$	2.38	3.07
Estimated Michaelis constants		
$K_m^{oxygen}(K_o)$ (μM)	370	1100
K_m^{amine} (μM)	76	14
Absolute V_{max} (nmol/mg protein/min)	3.8	1.1
Ratio K_o:absolute V_{max}	97:1	1050:1

Note: K_m and V_{max} values were calculated from determinations with 6 substrate concentrations in each case (values are means ± S.D.). The K_m^{oxygen} and K_m^{amine} and absolute V_{max} (V_{max} as both oxygen and amine concentrations tend to infinity) were estimated as described elsewhere,[37] assuming oxygen concentrations of 217 μM (air) and 1085 μM (oxygen). Values are calculated from the original data for References 11 and 35.

with respect to values found in other tissues, but this is consistent with the finding of Mitra and Guha[39] that the K_o values for MAO-B are higher than those for MAO-A.

For a simple double-displacement reaction, it can be demonstrated that the K_o value should be proportional to the k_{cat} (maximum molecular turnover number) value of the enzyme (see Reference 38). Such proportionality has been demonstrated for rat liver MAO,[38] although in those experiments the conclusions made were confused to some extent by the postulation that MAO shows half-sites reactivity, a conclusion which need not be made if it is assumed that the two forms of MAO are separate entities, but which interact to some extent with all monoamine substrates.[40] For platelet MAO, however, only the B form of the enzyme is present and thus the situation is simpler. Furthermore, since the concentration of available MAO-B active centers in the platelet is the same when tryptamine is used to assay for activity as when benzylamine is used,[11] a direct proportionality between the K_o value and the absolute V_{max} value found for a given monoamine substrate should be expected for a double-displacement reaction. The data shown in Table 2 would, however, suggest that this is not the case for human platelet MAO, since the ratio of K_o:V_{max} is very much higher when tryptamine is used as the amine substrate than when benzylamine is used. Although more detailed experiments with many more substrates are clearly necessary before this discrepancy can be satisfactorily explained, it is possible that the difference in K_o:V_{max} ratios found in Table 2 is simply because benzylamine follows a different reaction mechanism to that found with other monoamines, as suggested by studies in the ox liver[33] and human brain.[34]

III. ALTERATION OF PLATELET MAO ACTIVITY BY CONSTITUENTS OF PLASMA

In 1979, Yu and Boulton[41] reported the presence in human plasma of a factor that could, in a noncompetitive manner, increase the activity of human platelet MAO. The authors further suggested that this factor may play an important physiological role in the modulation of platelet MAO activity in man.[41] Two more recent studies have confirmed the activating properties of human plasma. At an assay pH of 7.8, human platelet-poor plasma has been shown to activate platelet MAO activity, in a noncompetitive manner, towards dopamine and tyramine, but not towards 2-phenylethylamine.[12] This activator was sensitive to dialysis. It was suggested that the substrate-selective activation could be explained without the need to invoke enzyme heterogeneity as follows: for a substrate where the rate-limiting step is the release of the imine intermediate, a facilitation of this release would result in a noncompetitive activation, whereas if the rate-limiting step was the release of hydrogen peroxide, then facilitation of the imine release would not change the reaction velocity.[12] The activity of MAO-B towards 2-phenylethylamine is higher than for tyramine and dopamine (see Figure 2), and it is thus possible that for 2-phenylethylamine the release of hydrogen peroxide is the rate-limiting step (for further discussion, see Reference 33).

At an assay pH of 7.4, however, a different pattern of activation was found: in this case, 2-phenylethylamine was activated by a decreased K_m of the enzyme for the substrate without any change in the V_{max} value.[42] The activating factor was not sensitive to dialysis, and appeared to have a molecular weight higher than 50,000.[42] Despite this large molecular weight, the activator appeared to be equally effective when MAO was assayed in intact platelets as when sonicated platelets were used.[42]

At first glance, the results from these two studies appear to be irreconcilable. However, the differences in the activating properties appear to be due to the different pH values used in the two studies. This is shown in Figure 4. At pH 7.8, there does not appear to be any activation of 2-phenylethylamine, whereas at pH 7.4 the K_m of the enzyme for the substrate is reduced. At pH values of 7.0 and lower there is a large effect on the V_{max} of the enzyme

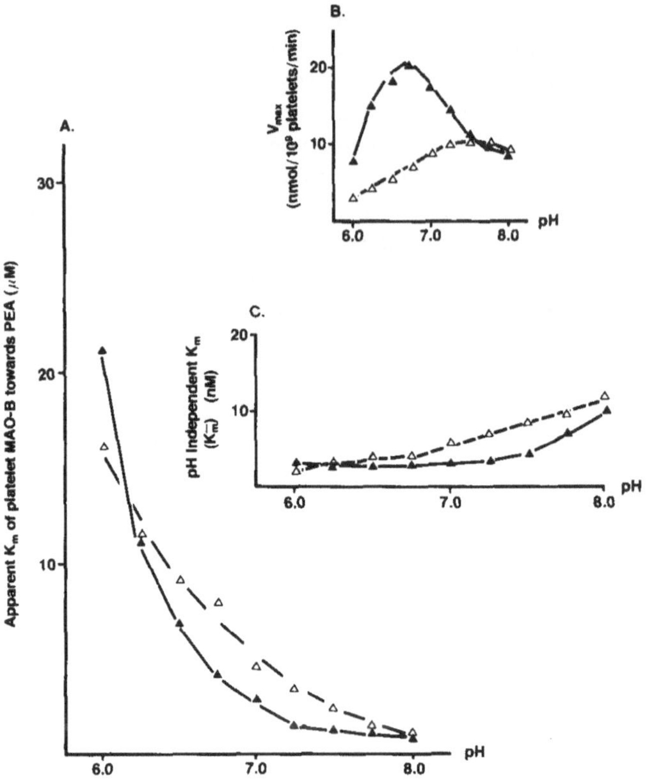

FIGURE 4. Influence of pH on the activity of human platelet MAO towards 2-phenylethylamine assayed in the absence (open triangles) and presence (closed triangles) of human platelet-poor plasma (diluted 1:50 v/v). The pH-independent K_m values ($K_{\bar{m}}$) were calculated from the individual K_m values at each pH by use of a modified Henderson-Hasselbach equation (see References 47 to 49) assuming a pK_a for 2-phenylethylamine of 9.88.[42]

(Figure 4). Thus, it is possible that the two studies measure two separate activating factors: one which is sensitive to dialysis[12] and one which is not.[42] This conclusion is consistent with the work of Demisch et al.,[43] who have described a number of plasma factors of different molecular weight that are capable of influencing platelet MAO activity. The enzyme activity can also be influenced by other factors such as calcium[44] and phosphatidylserine.[45] The physiological significance of these activators and inhibitors is not as yet elucidated, although changes in the activating properties of plasma from migraine patients[42] and from schizophrenic patients undergoing hemodialysis treatment[46] have been reported.

There is a considerable variation in the K_m values towards 2-phenylethylamine when assayed at different pH values (Figure 4), which is consistent with earlier studies in the human platelet with this substrate[36] and with tyramine.[12] However, the variation is rather small when the pH-independent K_m values ($K_{\bar{m}}$) are determined by the use of a modified Henderson-Hasselbach equation (see Figure 4). This is in agreement with earlier studies in the human liver[47,48] and pig brain[49] which demonstrated that MAO preferentially metabolized the nonionized form of the substrate.

IV. VARIATION OF THE ACTIVITY OF HUMAN PLATELET MAO IN HEALTHY INDIVIDUALS AND IN DISEASE

There has been a very large number of studies into the activity of human platelet MAO

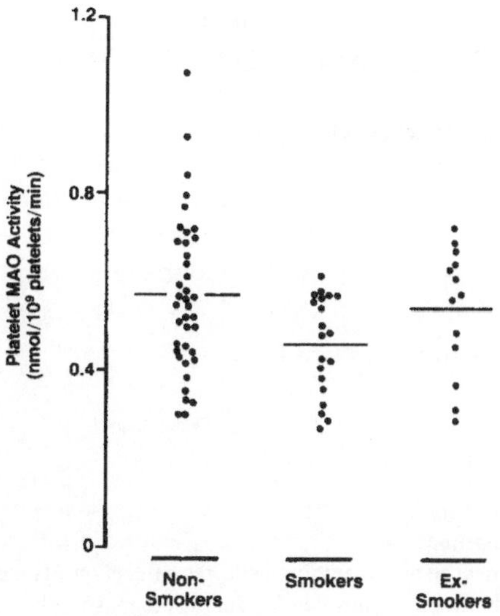

FIGURE 5. Scattergram of platelet MAO activity assayed with 50 μM tryptamine as substrate in platelet-rich plasma from 39 female nonsmokers, 20 female smokers, and 13 female exsmokers. The mean activity for the smokers was significantly lower ($p<0.05$, two-tailed t-test) than for the nonsmokers, whereas the mean for the exsmokers did not differ significantly from that for the nonsmokers.[62]

in different disease processes. Since these studies have been reviewed elsewhere,[6,50,51] they will only briefly be summarized here.

There is a very large variation of platelet MAO activities from individual to individual (see Figures 1 and 5 as examples). Studies with very large sample sizes have demonstrated that this scatter is unimodally distributed, but that women have a significantly higher mean activity than men.[52,53] The variation in activity from individual to individual appears to be due to differences in the concentrations of the available MAO-B active centers rather than due to different molecular properties of the enzyme.[11,28]

In 1976, Buchsbaum et al.[54] found that out of a group of 365 college students, the 19 individuals with the lowest platelet MAO activities had a rather different personality profile than the 17 individuals with the highest MAO activities. Those with the lowest values (''low MAO probands'') were found to have a higher incidence of psychiatric counselling than the high MAO probands.[54] In addition, there was a higher incidence of suicide and suicide attempts[54] and alcoholism[55] in the first-degree relatives of the low MAO probands than in those of the high MAO probands. Buchsbaum et al.[54] suggested that low platelet MAO activity may therefore act as a marker for vulnerability to certain psychiatric disorders. A 2-year follow up study of the college students[56] demonstrated that the low MAO probands reported more job instability and were $^1/_2$ year behind the high MAO probands in their studies. Several independent studies have found correlations between personality variables and platelet MAO activity: there, in particular, appears to be a negative correlation between platelet MAO activity and variables associated with sensation-seeking behaviour.[57,58] However, a recent study[59] was not able to demonstrate significant differences between low and high MAO probands with respect to a large number of personality variables.

One of the drawbacks of the study of platelet MAO activity in psychiatric disorders is that there are a large number of factors that can influence the MAO activity (see References 6 and 51). For example, the low platelet MAO activity thought to be associated with schizophrenic illnesses[60] appears to be an artifact of neuroleptic treatment.[61] Another factor that may influence the results is whether or not the patients were smokers: in two independent

Table 3
PLATELET MAO ACTIVITIES IN DISEASE

Diseases where an increased platelet MAO activity is found
- Nonalcoholic liver disease
- Megaloblastic anaemia
- Huntington's Disease
- Alzheimer's Disease
- Epilepsy
- Anxiety states

Diseases where a decreased platelet MAO activity is found
- Migraine headache
- Insulin-dependent diabetes
- Down's Syndrome
- Alcoholism
- Patients with alcoholic liver cirrhosis
- Patients with iron deficiency anaemia

Diseases where there is disagreement between authors as to the effects on platelet MAO activity
- Schizophrenia — Differences between studies can be due to diagnostic differences (see Reference 64 for discussion), although a considerable proportion of the studies where a decreased activity has been found may be due to neuroleptic effects,[6,61] since these are known to inhibit MAO-B activity in vitro[65]
- Depression — Initially suggested that there is a decreased platelet MAO activity associated with bipolar affective disorder.[66] This has only been confirmed in about half of the subsequent studies.[67] Relevant in this respect is the finding that tricyclic antidepressants do have MAO inhibitory actions both in vitro[68] and in vivo.[69] Diagnosis also important: in a recent article, it was suggested that people attempting suicide by active means have low platelet MAO activities, whereas individuals attempting suicide by passive methods do not.[70]

Note: It should be stressed that the above is not a complete list of the changes reported. Detailed discussions and references are given in References 6, 50, 51, and 67. It should also be stressed that there is a wide scatter of activities in both control groups and in the disease groups (e.g., Figure 5). Thus it is possible to find an alcoholic, for example, who has a platelet MAO activity higher than a nonalcoholic (see the scattergrams given in References 6 and 71 as examples). Thus the changes discussed above refer to groups taken as a whole and cannot be applied to individuals.

studies,[62,63] a decreased platelet MAO activity in female smokers has been observed (see Figure 5), although this is not found for male smokers.[63] Ex-smokers, on the other hand, have normal activities (Figure 5). These findings can be interpreted in two ways: either there is a constituent of cigarette smoke that affects (in a sex-linked manner) platelet MAO activity (so that the activity is lower in smokers than in nonsmokers, but normal in exsmokers), or alternatively that those who smoke have a different personality profile (and hence a lower platelet MAO activity) than those who do not smoke. In this latter case, the normal platelet MAO activity found in the ex-smokers is consistent with the suggestion that the individuals who are able to give up smoking do not have the same sort of personality as those who are not able to give up smoking. This example illustrates rather well the pitfalls associated with the study of platelet MAO and disease. Thus, despite the large documentation of the activities of platelet MAO in different disorders (summarized in Table 3), the notion that platelet MAO activity can serve as a predictor for certain psychiatric disease states must be treated with caution.

REFERENCES

1. **Marchbanks, R. M.,** Biochemistry of Alzheimer's disease, *J. Neurochem.*, 39, 9, 1982.
2. **Åsberg, M., Bertilsson, L., Rydin, E., Schalling, D., Thorén, P., and Träskman-Bendz, L.,** Monoamine metabolites in cerebrospinal fluid in relation to depressive illness, suicidal behaviour and personality, *Adv. Biosci.*, 31, 257, 1981.
3. **Stahl, S. M.,** The human platelet. A diagnostic and research tool for the study of biogenic amines in psychiatric and neurologic disorders, *Arch. Gen. Psychiat.*, 34, 509, 1977.
4. **Pletscher, A.,** Platelets as models for monoaminergic neurones, *Essays Neurochem. Neuropharmacol.*, 3, 49, 1978.
5. **Campbell, I. C., Marangos, P. J., Murphy, D. L., and Pearse, A. G. E.,** Neuron specific enolase (NSE) in human blood platelets: implications for the neuronal model, *Adv. Biosci.*, 31, 203, 1981.
6. **Fowler, C. J., Tipton, K. F., MacKay, A. V. P., and Youdim, M. B. H.,** Human platelet monoamine oxidase — a useful enzyme in the study of psychiatric disorders?, *Neuroscience*, 7, 1577, 1982.
7. **Fozard, J. R.,** Serotonin, migraine and platelets, *Progr. Pharmacol.*, 4, 135, 1982.
8. **Jérome, H. and Kamoun, P.,** Anomalie de la captation de la sérotonine par les plaquettes sanguines des sujets Trisomiques 21. I- étude des caractéristiques générales de la captation, *Bull. Soc. Chim. Biol.*, 50, 907, 1968.
9. **Buffoni, F.,** Biochemical pharmacology of amine oxidases, *Trends Pharmacol. Sci.*, 4, 313, 1983.
10. **Winter, H., Herschel, M., Propping, P., Friedl, W., and Vogel, F.,** A twin study on three enzymes (DBH, COMT, MAO) of catecholamine metabolism. Correlations with MMPI, *Psychopharmacology*, 57, 63, 1978.
11. **Fowler, C. J. and Wiberg, Å,** Molecular characteristics of human platelet monoamine oxidase, *Naunyn-Schmiedeberg's Arch. Pharmacol.*, 313, 77, 1980.
12. **Fowler, C. J. and Oreland, L.,** Human platelet monoamine oxidase — some biochemical findings, in *Monoamine Oxidase — Basic and Clinical Frontiers*, Kamijo, K., Usdin, E., and Nagatsu, T., Eds., Excerpta Medica, Amsterdam, Holland, 1982, 28.
13. **Lyles, G. A. and Callingham, B. A.,** Evidence for a clogyline-resistant monoamine metabolising activity in the rat heart, *J. Pharm. Pharmacol.*, 27, 682, 1975.
14. **Houslay, M. D. and Tipton, K. F.,** Inhibition of beef plasma amine oxidase by clorgyline, *Biochem. Pharmacol.*, 24, 429, 1975.
15. **Eriksson, M. and Fowler, C. J.,** Inhibition of monoamine oxidase and a clorgyline-resistant, semicarbazide-sensitive amine oxidase by mexiletine and related compounds, *Naunyn-Schmiedeberg's Arch. Pharmacol.*, 327, 273, 1984.
16. **Johnston, J. P.,** Some observations upon a new inhibitor of monoamine oxidase in brain tissue, *Biochem. Pharmacol.*, 17, 1285, 1968.
17. **Knoll, J. and Magyar, K.,** Some puzzling pharmacological effects of monoamine oxidase inhibition, *Adv. Biochem. Psychopharmacol.*, 5, 393, 1972.
18. **Donnelly, C. H. and Murphy, D. L.,** Substrate- and inhibitor-related characteristics of human platelet monoamine oxidase, *Biochem. Pharmacol.*, 26, 853, 1977.
19. **Kinemuchi, H., Fowler, C. J., and Tipton, K. F.,** Substrate specificities of the two forms of monoamine oxidase, in *Monoamine Oxidase and Disease; Prospects for Therapy with Reversible Inhibitors*, Strolin Benedetti, M., Dostert, P., and Tipton, K. F., Eds., Academic Press, London, in press.
20. **Fowler, C. J., Oreland, L., Marcusson, J., and Winblad, B.,** Titration of human brain monoamine oxidase -A and -B by clorgyline and 1-deprenyl, *Naunyn-Schmiedeberg's Arch. Pharmacol.*, 311, 263, 1980.
21. **Fowler, C. J. and Oreland, L.,** unpublished observations.
22. **Oreland, L., Arai, Y., Stenström, A., and Fowler, C. J.,** Monoamine oxidase activity and localisation in the brain and the activity in relation to psychiatric disorders, *Mod. Probl. Pharmacopsychiat.*, 19, 246, 1983.
23. **Greenawalt, J. W.,** Localisation of monoamine oxidase in rat liver mitochondria, *Adv. Biochem. Psychopharmacol.*, 5, 207, 1972.
24. **Barrand, M. A. and Callingham, B. A.,** Monoamine oxidase activities in brown adipose tissue of the rat: some properties and subcellular distribution, *Biochem. Pharmacol.*, 31, 2177, 1982.
25. **Solatunturi, E. and Paasonen, M. K.,** Intracellular distribution of monoamine oxidase, 5-hydroxytryptamine and histamine in blood platelets of rabbit, *Ann. Med. Exp. Biol. Fenn.*, 44, 427, 1966.
26. **Collins, G. G. S. and Sandler, M.,** Human blood platelet monoamine oxidase, *Biochem. Pharmacol.*, 20, 289, 1971.
27. **Yu, P. H., Bowen, R., Carlson, K., O'Sullivan, K., and Boulton, A. A.,** Comparison of biochemical properties of platelet monoamine oxidase in mentally disordered and healthy individuals, *Psychiat. Res.*, 6, 107, 1982.

28. **Craig, I. W., Powell, J. F., Brown, G. K., and Summers, K. M.,** Studies on human monoamine oxidase, in *Monoamine Oxidase — Basic and Clinical Frontiers*, Kamijo, K., Usdin, E., and Nagatsu, T., Eds., Excerpta Medica, Amsterdam, Holland, 1982, 18.
29. **Abeles, R. H. and Maycock, A. L.,** Suicide enzyme inactivators, *Acc. Chem. Res.*, 9, 114, 1976.
30. **Sourkes, T. L.,** Influence of hormones, vitamins and metals on monoamine oxidase activity, in *Monoamine Oxidase: Structure, Function and Altered Functions*, Singer, T. P., Von Korff, R. W., and Murphy, D. L., Eds., Academic Press, New York, 1979, 291.
31. **Youdim, M. B. H., Woods, H. F., Mitchell, B., Grahame-Smith, D. G., and Callender, S.,** Human platelet monoamine oxidase in iron-deficiency anaemia, *Clin. Sci. Molec. Med.*, 48, 289, 1975.
32. **Tipton, K. F.,** Monoamine oxidase, in *Handbook of Physiology, Vol. 6, Sect. 7, Endocrinology*, Blaschko, H., Sayers, G., and Smith, A. D., Eds., American Physiological Society, Washington, 1976, 677.
33. **Husain, M., Edmondson, D. E., and Singer, T. P.,** Kinetic studies on the catalytic mechanism of liver monoamine oxidase, *Biochemistry*, 21, 595, 1982.
34. **Roth, J. A. and Pearce, L. B.,** MAO-B: mechanism of deamination, d-amphetamine and alternative substrate inhibition as steady state kinetic probes, *L.Encéphale*, IX Suppl. 1, 61A, 1983.
35. **Fowler, C. J., Ekstedt, B., Egashira, T., Kinemuchi, H., and Oreland, L.,** The interaction between human platelet monoamine oxidase, its monoamine substrates and oxygen, *Biochem. Pharmacol.*, 28, 3063, 1979.
36. **Koide, Y., Koide, N., Ross, S., Sääf, J., and Wetterberg, L.,** Monoamine oxidase in human platelets — kinetics and methodological aspects, *Biochem. Pharmacol.*, 30, 2893, 1981.
37. **Fowler, C. J. and Callingham, B. A.,** Substrate-selective activation of rat liver mitochondrial monoamine oxidase by oxygen, *Biochem. Pharmacol.*, 27, 1995, 1978.
38. **Fowler, C. J. and Oreland, L.,** The nature of the substrate-selective interaction between rat liver mitochondrial monoamine oxidase and oxygen, *Biochem. Pharmacol.*, 29, 2225, 1980.
39. **Mitra, C. and Guha, S. R.,** Effects of oxygen tension on the activity of MAO types, in *Monoamine Oxidase — Basic and Clinical Frontiers*, Kamijo, K., Usdin, E., and Nagatsu, T., Eds., Excerpta Medica, Amsterdam, 1982, 125.
40. **Fowler, C. J. and Tipton, K. F.,** Deamination of 5-hydroxytryptamine by both forms of monoamine oxidase in the rat brain, *J. Neurochem.*, 38, 733, 1982.
41. **Yu, P. H. and Boulton, A. A.,** Activation of platelet monoamine oxidase by plasma in the human, *Life Sci.*, 25, 31, 1979.
42. **Sääf, J., Wahlund, L.-O., Ross, S. B., and Wetterberg, L.,** The properties and significance of the plasma activator of monoamine oxidase, in *Monoamine Oxidase and Disease: Prospects for Therapy with Reversible Inhibitors*, Strolin Benedetti, M., Dostert, P., and Tipton, K. F., Eds., Academic Press, London, 1984, 265.
43. **Demisch, L., Gebhart, P., Kaszmarczyk, P., von der Mühlen, H., and Bochnik, H. J.,** Low platelet MAO activity in psychiatric patients and plasma factors: no evidence for inhibitory influences on MAO in the circulating platelet population, *Biol. Psychiat.*, 16, 21, 1981.
44. **Demisch, L., Gebhart, P., Kaczmarczyk, P., and Roser, M.,** Factors that may regulate MAO activity, *L'Encéphale*, IX Suppl. 1, 25A, 1983.
45. **Buckman, T. and Eiduson, S.,** A potential natural inhibitor of MAO-B, *L'Encéphale*, IX Suppl. 1, 64A, 1983.
46. **Wahlund, L.-O., Smedby, Y., Sääf, J., Schulman, A., Ross, S. B., and Wetterberg, L.,** Influence of hemodialysis on monoamine oxidase kinetics in platelets from chronic schizophrenic patients, *Artif. Organs*, 7, 334, 1983.
47. **McEwen, C. M., Sasaki, G., and Lenz, W. R.,** Human liver mitochondrial monoamine oxidase, *J. Biol. Chem.*, 243, 5217, 1968.
48. **McEwen, C. M., Sasaki, G., and Jones, D. C.,** Human liver mitochondrial monoamine oxidase. II. Determination of substrate and inhibitor specificity, *Biochemistry*, 8, 3952, 1969.
49. **Williams, C. H.,** Monoamine oxidase. I. Specificity of some substrates and inhibitors, *Biochem. Pharmacol.*, 23, 615, 1974.
50. **Sandler, M., Reveley, M. A., and Glover, V.,** Human platelet monoamine oxidase activity in health and disease: a review, *J. Clin. Pathol.*, 34, 292, 1981.
51. **Demisch, L., Kaczmarczyk, P., and Gebhart, P.,** Methodological problems of using platelet MAO in psychiatric research, *Mod. Probl. Pharmacopsychiat.*, 19, 265, 1983.
52. **Murphy, D. L., Wright, C., Buchsbaum, M., Nichols, A., Costa, J. L., and Wyatt, R. J.,** Platelet and plasma amine oxidase activity in 680 normals: sex and age differences and stability over time, *Biochem. Med.*, 16, 254, 1976.
53. **Shaskan, E. G., Peszke, M. A., and Niederman, J. C.,** Studies of platelet monoamine oxidase activity in Epstein-Barr and Dengue virus infections, *Schizophrenia Bull.*, 6, 347, 1980.
54. **Buchsbaum, M. S., Coursey, R. D., and Murphy, D. L.,** The biochemical high-risk paradigm: behavioural and familial correlates of low platelet monoamine oxidase activity, *Science*, 194, 339, 1976.

55. **Puchall, L. B., Coursey, R. D., Buchsbaum, M. S., and Murphy, D. L.,** Parents of high risk subjects defined by levels of monoamine oxidase activity, *Schizophrenia Bull.*, 6, 338, 1980.
56. **Coursey, R. D., Buchsbaum, M. S., and Murphy, D. L.,** Two-year follow-up of subjects and their families defined as at risk for psychopathology on the basis of platelet MAO activities, *Neuropsychobiology*, 8, 51, 1982.
57. **Schooler, C., Zahn, T. P., Murphy, D. L., and Buchsbaum, M. S.,** Psychological correlates of monoamine oxidase activity in normals, *J. Nerv. Ment. Dis.*, 166, 177, 1978.
58. **Perris, C., Jacobsson, L., von Knorring, L., Oreland, L., Perris, H., and Ross, S. B.,** Enzymes related to biogenic amine metabolism and personality characteristics in depressed patients, *Acta Psychiat. Neurol. Scand.*, 61, 477, 1980.
59. **Propping, P. and Friedl, W.,** Platelet MAO activity and high risk for psychopathology in a German population, *Mod. Probl. Pharmacopsychiat.*, 19, 304, 1983.
60. **Murphy, D. L. and Wyatt, R. J.,** Reduced monoamine oxidase activity in blood platelets from schizophrenic patients, *Nature (London)*, 238, 225, 1972.
61. **DeLisi, L. E., Wise, C. D., Bridge, T. P., Rosenblatt, J. E., Wagner, R. L., Morihisa, J., Karson, C., Potkin, S. G., and Wyatt, R. J.,** A probable neuroleptic effect on platelet monoamine oxidase in chronic schizophrenic patients, *Psychiat. Res.*, 4, 95, 1981.
62. **Oreland, L., Fowler, C. J., and Schalling, D.,** Low platelet monoamine oxidase activity in cigarette smokers, *Life Sci.*, 29, 2511, 1981.
63. **Norman, T. R., Chamberlain, K. G., French, M. A., and Burrows, G. D.,** Platelet monoamine oxidase activity and cigarette smoking, *J. Affect. Dis.*, 4, 73, 1982.
64. **Eckert, B., Gottfries, C.-G., von Knorring, L., Oreland, L., Wiberg, Å., and Winblad, B.,** Brain and platelet monoamine oxidase in mental disorders. I. Schizophrenics and cycloid psychotics, *Progr. Neuropsychopharmacol.*, 4, 57, 1980.
65. **Roth, J. A.,** Effect of drugs on inhibition of oxidised and reduced form of MAO, in *Monoamine Oxidase: Structure, Function and Altered Functions*, Singer, T. P., Von Korff, R. W., and Murphy, D. L., Eds., Academic Press, New York, 1979, 153.
66. **Murphy, D. L. and Weiss, R.,** Reduced monoamine oxidase activity in blood platelets from bipolar depressed patients, *Am. J. Psychiat.*, 128, 1351, 1972.
67. **Belmaker, R. H., Bracha, H. S., and Ebstein, R. P.,** Platelet monoamine oxidase in affective illness and alcoholism, *Schizophrenia Bull.*, 6, 320, 1980.
68. **Edwards, D. J. and Burns, M. O.,** Effects of tricyclic antidepressants upon human platelet monoamine oxidase, *Life Sci.*, 15, 2045, 1974.
69. **Sullivan, J. L., Dackis, C., and Stanfield, C.,** *In vivo* inhibition of platelet MAO activity by tricyclic antidepressants, *Am. J. Psychiat.*, 134, 188, 1977.
70. **Gottfries, C.-G., von Knorring, L., and Oreland, L.,** Platelet monoamine oxidase activity in mental disorders. II. Affective psychoses and suicidal behaviour, *Progr. Neuropsychopharmacol.*, 4, 185, 1980.
71. **Agarwal, D. P., Philippu, G., Milech, U., Goedde, H. W., and Schrappe, O.,** Platelet monoamine oxidase activity in alcoholics, *Mod. Probl. Pharmacopsychiat.*, 19, 260, 1983.

Chapter 24

CONCLUDING REMARKS

E. Albert Zeller

In 1951, for the first time, the oxidative deamination of monoamine, diamines, and polyamines was reviewed together.[1] There I summarized the discovery of the degradation of tyramine,[2] histamine,[3] aliphatic monoamines,[4] adrenaline,[5,6] aliphatic diamines,[7] and spermidine.[8] In addition, the identity of adrenaline oxidase with aliphatic amine oxidase and tyramine oxidase[6,9,10] and of histaminase with diamine oxidase[7,11] was reported. During this initial period, only a few investigators were interested in amine oxidases. This situation changed radically when the first irreversible inhibitor of monoamine oxidase (MAO) — iproniazid — was discovered.[12,13] The new inhibitors made it possible to modulate tissue levels of transmitter amines and, thus, to analyze the action of these agents in countless biochemical and pharmacological experiments as described in a rapidly increasing number of papers.[14-16] A similar transition with respect to the interest in diamine and polyamine oxidases took place a few years later, soon after it was found that these compounds play an important role in controlling growth and differentiation processes,[17] as reviewed by Bachrach. At a very early time it already had been suggested that DAO "in several ways is connected with human reproduction", because we found it to occur at high activity in sperm plasma,[18] prostatic gland,[19] placenta,[20,21] and in blood plasma of pregnant women.[22]

After the slow beginning, the rapid development of our knowledge of amine oxidases was favored by the increasing availability of incomparatively more sensitive and specific analytical tools, by the improvements of isolation procedures for enzymes, and by the rapidly evolving concepts regarding enzyme mechanisms.

In these remarks, a few examples, chosen almost at random out of an immense body of informations, are given to mark some stations on the way from the origin to the present state of the art as it is given in this book. The major part of this contribution, however, will be devoted to a discussion of the biological function of amine oxidases.

The enzymology of polyamine oxidation began with some simple observations pertaining to the degradation of spermine and spermidine by diamine oxidase. We had observed that spermidine and several spermidine homologs were substrates of the classical hog kidney DAO[8] and that spermine is rapidly degraded by a DAO of *Mycobacterium smegmatis.*[23] Much later, the degradation and interconversion of polyamines turned out to be a rich fabric of interacting enzymes. An important step in the unravelling of this reaction consisted of the introduction of acetyl residues into polyamines. By this process substrates of copper amine oxidases became substrates of flavin-containing enzymes and brought classical diamine and monoamine oxidases into close interaction (Chapter 3, Figure 2).

It took a long time to reach a safe identification of nonproteinaceous components of amine oxidases.[24] However, the occurrence of an aldehyde residue (not belonging to a carbohydrate moiety) in DAO was postulated at an early date.[8,11] This conclusion was drawn from our observation that a series of carbonyl reagents, including the aldehyde reagent dimedone (dimethylcyclohexanedione) displayed inhibitory action toward DAO.[8,25] The carbonyl residue was later considered to be a part of pyridoxal phosphate.

To give one of many examples of having missed the boat, I mention the copper component of DAO: it used to be one of the first steps of characterizing a new oxidative enzyme by testing its sensitivity toward cyanides. When this was done for DAO, the latter was found to be inhibited. Soon it became clear that the inhibition differed from that of the classical action of cyanide on iron-porphyrins. To further characterize the cyanide target, several heavy metal reagents were tested, but they did not influence DAO activity. Instead of pushing this matter further, we related the sensitivity of DAO toward KCN with that toward carbonyl reagents and concluded that cyanide formed a cyanhydrin with the carbonyl moiety mentioned above. Thus, heavy metals such as copper and iron unfortunately were included as components of DAO.[8,25,26]

One of the first steps toward a better understanding of the role of the flavin moiety in the catalytic action of MAO came from the spectroscopic studies of Hellerman and associates on the action of an acetylenic inhibitor on this enzyme; these authors assumed that C-2 was the point of attack in the flavin nucleus.[27] Later, N-5 was demonstrated to be the *locus* in model studies[28,29] and with MAO.[30] Our knowledge regarding the interaction between FAD, substrates, suicidal substrates, and various types of inhibitors has grown tremendously during the past decade (see Chapter 21).

As indicated in the preceding paragraph, not all major lines of investigation can be traced back to the early exploration of amine oxidases. As another example there is Gorkin's observations, made in 1967 and later, that the specificity pattern of monoamine oxidases can be reversibly transformed into one resembling that of diamine oxidases. This change is accomplished by treating the enzyme with oxidative agents. Transformed MAO can deaminate hydrolytically AMP. One could consider the possibility of a temporary withdrawal of electrons which, after the release of the amino residue, are restored to the substrate molecule instead to the oxygen. Gorkin's concise review (see Chapter 19) summarizes a large body of phenomena which are more extensively presented in his recent book.[31]

In several sections of this book many interesting examples regarding the function of amine oxidases in physiological and pathological processes are presented. No systematic treatment of this topic, however, has been undertaken here or elsewhere. One reason for this state of affairs might be related to the experience that questions of this kind are rarely completely answerable. When we start with a system which produces or eliminates components required in a second system, the role of the first one can only be fully discussed with respect to the second one. In general, the second system will then interact with a third one, and we may end up with a near-infinite regress.

While these remarks are not the place for a comprehensive review of this question, I would like to present a few thoughts which may provide a loose conceptual framework for some of the statements made in this book.

The situation is relatively simple when we explore amine oxidases whose products are essential components of another system. Lysyl oxidase is a particularly clear case, because this enzyme provides mechanical strength to collagen and elastic fibers by contributing covalent bonds between structural subunits. When the enzyme fails to perform adequately, the defect becomes recognizable by mechanical failure within the connective tissue (Chapter 8).

For several aminoaldehydes formed from diames and polyamines, a role in the differentiation and maturation of cells has been proposed (Chapter 14). Aminoaldehydes formed from polyamines interact with DNA and are toxic to virus, bacteria, parasites, and tumor cells (Chapter 2).

When the oxidative deamination of tyramine and phenylethylamine by tyramine oxidase in liver — a forerunner of MAO — was discovered, so little was known about the occurrence and physiology of these and related compounds that the only function to be given to this enzyme was that of "effecting a rapid detoxification of excessive amounts of tyramine absorbed from the intestine".[2] The same role had to be ascribed to the oxidation of putrescine and cadaverine in the intestinal mucosa.[11] When more and more biogenic amines of high biological activity (e.g., transmitter amines) and the distribution of the corresponding amine oxidases became better known, the concept of the scavenger function of the latter had to be limited to special situations. In view of the localization of MAO in adrenergic synapses, this enzyme gained for a while the reputation of being an impulse terminator, to be compared with that of cholinesterase in neuromuscular junctions, until it was shown that not MAO, but the re-uptake of the monoamine is the major process of synaptic impulse termination.

Is there any general task to be accomplished by amine oxidases outside those mentioned above? I begin my answer to this question with some statements made by four authors of this book. Bachrach (Chapter 2) proposes that "amine oxidases catalyze the oxidation of amines or polyamines and thereby regulate their cellular levels. Growth and differentiation processes are closely related to cellular polyamine levels, hence the vital processes may be regulated by modulating cellular amine levels". Similarly, Maslinski et al. (Chapter 14) compare DAO with "a sluice gate which adjusts inflow and outflow of putrecine, polyamines, and histamine, and which controls their levels and levels of their metabolites in cells". For a more complex system, Seiler (Chapter 3) suggests that "all available examples of enhanced acetylation rates ...can be interpreted as a means of the cell to get rid of excesive polyamines, whereby excessive indicates not necessarily a supernormal concentration, but only an amount above that which is physiologically needed under the specific circumstances". Finally, we learn from Perin, Sessa, and Desiderio (Chapter 16) that DAO "is probably implicated in the regulation of putrescine levels as a part of multienzymatic system".

To provide a solid basis for the evaluation of propositions of this kind, a model has to be developed which will permit quantitative analysis of the pertinent phenomena. We begin with the following assumption: whenever the biogenic amines themselves (and not their metabolites) produce a physiological or pathological reaction, the tissue concentration of the amine determines the intensity of the response. Not only the rate of formation and inactivation of our amines, but also their elimination by sesquestration into storage granules, by exocytosis, and by conjugation have to be considered. Among the major factors in establishing tissue concentrations of a given amine, I believe, are the enzymes that are most immediately involved in its formation and degradation (e.g., aromatic amino acid decarboxylases and amine oxidases). In some instances, the rate-limiting reactions of the sequences leading to and from the amines (e.g., tyrosine hydroxylase for catecholamine formation) have to replace decarboxylase and amine oxidase.

Amino acid decarbosylase and amine oxidase form a coupled enzyme system. The kinetics of coupled enzymes has been thoroughly worked out[32] because they are of considerable practical importance: whenever the activity of one enzyme is measured by the addition of a second enzyme whose rate is assessed we obtain coupled enzyme systems. We became interested in the latter because, after a lag phase, they reach a steady state with respect to the concentration of the intermediate S_2 which can be maintained for a considerable length of time.

$$S_1 \xrightarrow{\text{enzyme 1}} S_2 \xrightarrow{\text{enzyme 2}} S_p$$

If certain conditions are fulfilled, the steady state concentration of the intermediate S_2 can be calculated from the kinetic data of the individual enzymes. Thus, a coupled enzyme system may serve as a "buffer" for the maintenance of S_2 concentration.

If we combine an amino acid with aromatic L-amino acid decarboxylase and with MAO, we can determine the concentration of the intermediate monoamine and find out when, for how long, and at what level the steady state evolves. We started with a system consisting of the aromatic amino acid decarboxylase, MAO, and L-phenylalanine and measured phenylethylamine concentrations for 30 min. The two enzymes were those of beef liver homogenate, and the activity of MAO was modulated with the MAO-B inhibitor deprenyl. The concentration of phenylethylamine rose at first and leveled off after several minutes, depending on the MAO activity, and remained at this level for a few more minutes. It seems, therefore, that we may be able to develop an experimental set up for the study of the influence of enzyme activities on amine levels.[33] In the end, MAO may appear not to serve primarily as a destroyer of amines, but rather as a part of a buffer system which establishes and maintains certain monoamine levels in tissues.

The exploration of lysosomal function is one of the topics which may benefit from the application of these concepts as the following observations suggest: histamine appears to increase the fragility of lysosomes[34] while spermine and spermidine stabilize these cell organelles.[35] Most work so far has been dedicated to the study of the effect of catecholamines on the integrity of lysosomes; the action of the amines appear to be mediated by c-AMP.[36] One can expect that the amine concentrations *in loco* will determine the intensity of the lysosomal respones. Since the lysosomal apparatus is involved in countless reactions, we can foresee a substantial enhancement of our knowledge of the biofunction of amine oxidases when this line of investigation is further pursued. Recently, we have started to test this conjecture experimentally.

REFERENCES

1. **Zeller, E. A.,** Oxidation of amines, in *The Enzymes,* Sumner, J. B. and Myrback, K., Eds., Academic Press, New York, 1951, chap. 59.
2. **Hare, M. L. C.,** Tyramine oxidase. I. A new enzyme system in liver, *Biochem. J.,* 22, 968, 1928.
3. **Best, C. H. and McHenry, E. S.,** The inactivation of histamine, *J. Physiol.,* 70, 349, 1930.
4. **Pugh, C. E. M. and Quastel, J. H.,** Oxidation of aliphatic amines by brain and other tissues, *Biochem. J.,* 31, 286, 1937.
5. **Richter, D.,** Adrenaline and amine oxidase, *Biochem. J.,* 31, 2022, 1937.
6. **Kohn, H. I.,** Tyramine oxidase, *Biochem. J.,* 31, 1963, 1937.
7. **Zeller, E. A.,** Ueber den enzymatischen Abbau von Histamin und Diaminen, *Helv. Chim. Acta,* 21, 880, 1938.
8. **Zeller, E. A.,** Zur Kenntnis der Diamin-oxydase, *Helv. Chim. Acta,* 21, 1645, 1938.
9. **Blaschko, H., Richter, D., and Schlossman, H.,** The oxidation of adrenaline and other amines, *Biochem. J.,* 31, 2187, 1937.
10. **Pugh, C. E. M. and Quastel, J. H.,** Oxidation of amines by animal tissues, *Biochem. J.,* 31, 2306, 1937.
11. **Zeller, E. A.,** Diamin-oxydase, *Adv. Enzymol.,* 2, 93, 1942.
12. **Zeller, E. A. and Barsky, J.,** *In vivo* inhibition of liver and brain monoamine oxidase by 1-isonicotinyl-2-isopropyl hydrazine, *Proc. Soc. Exp. Biol. Med.,* 81, 459, 1952.
13. **Zeller, E. A., Barsky, J., and Berman, E. R.,** Amine oxidases. XI. Inhibition of monoamine oxidase by 1-isonicotinyl-2-isopropylhydrazine, *J. Biol. Chem.,* 214, 267, 1955.
14. **Griesemer, E. A., Barsky, J., Dragstedt, C. A., Wells, J. A., and Zeller, E. A.,** Potentiating effect of iproniazid on the pharmacological action of sympathomimetric amines, *Proc. Soc. Exp. Biol. Med.,* 84, 699, 1953.
15. **Zeller, E. A., Consulting Ed.,** Amine oxidase inhibitors, *Ann. N.Y. Acad. Sci.,* 80, 551, 1959.
16. **Zeller, E. A., Consulting Ed.,** New reflections on monoamine oxidase inhibition, *Ann. N.Y. Acad. Sci.,* 107, 809, 1963.
17. **Russell, D. H.,** *Polyamines in Normal and Neoplastic Cells,* Raven Press, New York, 1973.
18. **Zeller, E. A.,** Ueber das Vorkommen der Diamin-oxydase in menschlichen Sperma, *Helv. Chim. Acta,* 24, 117, 1941.
19. **Zeller, E. A. and Joël, C. A.,** Beitraege zur Fermentchemie des maennlichen Geschlectsapparates, *Helv. Chim. Acta,* 24, 968, 1941.
20. **Zeller, E. A., Schaer, B., and Staehlin, S.,** Weitere Beitraege zur Kenntnis der Diamin-oxydase (Histaminase), *Helv. Chim. Acta,* 22, 837, 1939.
21. **Danforth, D. N.,** Placental histaminase, *Am. J. Physiol.,* 119, 294, 1937.
22. **Zeller, E. A. and Birkhaeuser, H.,** Ueber eine in vitro-Schwanger-schaftsereaktion, *Schweiz. Med. Wschr.,* 70, 975, 1940.
23. **Roulet, F. and Zeller, E. A.,** Ueber die Diamin-diamin-oxydase der Smegmabazillen, *Helv. Chim. Acta,* 28, 1326, 1945.
24. **Blaschko, H.,** The natural history of amine oxidases, *Rev. Physiol. Biochem. Pharmacol.,* 70, 83, 1974.
25. **Zeller, E. A.,** Ueber die Beeinflussung der Diamin-oxydase durch Kaliumcyanid, *Helv. Chim. Acta,* 23, 1418, 1940.

26. **Zeller, E. A., Stern, R., and Wenk, M.,** Ueber die Diamin-Diamin-oxydase-Reaktion, *Helv. Chim. Acta*, 23, 3, 1940.
27. **Hellerman, L. and Erwin, V. G.,** Mitochondrial monoamine oxidase. II. Action of various inhibitors for the bovine kidney enzyme. Catalytic mechanism, *J. Biol. Chem.*, 243, 5234, 1968.
28. **Zeller, E. A., Gaertner, B., and Hemmerich, P.,** 4a,5-Cycloaddition reactions of acetylenic compounds at the flavoquinone nucleus as mechanism of flavoprotein inhibition, *Z. Naturforsch.*, 27, 1050, 1972.
29. **Gaertner, B., Hemmerich, P., and Zeller, E. A.,** Structure of flavin-adducts with acetylenic (photo)substrates: chemistry of monoamine oxidase and lactase oxidase inhibition, *Eur. J. Biochem.*, 63, 211, 1976.
30. **Zeller, E. A. and Hsu, M.,** On the mechanism of inhibition of monoamine oxidase, in *Frontiers in Catecholamine Research*, Usdin, E. and Snyder, S., Eds., Pergamon Press, New York, 1973, 153.
31. **Gorkin, V. Z.,** *Amine Oxidases in Clinical Research*, Pergamon Press, Oxford, 1983.
32. **Roberts, D. V.,** *Enzyme Kinetics*, Cambridge University Press, Cambridge, 1977, chap. 9.
33. **Huprikar, S. V. and Zeller, E. A.,** Unpublished data, 1984.
34. **Yoffe, J. R.,** Effects of inflammatory agents on endothelial lysosomal fragility and their inhibition by antiinflammatory drugs, *Br. J. Pharmacol.*, 72, 603, 1981.
35. **Powell, J. H. and Reidenberg, M. M.,** *In vitro* response of rat and human kidney lysosomes to aminoglycosides, *Biochem. Pharmacol.*, 31, 3447, 1982.
36. **Ignarro, L. J.,** Regulation of lysosomal enzyme release by prostaglandins, autonomic neurohormones and cyclic nucleotides, *Front. Biol.*, 43, 481, 1975.

INDEX

A

C

D

E

F

G

H

I

M

P

Q

R

S

T

U

V

W

Y

Z